# Reading Essentials

## An Interactive Student Workbook

earth.msscience.com

New York, New York Columbus, Ohio Chicago, Illinois Peoria, Illinois Woodland Hills, California

# To the Student

In today's world, knowing science is important for thinking critically, solving problems, and making decisions. But understanding science sometimes can be a challenge.

***Reading Essentials*** takes the stress out of reading, learning, and understanding science. This book covers important concepts in science, offers ideas for how to learn the information, and helps you review what you have learned.

In each chapter:

- **Before You Read** sparks your interest in what you'll learn and relates it to your world.
- **Read to Learn** describes important science concepts with words and graphics. Next to the text you can find a variety of study tips and ideas for organizing and learning information:
    - The **Study Coach** offers tips for getting the main ideas out of the text.
    - **Foldables™ Study Organizers** help you divide the information into smaller, easier-to-remember concepts.
    - **Reading Checks** ask questions about key concepts. The questions are placed so you know whether you understand the material.
    - **Think It Over** elements help you consider the material in-depth, giving you an opportunity to use your critical-thinking skills.
    - **Picture This** questions specifically relate to the art and graphics used with the text. You'll find questions to get you actively involved in illustrating the concepts you read about.
    - **Applying Math** reinforces the connection between math and science.
- Use **After You Read** to review key terms and answer questions about what you have learned. The **Mini Glossary** can assist you with science vocabulary. Review questions focus on the key concepts to help you evaluate your learning.

See for yourself. ***Reading Essentials*** makes science easy to understand and enjoyable.

The McGraw·Hill Companies



Send all inquiries to:
Glencoe/McGraw-Hill
8787 Orion Place
Columbus, OH 43240

ISBN 0-07-866970-7
Printed in the United States of America
8 9 10 024 09 08 07

# Table of Contents

# The Nature of Science

## section 1 Science All Around

### Before You Read

What do you think of when you hear the words *Earth science?* In the space below, list some topics that you would like to learn about in Earth science.

______________________________________________

______________________________________________

______________________________________________

**What You'll Learn**

- how scientific methods are used
- what science and Earth science are
- the parts of a scientific experiment

### Read to Learn

**Study Coach**

**Summarize** As you read this section, stop after each paragraph and summarize the main idea in your own words.

#### Mysteries and Problems

Scientists are often like detectives trying to solve a mystery. In 1996, Japanese scientists discovered reports from almost 300 years ago when a chain of huge, fast waves, called a tsunami, had smashed the coast of the island of Honshu. What caused the tsunami? This was the mystery the scientists wanted to solve.

#### How do scientists look for answers?

The scientists suspected that an earthquake along the Pacific coast of North America might have caused the tsunami in Japan. The map on the next page shows the northwest coast of the United States. Here, one section of Earth's outer layer, called a plate, is sinking beneath another plate. This is called a subduction zone. In the Cascadia subduction zone, the Juan de Fuca Plate is sinking under the North American Plate. Earthquakes are common in areas like this. But it would take a stronger earthquake than ever recorded to send waves rolling all the way across the Pacific Ocean to Japan. How would scientists find evidence for such a large earthquake?

## Picture This

**1. Identify** Mark on the map the area where earthquakes might occur.

**Pacific Earthquake Zone**

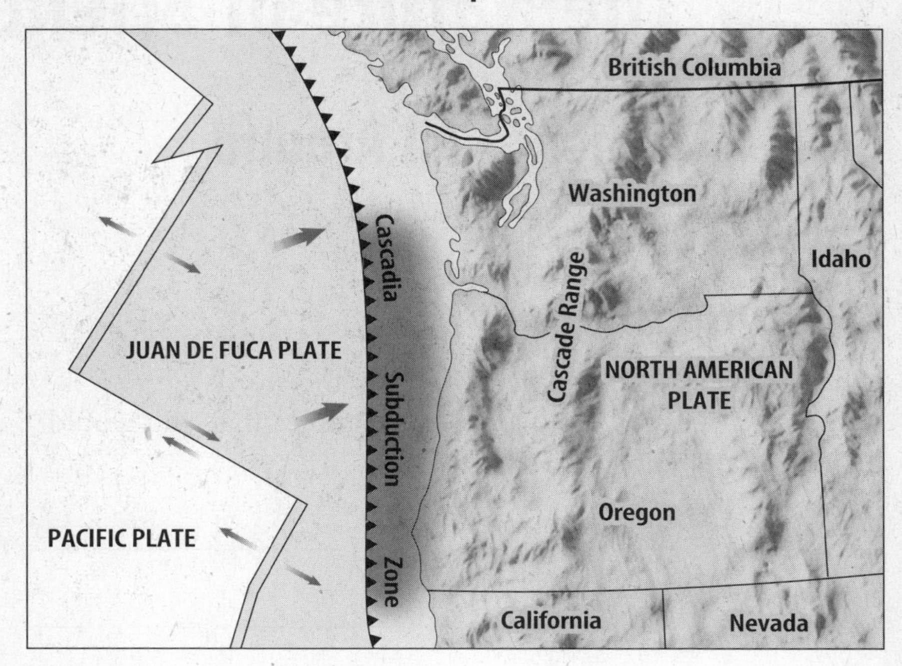


## What evidence did scientists find?

The scientists were able to find evidence of a long-ago earthquake along the coast of Oregon and Washington. Much of the coast had sunk. Thousands of trees had been covered with water. However, the scientists still needed to figure out exactly when the earthquake occurred.

## What is a hypothesis?

One scientist thought of a way to determine when the earthquake happened. He made an educated guess called a **hypothesis.** He guessed that tree rings of the drowned trees could provide clues about the date of the earthquake.

Each year, a tree makes a new ring of tissue in its trunk. Scientists used data from carbon dating and examined the tree rings. They were able to pinpoint that the trees had died or were damaged after the summer of 1699 but before the spring of 1700. This evidence meant that the earthquake and the tsunami happened at nearly the same time.

## Why is solving mysteries important?

The trees helped explain what caused the powerful tsunami in Japan. They also provided a warning for people living in the Pacific Northwest. Strong earthquakes are possible in that area. Scientists warn that it is only a matter of time before another huge earthquake strikes.

## Think it Over

**2. Recognize Cause and Effect** How does solving this mystery help people in the future?

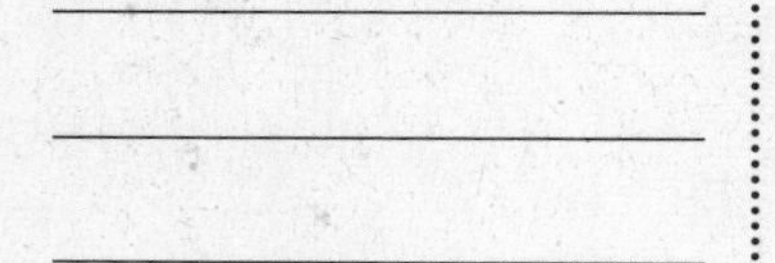

## Scientific Methods

Scientists solve mysteries by using scientific methods. **Scientific methods** are procedures or steps used to solve a problem. The steps used in a scientific method are shown below. When you used methods like these, you are solving problems in a scientific way.

| Steps | Example |
|---|---|
| Identify a problem | What caused the tsunami in Honshu in 1700? |
| Gather information | Scientists found evidence of a long-ago earthquake along the coast of Oregon and Washington. Much of the coast had sunk. |
| Make a hypothesis | The rings of the drowned trees and carbon dating would show that the earthquake happened about the same time as the tsunami. |
| Test the hypothesis | Scientists studied the growth rings and age of drowned trees. |
| Analyze the results | The results showed that the trees had died or been damaged between 1699 and 1700—the same time as the tsunami. |
| Draw conclusions | Three hundred years ago, the earthquake along the coast of Oregon and Washington caused the tsunami in Honshu, Japan. |

### Picture This

3. **Sequence** List the first three steps of the scientific method used here.

______________________

______________________

______________________

## Science

**Science** is a way of observing, studying, and thinking about things in your world to gain knowledge. When you see something that you cannot explain, you ask questions about it. For example, you might see that the sky is blue during the day, but red at sunset and sunrise. You might ask yourself why the color changes. Or, you might see an unusual rock formation and wonder why and how it was formed. You might ask why some of the rocks are smooth while others are rough.

Science involves trying to answer questions and solve problems to better understand the world. Whenever you try to find out how and why things look and behave the way they do, you are doing science. ☑

### What is Earth science?

**Earth science** is the study of Earth and space. Earth scientists study topics like rocks, minerals, soil, volcanoes, earthquakes, and fossils. They also study maps, mountains, climates, weather, ocean water, and objects in space. By using tests and investigations, these scientists have discovered much of the information you will study. There are still many unanswered questions about Earth science.

### Reading Check

4. **Explain** What is science?

______________________

______________________

______________________

______________________

FOLDABLES™

**A Understand Main Ideas** Make a four-tab Foldable, as shown below. As you read, write what you learn about independent variables, dependent variables, constants, and controls.

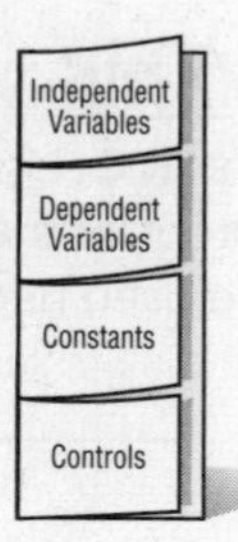


**Reading Check**

5. **Identify** What is the dependent variable in the dishwashing soap experiment?

______________________

______________________

## Working in the Lab

An important part of science is testing, or experimenting. Good experiments take careful planning. For instance, imagine that you want to know which brand of dishwashing soap cleans best. You may hypothesize that one brand is better than another. Or, you may hypothesize that all brands are the same. Next, you plan an experiment that will test your hypothesis. You choose which soaps you will test and how much soap to use. You figure out the best water temperature for your experiment and how many dishes you will wash. Finally, you decide how much grease to put on the dishes and the brand of paper towel you will use to dry the dishes. All of these factors can affect the outcome of the experiment.

### What are variables?

The factors that can change in an experiment are called **variables.** Test one variable at a time. The variable you want to test is the brand of dishwashing soap. This is called the independent variable. The **independent variable** is the one variable that you will change during the experiment.

### What are constants?

The variables that do not change in the experiment are called **constants.** Some constants in your experiment include the number of dishes you wash and the temperature of the water. You would wash the same number of dishes with each brand of soap. You would use the same water temperature each time.

### What is measured in an experiment?

You also need a way to measure how clean each dish is. You could find out how much grease is left on the dish by wiping it with a paper towel. How clean the dishes are is called the dependent variable. A **dependent variable** is the variable being measured in an experiment. ☑

### What is a control?

Most experiments need a control. A **control** is the standard to which your results can be compared. In this experiment, the control would use the same number of dishes, the same water temperature, and the same kind of towel to wipe the dishes. The control would leave out the dishwashing soap. You could then compare the results of the control to the results you got from using the different soaps.

## Why are experiments repeated?

You have to repeat your tests many times to gather data. Each brand should be tested the same number of times. Also, there should be a large number of samples. If one brand of soap comes out best five times in five tries, then you can be more confident in your conclusions. If you used 20 plates in each test, your total sample for each dishwashing liquid would be 100 plates. If something in a scientific experiment occurs only once, you cannot base a scientific conclusion on it. ☑

## How do we arrive at conclusions?

After you plan your experiment, you can begin your tests. You make observations and record data as you work. Your final step is to draw your conclusions. You draw conclusions by analyzing your results and by trying to understand what they mean.

When you are making and recording observations, be sure to include any unexpected results. Many discoveries have been made when experiments produced unexpected results.

# Technology

**Technology** is the use of scientific discoveries for practical purposes. Technology uses the knowledge of science to help people and to meet their needs. From the beginning of time, science and technology have worked together to help shape the world.

## How is technology used?

Advances in technology have resulted in many familiar inventions. These include everyday products such as paper, can openers, and rubber boots. Technology also includes calculators and computers that process information.

## What is transferable technology?

Technology designed for one situation often can be transferred to solve other problems. Some types of technology invented for use in outer space now are used here on Earth. Robotic parts were developed for spacecraft and satellites. But they also have become valuable in such areas as manufacturing and communications. The same is true for radar and sonar. These technologies were first created for the military. They are used now for studying weather and medicine. ☑

**Reading Check**

**6. Determine** Why are experiments repeated?

______________________

______________________

**Reading Check**

**7. Identify** Give two examples of transferable technology.

______________________

______________________

## After You Read

### Mini Glossary

**constant:** a variable that does not change in an experiment
**control:** the standard to which your results can be compared
**dependent variable:** the variable being measured in an experiment
**Earth science:** a way of looking and thinking about Earth and space
**hypothesis:** an educated guess; a likely explanation
**independent variable:** the variable in an experiment that changes
**science:** a way of looking at and thinking about the world
**scientific methods:** the steps scientists use to solve a problem
**technology:** the use of science for practical purposes
**variable:** a factor that can change in an experiment

1. Review the terms and their definitions in the Mini Glossary. Write a sentence using two terms that are part of scientific experiments.

2. The flowchart below shows steps in a scientific method. List the part of the dishwashing soap experiment that goes with each step. Write your answers below the matching step.

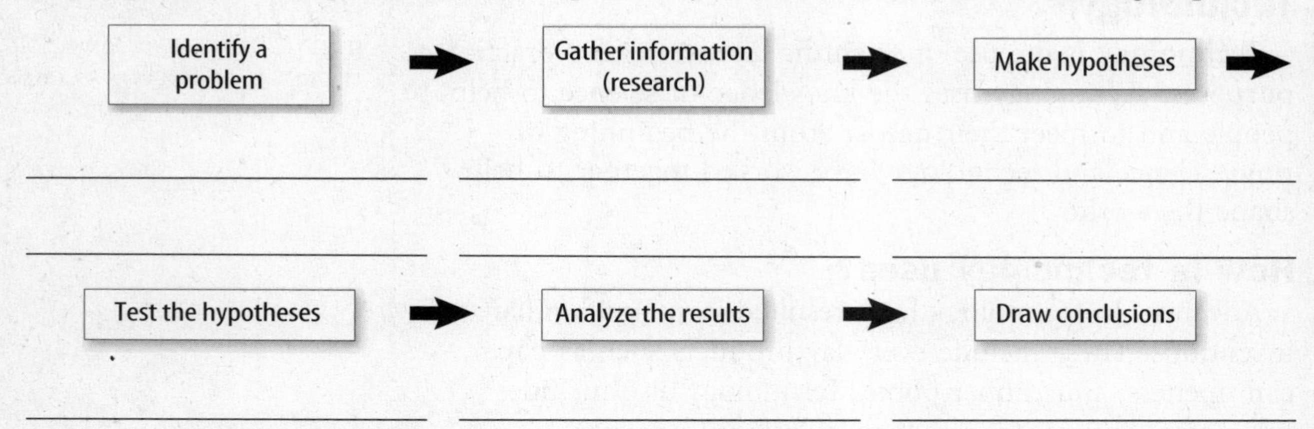


3. How did summarizing the main ideas in each paragraph help you to understand this section?

Visit **earth.msscience.com** to access your textbook, interactive games, and projects to help you learn more about science all around.

# The Nature of Science

## section 2 Scientific Enterprise

### Before You Read

Think of a question you have about the world around you. Write your question on the lines below. What steps could you take to find the answer to your question?

_______________________________________________

_______________________________________________

**What You'll Learn**

- how science is changing all the time
- what scientific theories and scientific laws are
- what science cannot answer

### Read to Learn

## A Work in Progress

People have always been curious about the world around them. They are especially curious about natural events such as storms, volcanoes, earthquakes, and comets. Early people made up stories, called myths, to explain what they observed. They believed that gods were responsible for storms, volcanoes, earthquakes, and comets.

**Mark the Text**

**Identify the Main Point** Highlight the main point of each paragraph. Use a different color to highlight a detail or example that helps explain the main point.

### Why are observations recorded?

Many of these early people wrote about events they saw. Six thousand years ago, farmers in Egypt noticed that the Nile River flooded every summer. Their crops had to be planted at just the right time to use the river's water. The farmers also noticed something else. The same bright star shone at dawn before the flood each year. The Egyptians used this information to develop a calendar. It was based on the appearance of that star every 365 days.

Later, civilizations created tools to measure with. As the tools were improved, people were able to make more accurate observations. They spent time trying to understand why things happened. They made inferences, or conclusions, to help explain things. They developed hypotheses and tested them. The results of these experiments helped them to learn even more.

**FOLDABLES™**

**B Organize Information** Make a Foldable from three sheets of notebook paper, as shown below. As you read, write what you learn about scientific enterprise.

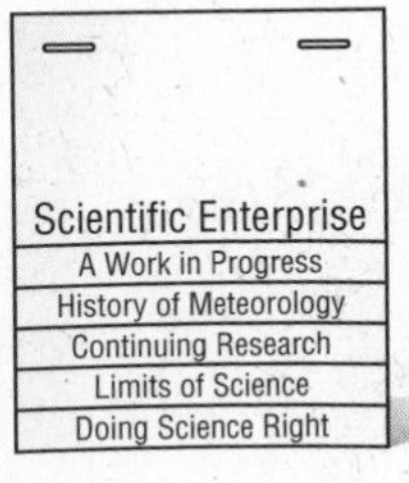

## The History of Meteorology

Today, scientists have information because of observations and conclusions made by scientists in earlier ages. Meteorology, the study of the weather, is one area of Earth science that developed over time.

**Reading Check**

**1. Define** What is meteorology?

### What are weather instruments used for?

The rain gauge was probably one of the first weather instruments. It measured the amount of rainfall. By the 1600s, scientists began to use other instruments. They used barometers to measure air pressure and thermometers to measure temperature. Other tools measured water vapor and wind speed. Modern versions of these tools are used in weather stations like the one shown in the figure below.

**Automated Weather Station**

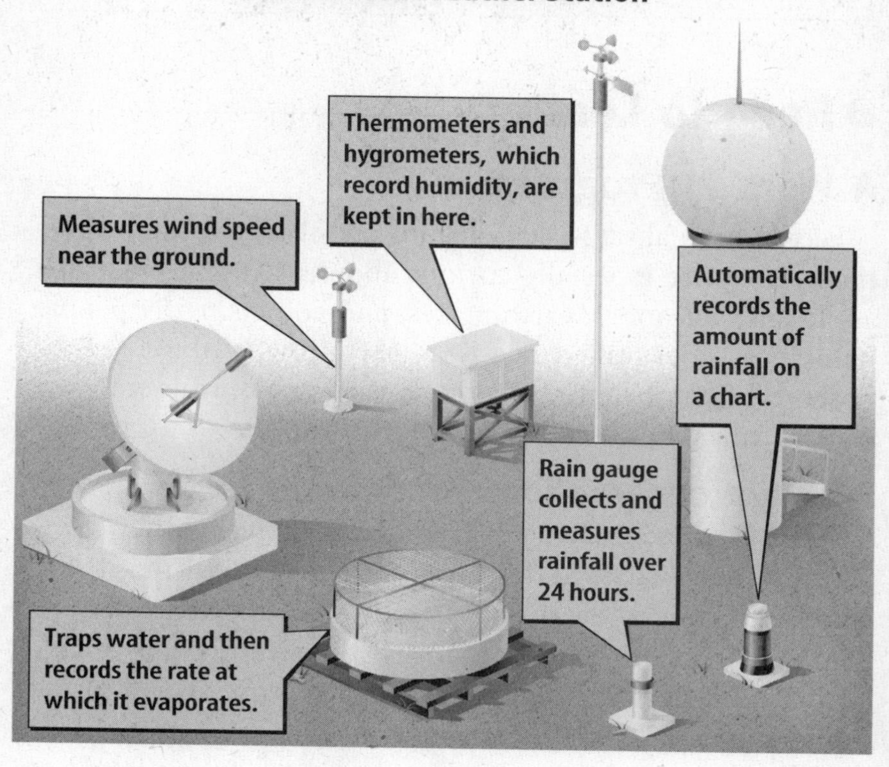


**Picture This**

**2. Apply** What does a hygrometer measure?

### When did weather prediction begin in the U.S.?

Benjamin Franklin was the first American to suggest predicting the weather. He had read newspaper accounts about storms across the country. From this information, he concluded that storms moved from west to east. His idea was that people could observe where a storm was moving and warn others that it was coming.

**Weather Reports** By 1849, there were many locations for observing the weather. Weather reports from these sites were sent by telegraph to the Smithsonian Institution in Washington, D.C. These data were recorded and a weather map was made. A weather report was then sent to *The Washington Evening Post* to be published.

### How has weather reporting changed?

By the late 1800s, there were more than 350 locations for observing and reporting the weather around the United States. By 1923, weather forecasts could be heard on 140 radio stations. In 1970, these reporting sites became part of the National Weather Service.

Today, instruments record weather information automatically. Satellites, weather balloons, and radar help predict weather. Reports also come from weather stations located on ships and aircraft. ☑

Today, it is easy to find out the weather anywhere in the world. You can watch a television weather channel, listen to a radio, or check a Web site. Weather information and warnings are important to people in areas where tornadoes, hurricanes, or severe weather conditions occur.

## Continuing Research

What scientists know is always changing. Better instruments are being developed. Methods of testing are improved. As science changes, scientists learn more about nature. They learn more about Earth's interior, its oceans, and its environment. One day, you might make a scientific discovery that changes people's understanding of the world.

### What are scientific theories?

A **scientific theory** is an explanation about the behavior of something in nature. A theory is based on tests and experiments. Remember, scientists make and test hypotheses. If the hypotheses are good, the data scientists collect will support them. Scientists use the results of many tests to develop a scientific theory.

Here's how one hypothesis became a scientific theory. Early scientists hypothesized that comets were made up of many different materials swirling together. In 1949, astronomer Fred L. Whipple proposed a new hypothesis. He said a comet was more like a dirty snowball. It was made up only of ice and dust particles.

**Reading Check**

3. **Determine** What tools are used to predict weather today?

______________________

______________________

______________________

**Think it Over**

4. **Analyze** Why do scientists use the results from many tests to develop a scientific theory?

______________________

______________________

______________________

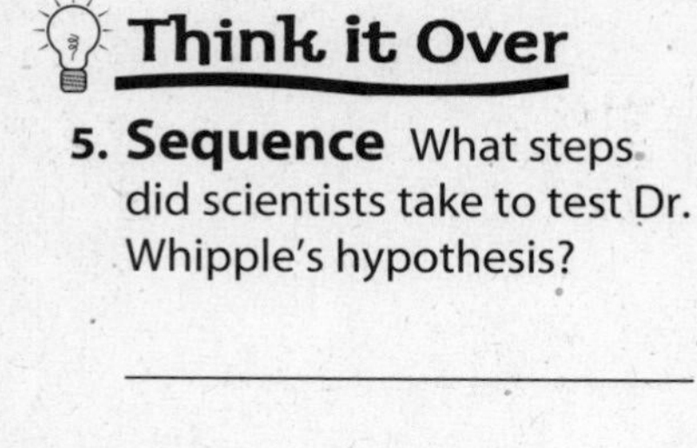

**5. Sequence** What steps did scientists take to test Dr. Whipple's hypothesis?

**Reading Check**

**6. Define** What is ethics?

## How does a hypothesis become a scientific theory?

Dr. Whipple's hypothesis was tested for many years. Comets were observed through giant telescopes. Finally, in 1986, a group of scientists looked closely at Halley's comet. They collected data. They also looked at the observations of other scientists. All the observations and data supported Dr. Whipple's original hypothesis. It has become an accepted scientific theory.

## What are scientific laws?

A **scientific law** is a rule that describes the behavior of something in nature. It usually describes what will happen, but it does not explain why.

Newton's first law of motion is a scientific law. This law states that an object, such as a marble or a spacecraft, will keep moving or remain at rest until it is acted on by an outside force. In other words, an object will move in the same direction and at the same speed until an outside force changes its direction or speed, including stopping it. Or an object will remain still until an outside force moves it. This scientific law describes the behavior of something by telling what will happen to it, but not why it happens.

# Limits of Science

Science does not have answers to all the questions in the universe. Science is limited in what it can explain. For a question or problem to be studied using scientific methods, there must be variables that can be observed, measured, and tested. Questions that deal with ethics or belief systems cannot be answered by science. **Ethics** is a system of understanding what is good and what is bad. Belief systems deal with religious beliefs. Questions that science cannot answer might include: Do humans have more value than other life forms? Should animals be used for medical testing? ☑

# Doing Science Right

Questions about ethics cannot be answered by science, but there are still ethical ways of doing science. This means that tests will be performed and conclusions drawn in a fair way. A scientist who is objective is one whose mind is open to all possible answers and outcomes.

## How can scientists be objective?

Scientists know that **bias,** or their own personal opinions, can affect their observations. They design and carry out experiments with this idea in mind. They do not decide ahead of time how the experiment will turn out or what the results will be. They make sure that the conditions are the same for testing each variable.

## How can scientists be ethical and open?

Ethical scientists keep detailed notes about their experiments. They base their conclusions on their notes and on their tests and measurements. They know that scientific knowledge grows when people work together. They share their results with other scientists. This allows other scientists to examine and evaluate their work. Most of the science we know today is a result of shared information.

Like the student in the figure below, scientists often use tables to record their data in a clear and organized way. Tables also help scientists share results with other scientists.

Aaron Haupt

## What is fraud?

Behaving in a way that is not open and ethical is fraud. Scientific fraud includes dishonest acts or statements, such as making up measurements, changing results, or taking credit for work that others have done.

**Think it Over**

7. **Think Critically** Why is recording data in a table important in a scientific process?

______________________

______________________

______________________

**Think it Over**

8. **Infer** How do you think fraud hurts science?

______________________

______________________

## After You Read

### Mini Glossary

**bias:** a person's own opinion

**ethics:** a system of understanding what is good and bad

**scientific law:** a rule that describes the behavior of something in nature

**scientific theory:** an explanation or model based on many tests or experiments

1. Review the terms and their definitions in the Mini Glossary. Circle one term. Then, in the space provided, explain what the term means in your own words.

_______________________________________________

_______________________________________________

2. Use the table below to review questions that science can and cannot answer. Place the letter of each of the questions from the list that follows in the correct column.

| Science Can Answer | Science Cannot Answer |
|---|---|
| | |
| | |
| | |

A. What are comets?
B. Should animals be used for testing?
C. Should people become vegetarians?
D. When will the hurricane reach land?
E. What do plants need to grow?
F. Should the government regulate hunting?

3. Before you read this chapter, you thought of a question you have about the world around you. Do you think science could answer this question? Why or why not?

_______________________________________________

_______________________________________________

Visit **earth.msscience.com** to access your textbook, interactive games, and projects to help you learn more about the changes in and limits of science.

# Matter

## section 1 Atoms

### Before You Read

Choose an object and think about the material it is made of. If you could break down that material into smaller and smaller units, what would the smallest possible particle be?

______________________________________________

______________________________________________

______________________________________________

**What You'll Learn**

- the different states of matter
- the internal structure of an atom
- about isotopes of an element

### Read to Learn

## The Building Blocks of Matter

What do the things you see, the air you breathe, and the food you eat have in common? They are matter. **Matter** is anything that has mass and takes up space. Heat and light are not matter because they have no mass and do not take up space. Look around the room. If all the objects you see are matter, why do they look so different from one another?

### What makes up matter?

Matter, in its many forms, is all around you. You can't see all matter as clearly as you can see water, a clear liquid, or a rock, a colorful solid. You can't see air because it contains colorless gases.

**Atoms** are the tiny particles that make up all matter. The forms or properties of one type of matter are different from those of another type of matter because of the atoms they contain. The structures of different atoms and the way they join together determine all the properties of matter that you can observe.

Study Coach

**Outline** Make an outline to organize the information in this chapter. The numbered headings should be the main ideas. Under each heading, list the details or examples that help explain the main idea.

FOLDABLES™

**A Organize Information** Use two sheets of paper to make a layered-book Foldable. Use it to describe atoms and the parts of an atom.

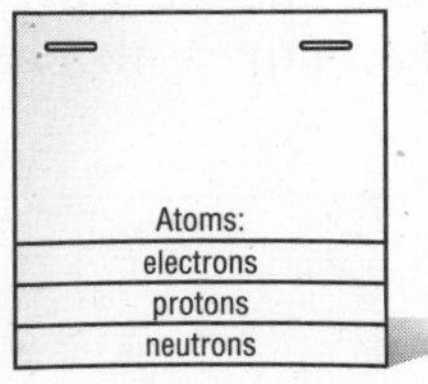

## How is matter joined together?

The figure below shows only two types of atoms, represented by two colors. Matter is joined together much like the blocks shown in the figure. The building blocks of matter are atoms. The types of atoms in matter and how they attach to each other give matter its properties. There are over 90 types of atoms, which may combine to form a huge variety of matter.

### Picture This

**1. Infer** How could this model help explain the variety of matter?

______________________

______________________

______________________

## What are elements?

When atoms combine, they form many different types of matter. Your body is made of many kinds of atoms combined in different ways. These atoms form the proteins, DNA, tissues, and other matter that make you the person you are. Most other objects that you see also are made of many different types of atoms. But some substances are made of only one type of atom. **Elements** are substances that are made of only one type of atom. Elements cannot be broken down into simpler substances by normal chemical or physical means. A table of elements, called the periodic table of the elements, contains a complete listing of all elements.

Elements combine to make a variety of items you depend on every day. They also combine to make up the minerals that compose Earth's crust. Some minerals, however, are made up of only one element. These minerals, which include copper and silver, are called native elements. Copper is often used to make wire. Silver is commonly used in tableware such as forks, knives, and spoons.

### Think it Over

**2. Apply** The element copper is used to make wire, and the element silver can be found in tableware. Name one other use for each of these elements.

______________________

______________________

# Modeling the Atom

How can you study things that are too small to be seen with the unaided eye? When something is too large or too small to observe directly, models can be used. A model also can describe tiny objects, such as atoms, that otherwise are difficult or impossible to see. ☑

## What is the history of the atomic model?

Over 2,000 years ago, the Greek philosopher Democritus (dih MAH kruh tuss) proposed that matter is made of small particles. He called these particles atoms and said that different types of matter were composed of different types of atoms. More than 2,000 year later, John Dalton expanded on these ideas. He thought that all atoms of an element contain the same type of atom.

## What are protons and neutrons?

In the early 1900s, the current model of the atom was developed. Three basic particles make up an atom—protons, neutrons (NOO trahnz), and electrons. **Protons** are particles that have a positive electric charge. **Neutrons** have no electric charge. Both particles are found in the nucleus—the center of an atom. With no negative charge to balance the positive charge of the protons, the charge of the nucleus is positive.

## What are electrons?

**Electrons** are particles with a negative charge that exist outside of the nucleus. In 1913, Niels Bohr, a Danish scientist, proposed that an atom's electrons travel in circles, or orbits, around the nucleus. He also thought that the electron's energy in an atom depends on its distance from the nucleus. Electrons that travel in paths closer to the nucleus have lower energy, and electrons that orbit farther from the nucleus have higher energy.

Today, as pictured in the figure to the right, scientists use a model that shows that electrons do not travel in orbitlike paths, but resemble a cloud surrounding the nucleus. Electrons can be anywhere within the cloud, but evidence suggests that they are located near the nucleus most of the time.

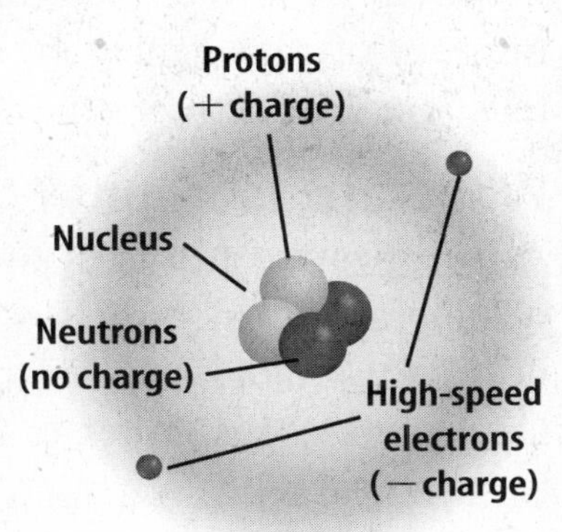


**Reading Check**

3. **Identify** What can you use to study something that is too small to see with the human eye?

_______________

**Picture This**

4. **Identify** Circle the protons and neutrons. Put an X next to each electron.

## Counting Atomic Particles

You now know where protons, neutrons, and electrons are located, but how many of each are in an atom? The number of protons in an atom depends on the element. All atoms of the same element have the same number of protons. For example, all iron atoms—whether in train tracks or breakfast cereal—contain 26 protons. All atoms with 26 protons are iron atoms. The number of protons in an atom is equal to the **atomic number** of the element. Elements in the periodic table are organized by their atomic number. This number can be found above the element symbol on the periodic table. As you go from left to right on the periodic table, the atomic number of the element increases by one.

### How many electrons are in an atom?

Some atoms are neutral, which means they don't have a positive or a negative charge. In a neutral atom, the number of protons is equal to the number of electrons. This makes the overall charge of the atom zero.

Atoms of an element can lose or gain electrons and still be the same element. When this happens, the atom is no longer neutral. Atoms with fewer electrons than protons have a positive charge. Atoms with more electrons than protons have a negative charge.

**Think it Over**

5. **Infer** If an atom is neutral, the atomic number is equal to the number of electrons. True or false?

_______________

### How many neutrons are in an atom?

Atoms of the same element always have the same number of protons, but they can have different numbers of neutrons. The number of neutrons in an atom isn't found on the periodic table. To figure out the number of neutrons in an atom, you need to know the atom's mass number. The **mass number** of an atom is equal to the number of protons plus the number of neutrons. To figure out the number of neutrons, subtract the number of protons (the atomic number) from the mass number. For example, if the mass number of nitrogen is 14, subtracting its atomic number, 7, tells you that nitrogen has 7 neutrons.

**Applying Math**

6. **Subtract** The atomic number of oxygen is 8. Its mass number is 16. How many neutrons does an atom of oxygen have?

_______________

### What are isotopes?

**Isotopes** are atoms of the same element that have different numbers of neutrons. Isotopes can be useful. For example, doctors use radioactive isotopes to treat certain types of cancer. Geologists use isotopes to date some rocks and fossils.

# After You Read

## Mini Glossary

**atom:** tiny building block of matter, made up of protons, neutrons, and electrons

**atomic number:** the number of protons in an atom

**electron:** particle with a negative charge

**element:** substance that is made up of only one type of atom

**isotopes:** atoms of the same element that have different number of neutrons

**mass number:** the number of protons plus the number of neutrons in an atom

**matter:** anything that has mass and takes up space

**neutron:** particle that has no electric charge

**proton:** particle that has a positive electric charge

1. Review the terms and their definitions in the Mini Glossary. Use the terms *electron, proton,* and *neutron* to describe the makeup of an atom.

_______________________________________________

_______________________________________________

_______________________________________________

2. Complete the concept map below to describe what you learned about atoms. Use the words *nucleus, electrons, cloud, neutrons,* and *positive.*

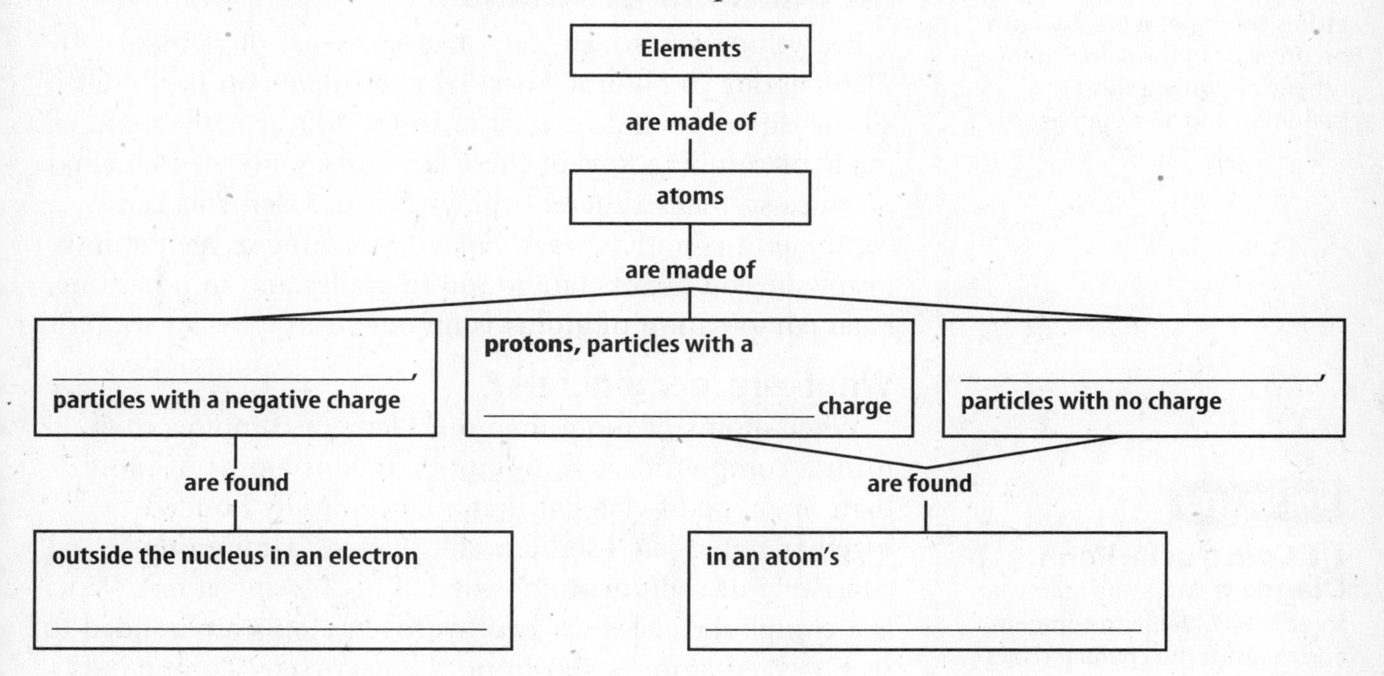


Science Online Visit earth.msscience.com to access your textbook, interactive games, and projects to help you learn more about atoms.

End of Section

# Matter

## section 2 Combinations of Atoms

### What You'll Learn

- ways atoms combine to form compounds
- the differences between compounds and mixtures

### Before You Read

On its own, the element chlorine is poisonous. But when combined with sodium, it forms table salt. Why do you think the properties of chlorine change this way?

_______________

_______________

_______________

Study Coach

**Make Flash Cards** to describe the different ways that atoms combine. Write the term on one side of the card. On the other side, write a short definition and an example.

### Read to Learn

#### Interactions of Atoms

Everything you touch, eat, or use is made of elements. There about 90 naturally occurring elements on Earth. Of all the different kinds of matter in the universe, most are made of combinations of these elements. Only 90 elements produce so many different things because elements can combine in countless ways. For example, the oxygen atoms in the air you breathe are found in apples and in limestone. Each combination of atoms is unique.

#### What are compounds?

When atoms of more than one element combine, they form a compound. A **compound** contains atoms of more than one type of element that are chemically bonded together. Table salt—sodium chloride—is a compound consisting of sodium atoms bonded to chlorine atoms. Water is a compound in which two hydrogen atoms are bonded to each oxygen atom as shown on the next page. Compounds are represented by chemical formulas. Every formula shows the ratios and types of atoms in the compound. For example, the chemical formula for water is $H_2O$.

FOLDABLES™

**B Construct a Venn Diagram** Make the following three-tab Foldable to compare and contrast the characteristics of compounds and mixtures.

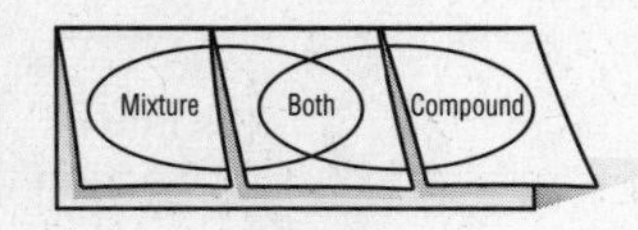

## Can compounds have different properties than the elements that form them?

The properties of compounds often are very different from the properties of the elements that form them. Sodium is a soft, silvery metal, and chlorine is a greenish, poisonous gas. Yet the compound they form is sodium chloride (NaCl), the table salt you use to season food. Under normal conditions on Earth, the hydrogen and oxygen that form water are gases. Water can be solid ice, liquid water, or gas, as shown in the figure above.

## What are chemical properties?

A property that describes a change that occurs when one substance reacts with another is called a chemical property. For example, one chemical property of water is that it changes to hydrogen gas and oxygen gas when an electric current passes through it. The chemical properties of a substance depend on what elements are in that substance and how they are arranged. Iron atoms in the mineral biotite react with water and oxygen to form iron oxide, or rust. Yet when it is mixed with chromium and nickel in stainless steel, iron resists rusting.

# Bonding

The forces that hold the atoms together in compounds are called chemical bonds. These bonds form when atoms share or exchange electrons. Remember that electrons are found in clouds outside the nucleus. Electrons are organized into energy levels. The highest energy levels are farther from the nucleus. Only the electrons with the highest energy can form bonds.

An atom is stable when it has eight electrons in its outer energy level. If an atom has exactly eight electrons in its outermost level it probably won't form bonds. If it has more than eight electrons, the extra electrons will form a new, higher energy level. An atom with fewer than eight electrons in its outermost level is unstable and is more likely to combine with other atoms. ☑

### Picture This

1. **Identify** Put an X on the oxygen atom and circle the hydrogen atoms in one water molecule in the figure.

### ✔ Reading Check

2. **Calculate** If an atom has six electrons in its outer shell, how many more electrons does it need to be stable?

______________________

## What is a covalent bond?

Atoms combine to form compounds in two ways. When atoms combine by sharing electrons in their outermost energy levels, they form a covalent bond. A **molecule** is a group of atoms connected by covalent bonds. In the figure below on the left, two atoms of hydrogen share electrons with one atom of oxygen to form a molecule of water. Each of the hydrogen atoms has one electron in its outermost level, and the oxygen atom has six electrons in its outermost level. This arrangement causes hydrogen and oxygen atoms to bond together. Each of the hydrogen atoms becomes stable by sharing one electron with the oxygen atom. The oxygen atom becomes stable by sharing two electrons with the two hydrogen atoms.

**Picture This**

**3. Interpret** Look at the diagram. How many atoms make up one molecule of water? How many electrons are in one molecule of water?

_______________

_______________



## What is an ionic bond?

Not all atoms combine by sharing electrons. Atoms also combine if they become positively or negatively charged. This type of bond is called an ionic bond.

Atoms can be neutral, or they can lose or gain electrons. When an atom loses electrons, it has more protons than electrons. This loss of electrons gives the atom a positive charge. When an atom gains electrons, it has more electrons than protons. This gain of electrons gives it a negative charge. **Ions** are atoms that are electrically charged. ☑

**Reading Check**

**4. Explain** What happens to an atom's charge when an atom has more protons than electrons?

_______________

_______________

Ions are attracted to each other when they have opposite charges. This is similar to the way magnets behave. Atoms with positive charges are attracted to atoms with negative charges. When ions join together, they create a compound that is electrically neutral. When the mineral halite forms, a sodium (Na) atom loses an outer electron and becomes an ion with a positive charge. If this positively charged ion gets close to a negatively charged chlorine (Cl) atom, they attract each other, as in the figure above. They bond to form the electronically neutral compound table salt.

**Metallic Bonds** Metallic bonds are found in metals such as copper, gold, aluminum, and silver. In this type of bond, electrons move freely from one positively charged ion to another. This movement of electrons, or conductivity, allows metals like copper to pass an electric current easily.

**Hydrogen Bonds** Hydrogen bonds can form without the interactions of electrons. A polar molecule, such as water shown below, has a positive end and a negative end. This happens because the atoms do not share electrons equally. When the positive end of one molecule is attracted to the negative end of another, a weak hydrogen bond forms, as shown by the water molecules below.

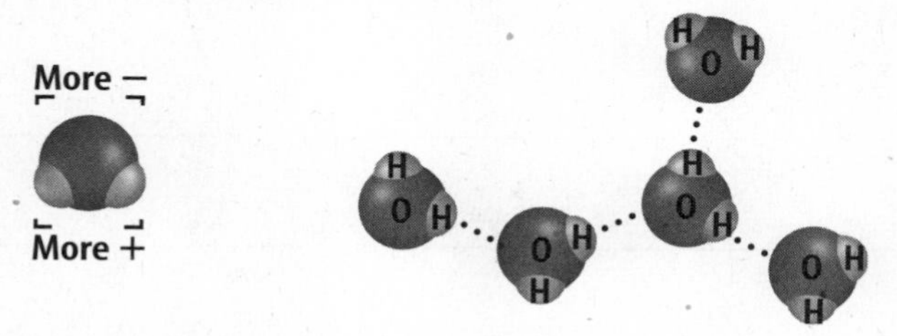


The hydrogen bonds that form between water molecules give water some of its unique properties. Cohesion is the attraction between water molecules that allows raindrops to form. Hydrogen bonds cause water to exist as a liquid at room temperature. As water freezes, hydrogen bonds force water molecules apart into a structure less dense than liquid water.

## Mixtures

A **mixture** is composed of two or more substances that are not chemically combined. There are two types of mixtures. The components of a **heterogeneous mixture** are not mixed evenly, and each component retains its own properties. The components of a **homogeneous mixture**, or a **solution**, are mixed evenly. The properties of the components of this type of mixture often are different from the properties of the mixture.

## Separating Mixtures and Compounds

The parts of a mixture can be separated physically. When the water evaporates from salt water, the salt remains. Parts of a compound must be separated chemically. Most compounds require several steps to separate them into their original compounds.

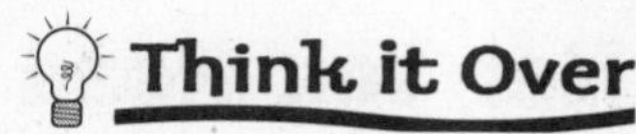


5. **Infer** What type of bond makes it possible for the wire in an extension cord to pass current?

______________________

### Picture This

6. **Identify** Put plus signs at the positive ends of the water molecules and minus signs at the negative ends of the water molecules.

# After You Read

## Mini Glossary

**compound:** atoms of more than one type of element that are chemically bonded together

**heterogeneous mixture:** a mixture that is not evenly mixed throughout and each component retains its own properties

**homogeneous mixture:** a mixture that is evenly mixed and the individual components cannot be seen

**ion:** electrically charged atom whose charge results from an atom losing or gaining electrons

**mixture:** composed of two or more substances that are not chemically bonded

**molecule:** a group of atoms bonded together by covalent bonds

**solution:** a homogeneous mixture

1. Review the terms and their definitions in the Mini Glossary. Are all mixtures solutions? Write one sentence to explain your answer.

_______________________________________________

_______________________________________________

2. Complete the concept chart using the terms *chlorine, compound, salt water, sodium, sodium chloride, solution, water.*

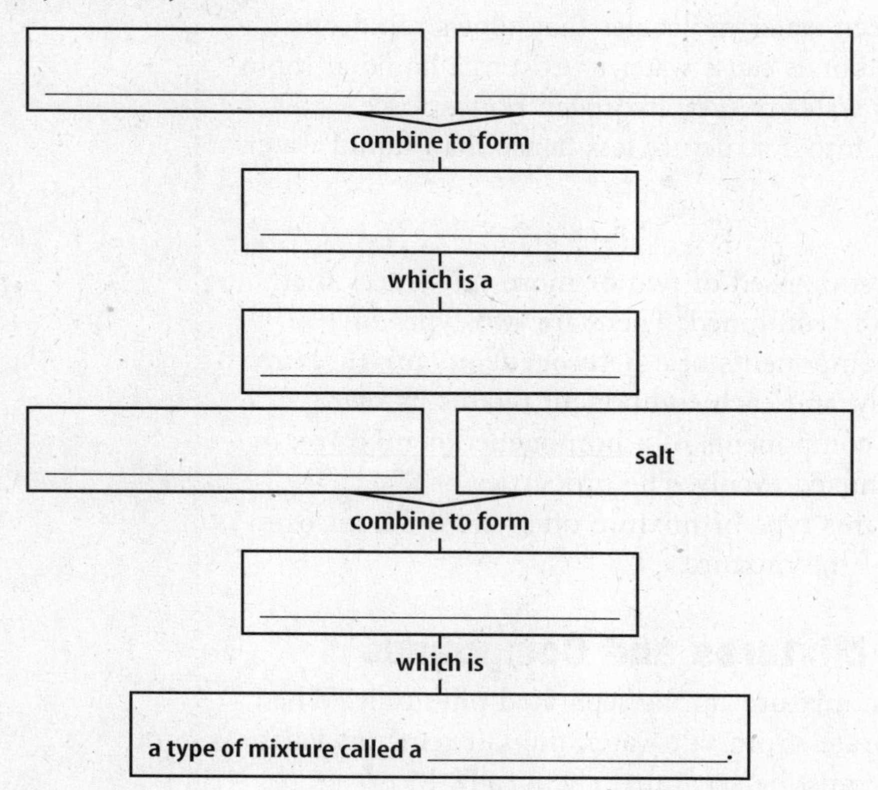

Science Online Visit **earth.msscience.com** to access your textbook, interactive games, and projects to help you learn more about combinations of atoms.

# Matter

## section 3 Properties of Matter

## Before You Read

Water can exist in three states of matter. Name them.

_______________________________________________

_______________________________________________

### What You'll Learn

- the physical properties of matter
- what causes matter to change state
- the four states of matter

## Read to Learn

### Physical Properties of Matter

In addition to the chemical properties of matter that you have already studied, matter has other properties that can be described. You might describe a pair of jeans as blue, soft, and about 80 cm long. This description can be made without changing the jeans in any way. The properties that you can observe without changing a substance into a new substance are physical properties.

#### What is density?

One physical property used to describe matter is density. **Density** is a measure of the mass of an object divided by its volume. You can think of density as the amount of matter in an object divided by the space the object takes up. Generally, this measurement is given in grams per cubic centimeter ($g/cm^3$). For example, the average density of liquid water is about 1 $g/cm^3$. So, 1 $cm^3$ of pure water has a mass of about 1 g.

An object that is more dense than water will sink in water. On the other hand, an object that is less dense than water will float. When oil spills occur on the ocean, the oil floats on the surface of the water and washes up on beaches. Because oil is less dense than the water and floats, even a small spill can spread out and cover large areas.

Study Coach

**Mark the Text** As you read, underline the answers to the question in each heading to help you focus on the main ideas.

FOLDABLES™

**C Organize Information** Make two half-sheet Foldables to take notes on the properties and states of matter.

Properties of Matter:

States of Matter:

## States of Matter

On Earth, matter occurs in four physical states. These four states—solid, liquid, gas, and plasma—are shown in the figure below. You might have had solid toast and liquid juice for breakfast this morning. You breathe air, which is a gas. A lightning bolt during a storm is an example of matter in its plasma state.

Solid Liquid Gas Plasma

### Picture This

1. **Identify** Name the four physical states of matter found on Earth.

_______________

_______________

_______________

_______________

**Solids** Matter is solid because its particles are in fixed positions relative to each other. The individual particles may vibrate, but they don't switch positions. Solids have a definite shape and take up a definite volume. A completely assembled puzzle can be a model for a solid. The pieces can move a little, but they cannot change positions.

**Liquids** Particles in a liquid are attracted to each other but are not in fixed positions. This is because liquid particles have more energy than solid particles. This energy allows them to move around and change positions with each other. Particles in a liquid fit the shape of the container they are in. You can pour a liquid into any container and it will flow until it takes the shape of the container.

**Gases** Particles that make up gases have even more energy than particles in solids or liquids. This energy can break any attractions between the particles allowing them to move freely and independently. Unlike solids and liquids, gases spread out evenly to fill the entire container in which they are placed.

### Think it Over

2. **Compare** In which state of matter do the particles have the *least* amount of energy?

_______________

**Plasma** Although plasma is probably unfamiliar to most people, it is the most common state of matter in the universe. This state is associated with high temperatures. Stars like the Sun are composed of matter in the plasma state. On Earth, plasma is found in lightning bolts. Plasma is made of ions and electrons. It forms when high temperatures cause some of the electrons to escape from an atom's electron cloud. These particles then move freely outside of the electron cloud.

## Changing the State of Matter

One way that matter changes state is through changes in temperature. Water begins to change from a liquid to a solid at 0°C. The temperature at which matter changes from a liquid to a solid is called the freezing point. Matter changes from a liquid to a gas at its boiling point. The boiling point for water is 100°C. Water is the only substance that occurs naturally on Earth as a solid, a liquid, and a gas.

The attraction between particles of a substance and their rate of movement are factors that determine the substance's state of matter. When thermal energy is added to a substance, the molecules move faster. This movement can cause the matter to change form. Adding thermal energy to ice causes the molecules to move more freely and the ice melts. ☑

Matter also changes state because of changes in pressure. Decreasing pressure lowers the boiling point of liquids. Solids tend to melt at lower temperatures when pressure is increased.

**Reading Check**

3. **Predict** What happens to the water molecules in ice when thermal energy is added?

______________________

______________________

## Changes in Physical Properties

Chemical properties of matter do not change when the matter changes state, but some of its physical properties change. The density of water changes as water changes state. Ice floats in water because it is less dense than liquid water. However, most materials are denser in their solid state than in their liquid state. Some physical properties of substances don't change when they change state. For example, water is colorless and transparent in each of its states.

## Matter on Mars

Matter in one state often can be changed to another state by adding or removing thermal energy. Today, there is no or very little water on Mars. What could explain the huge water-carved channels that formed on Mars long ago? Mars's liquid water might have changed to the solid state long ago when the planet cooled. Scientists believe that some of the water soaked into the ground and froze, forming permafrost. Some water might have frozen to form Mars' polar ice caps. Even more water might have evaporated and escaped to space.

**Think it Over**

4. **Explain** How did the huge channels that have been observed on Mars form?

______________________

______________________

______________________

# After You Read

## Mini Glossary

**density:** measurement of the mass of an object divided by its volume

1. Review the term and its definition in the Mini Glossary. Use the term *density* in a sentence to explain why ice floats in water.

_______________________________________________

_______________________________________________

2. Complete the table below to describe the different states of matter.

| STATE | DESCRIPTION | EXAMPLE |
|---|---|---|
| | Particles are packed tightly and don't change positions. | |
| Liquid | Particles have more ____________ than solid particles.<br>Particles move around and change positions.<br>Particles fit the ____________ of a container. | |
| | Particles have more ____________ than liquid particles.<br>Particles move freely and spread out to fill a container. | Air |
| Plasma | Associated with ____________ temperatures<br>Composed of ions and ____________ that escape the electron cloud. | |

3. Before you read this section, you marked key parts of the text. How did this help you understand the information in this section?

_______________________________________________

_______________________________________________

Science Online Visit **earth.msscience.com** to access your textbook, interactive games, and projects to help you learn more about properties of matter.

# Minerals

## section 1 Minerals

### Before You Read

Think about a diamond that you have seen. Describe what the diamond looked like on the lines below.

_______________________________________________

_______________________________________________

**What You'll Learn**

- what all minerals have in common
- how minerals form

### Read to Learn

#### What is a mineral?

You use minerals every day. In fact, minerals are all around you. The diamond you wrote about is a mineral. The glass in your windows is made from a mineral. The pencil lead, or graphite, you write with is also a mineral.

A **mineral** is a solid, non-living substance that is found in nature. A mineral is made up of atoms that are arranged in a certain, set way.

**Study Coach**

**Two-Column Notes** As you read, organize your notes in two columns. In the left-hand column, write the main idea of each paragraph. Next to it, in the right-hand column, write details about it.

#### How are minerals alike?

There are about 4,000 different minerals on Earth. All minerals have four things in common. First, all minerals are formed by natural processes. These processes occur on or inside the Earth with no help from humans. For example, diamonds are formed by the pressure deep within Earth. Second, minerals are inorganic, or not living. They are made from non-living substances, like the material that makes up a diamond. Third, minerals are made up of one element or a combination of elements. The finest diamonds, for example, are made up of one element—carbon. The mineral halite is made up of two elements—sodium and chlorine. Fourth, all minerals are solids. A solid keeps a set shape. Minerals are a special type of solid because their atoms are arranged in a specific way.

**FOLDABLES™**

**A Identify** Make a four tab Foldable to identify the four characteristics of minerals.

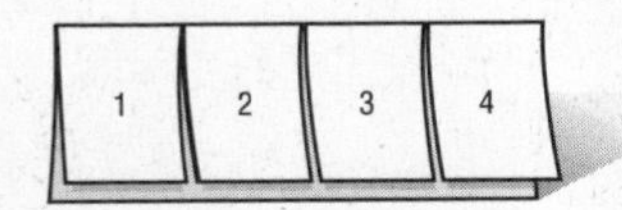

## Picture This

**1. Draw** In the box below, draw at least 10 dots that show how the atoms in a crystal might be arranged. Each dot is one atom.

**Reading Check**

**2. Determine** How are the atoms arranged in a crystal?

______________________

______________________

## How are atoms in minerals arranged?

Atoms in a mineral are arranged in an orderly pattern that repeats itself over and over again. This repeated pattern of atoms is called a crystalline pattern. For example, the atoms in the mineral graphite are arranged in layers. All true minerals are crystalline solids. But not all solids are minerals. Opal, for example, is not a true mineral because its atoms are not arranged in a definite repeating pattern. In the first box below, the atoms are scattered. In the second box the atoms are arranged in a crystalline pattern.

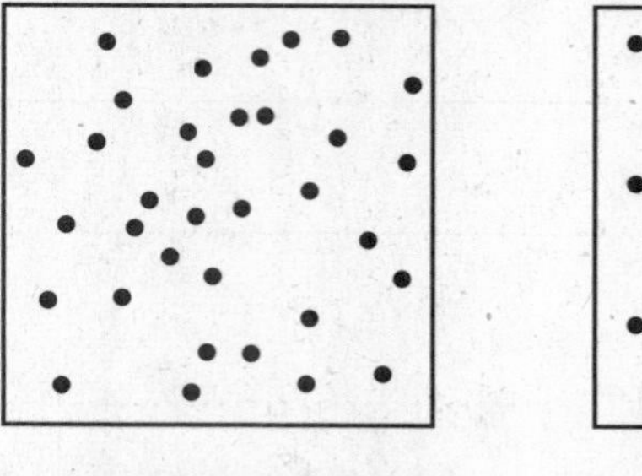

Scattered Pattern

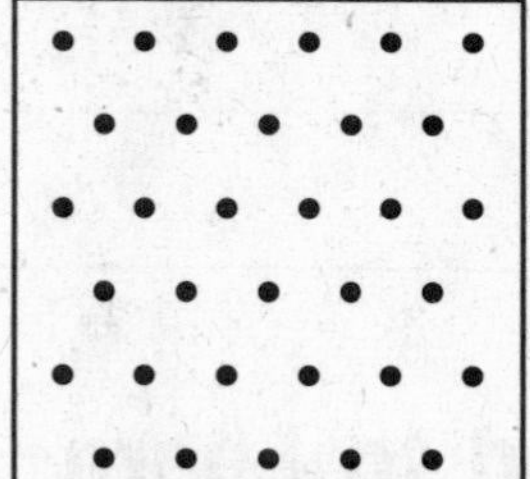

Crystalline Pattern

# The Structure of Minerals

All minerals are crystals. A **crystal** is a solid that has atoms arranged in an orderly, repeating pattern. For example, a diamond is a crystal. A diamond has many flat sides, or faces. The flat faces are evidence for the orderly arrangement of atoms in the diamond. ☑

## What affects the shape of crystals?

Like diamonds, the crystalline form of quartz can also be seen. The clear quartz crystal shown in the figure on the left on the next page has atoms arranged in flat layers. The clear quartz crystals formed in an open space and their crystalline form can be seen easily.

Sometimes crystal shapes are not seen so easily. The second figure shows a block of rose quartz. This rose quartz looks uneven on the outside. Yet it is a mineral with atoms arranged in a repeating pattern. The rose quartz crystals look this way because they formed in a tight space.

Crystals form in many ways. There are two ways in which crystals form most often from magma and from solution.

**Clear quartz**

______________________

**Rose quartz**

______________________

## Picture This

3. **Identify** Under each piece of quartz, write *tight space* or *open space* to show where each mineral formed.

## How do crystals form from magma?

**Magma** is the melted rock inside Earth. When magma rises to Earth's surface, it cools. As the magma cools, the atoms in the magma start to move closer together. The atoms begin to combine to form compounds. As they combine, some of the atoms of different compounds form orderly, repeating patterns. When they are completely cooled, these atoms have formed crystals. ☑

When magma cools slowly, large crystals form. The crystal form can be seen easily in these minerals. When magma cools quickly, smaller crystals form. These small mineral crystals are much harder to see.

Magma is made of different elements. The type and amount of the elements that make up magma partly determine which minerals will form.

**Reading Check**

4. **Define** What is magma?

______________________

______________________

## How do crystals form from solution?

When minerals dissolve in water, the result is called a solution. In a solution, the mineral's atoms are spread evenly through the water. As the water evaporates, ions come together to form crystals. For example, if you place a bowl of very salty water in the sun, the water soon evaporates. When the water has evaporated, salt crystals are left at the bottom of the bowl.

Crystals also may form when there is too much of a mineral in a solution. Because there is so much of the mineral, the ions come together and begin to form crystals in the solution. Minerals can form from a solution in this way without the need for evaporation.

## Think it Over

5. **Explain** What is one of the ways that crystals form from solution?

______________________

______________________

______________________

**Applying Math**

**6. Calculate** Look at the graph. Elements other than silicon and oxygen make up about what percent of Earth's crust? Show your work.

______________________

______________________

______________________

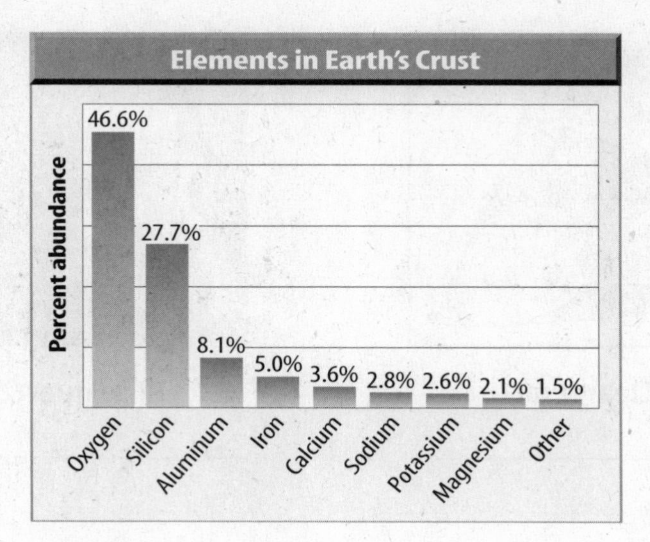


## Mineral Compositions and Groups

There are ninety elements that make up Earth's crust. About 98 percent of the crust is made up of only eight of these elements. These common elements are shown above.

There are thousands of known minerals. However, only a few dozen minerals are common. These common minerals are made up mostly of the eight most common elements in Earth's crust.

### How are minerals grouped?

Minerals are often grouped by the elements they are made of. Silicon and oxygen are the two most abundant elements in Earth's crust. They form the basic building blocks of most minerals.

The most common rock-forming minerals in Earth's crust are called silicates. A **silicate** is a mineral made up of the element silicon, the element oxygen, and usually one or more other elements. For example, quartz, a silicate, is a common rock-forming mineral in Earth's crust made of silicon and oxygen. Feldspar, another silicate, is made of silicon, oxygen, aluminum, and either potassium, sodium, calcium, or barium. ☑

**Reading Check**

**7. Determine** What two elements always make up silicates?

______________________

______________________

Another group of rock-forming minerals is the carbonates. Carbonates contain carbon and oxygen. They can include other elements. For example, calcite is a carbonate made of carbon, oxygen, and calcium. Dolomite is a carbonate made of carbon, oxygen, magnesium, and calcium. Other mineral groups are also defined according to the elements from which they are made.

# After You Read

## Mini Glossary

**crystal:** solid in which the atoms are arranged in an orderly, repeating pattern

**magma:** hot, melted rock material beneath Earth's surface

**mineral:** solid material made in Earth, whose atoms are arranged in a set, repeating pattern

**silicate:** mineral that contains silicon and oxygen and usually one or more other elements

1. Review the terms and their definitions in the Mini Glossary. Then write a sentence that explains why all minerals are crystals. Use at least two terms in your sentence.

_______________________________________________

_______________________________________________

2. Fill in the boxes to tell how some crystals form.

| ______________________ **cools** |
|---|
| ______________________ **move closer together** |
| **compounds form** ______________________ |

3. You organized your notes in two columns, one for main ideas and one for details. How did this strategy help you learn the information?

_______________________________________________

_______________________________________________

_______________________________________________

Visit **earth.msscience.com** to access your textbook, interactive games, and projects to help you learn more about minerals.

# Minerals

## section 2 Mineral Identification

**What You'll Learn**
- the physical properties of minerals
- how to use physical properties to identify minerals

### Before You Read

Someone hands you a gold ring and a lump of rock salt. Both gold and rock salt are minerals. On the lines below, describe the differences between the gold ring and rock salt.

______________________________________________

______________________________________________

______________________________________________

**Mark the Text**

**Highlight Key Terms** Highlight the key terms and their meanings as you read this section.

### Read to Learn

## Physical Properties

You can tell one person from another because people look different. People or things look different because they have different physical properties. Height, hair color, eye color, and face shape are some of a person's physical properties. Minerals have physical properties, too. You can identify minerals by their different physical properties.

### What do different minerals look like?

Different minerals may look different in several ways. Some minerals may have smooth surfaces, while others may be rough. Different minerals may have different colors. A mineral's appearance is one way to begin to identify a mineral.

Looking only at a mineral's color can trick you. For example, gold is gold colored. Gold is valuable. Pyrite is a mineral that also has a bright gold color but has little value. In fact, pyrite is often called fool's gold.

To identify a mineral correctly, you need to look at other physical properties of the mineral. Some other physical properties to study are a mineral's hardness, how a mineral breaks, and its color when it is crushed into a powder.

**FOLDABLES™**

**B Identification** Make a Foldable from 3 half sheets of notebook paper. Use it to list and explain the physical properties of minerals.

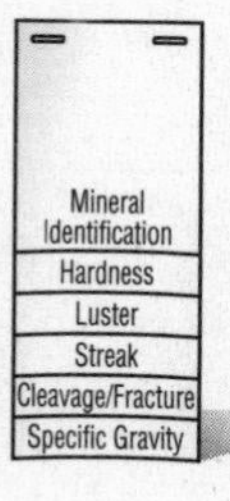

## How is the hardness of a mineral measured?

A mineral's **hardness** is a measure of how easily the mineral can be scratched. Diamond is the hardest mineral known. A diamond can be scratched only by another diamond. Diamonds are so hard that they are used to cut through other materials. They are used often as the blades of saws that cut through rock.

A mineral's hardness has nothing to do with how easily a mineral will break. Some minerals shatter easily, yet they still may be very hard.

## What is Mohs scale of hardness?

The Mohs scale compares the hardness of different minerals. In 1824, a scientist named Friedrich Mohs tested the hardnesses of many minerals. He then listed the minerals in order. Talc, the softest mineral, was given the number 1. Diamond, the hardest mineral, was given the number 10. The hardness of each mineral was compared with the hardnesses of talc and diamond. Each mineral was given a number. The higher its number, the harder a mineral is. ☑

**Mineral Hardness**

| Mohs Scale | Hardness | Hardness of Common Objects |
|---|---|---|
| Talc (softest) | 1 | |
| Gypsum | 2 | fingernail (2.5) |
| Calcite | 3 | piece of copper (2.5 to 3.0) |
| Fluorite | 4 | iron nail (4.5) |
| Apatite | 5 | glass (5.5) |
| Feldspar | 6 | steel file (6.5) |
| Quartz | 7 | streak plate (7.0) |
| Topaz | 8 | |
| Corundum | 9 | |
| Diamond (hardest) | 10 | |

The table above will help you understand the Mohs scale of hardness. The table shows that topaz has a hardness of 8. Quartz has a hardness of 7. Topaz is harder than quartz. That means topaz can scratch quartz, but quartz cannot scratch topaz. The table also lists the hardnesses of some common objects.

**Reading Check**

1. **Apply** What does the Mohs scale measure?

______________________

**Picture This**

2. **Interpret Data** Look at the table on this page. List three minerals that feldspar can scratch.

______________________

______________________

______________________

## How is the Mohs scale used?

Here's how the Mohs scale helps identify minerals. Suppose you are given a white mineral to identify. You know it is either fluorite or quartz. Your fingernail does not scratch it. But, you can scratch it with an iron nail. So the unknown mineral must have a hardness between 2.5 (your fingernail) and 4.5 (an iron nail). Now look at the Mohs scale. Because quartz has a hardness of 7 and fluorite has a hardness of 4, the unknown mineral must be fluorite.

## What is luster?

An important physical property that is used to identify minerals is called luster. **Luster** is the way a mineral reflects light. There are two types of luster. A mineral that shines like a bright piece of metal has a metallic luster. The mineral graphite has a metallic luster. ☑

**Reading Check**

3. **Explain** Why does graphite have a metallic luster?

______________________

______________________

A mineral that does not shine like metal has a nonmetallic luster. A nonmetallic luster may appear to be glassy, pearly, or dull. The mineral fluorite has nonmetallic luster.

## What is gravity?

Picture a one-inch cube of lead and a one-inch cube of wood. Both cubes are the same size. Yet the cube of lead weighs more than the cube of wood. Lead has a greater specific gravity than wood. The **specific gravity** of a mineral is the ratio of its weight compared with the weight of an equal volume of water. A mineral's specific gravity is given as a number. ☑

**Reading Check**

4. **Identify** Specific gravity is the weight of a mineral compared to the weight of an equal volume of what?

______________________

Specific gravity can be used to tell the difference between minerals that look similar. Gold and pyrite look alike. Gold has a specific gravity of 19. That means that gold is 19 times heavier than water. Pyrite has a specific gravity of 5. It is 5 times heavier than water. If you held equal-sized cubes of gold and pyrite, the gold would feel much heavier than the pyrite. The term *heft* can be used to describe this difference in weight. Gold has greater heft than pyrite.

## How is streak used to identify minerals?

A mineral's **streak** is the color of the mineral in a powdered form. To find a mineral's streak, rub the mineral across a white, unglazed porcelain tile, and look at the streak of powered mineral left behind. For example, gold has a yellow streak and pyrite has a greenish-black streak.

## When can a streak test be used?

A streak test works only when the mineral being tested is softer than the porcelain tile. A softer mineral will leave a streak. A harder mineral will leave a scratch, not a streak. A streak test cannot be used to identify a diamond because the diamond is much harder than the tile.

## How do minerals break?

Different minerals break apart in different ways. How a mineral breaks is another way to identify it. **Cleavage** is the physical property of minerals that break along smooth, flat surfaces. The way atoms are arranged in the minerals partly determines cleavage. The mica shown in the figure below breaks along flat cleavage surfaces. If you took a layer cake and separated its layers, you would show that the cake has cleavage. ☑

Not all minerals have cleavage. **Fracture** is the property that causes minerals to break with rough, jagged surfaces. Quartz, a mineral with fracture, is shown in the figure below. If you grabbed a chunk out of the side of that cake, it would be like breaking a mineral that has a fracture.

**Reading Check**

**5. Observe** How does a mineral with cleavage look when it breaks apart?

______________________

______________________

**Mica**

**Quartz**

**Picture This**

**6. Identify** Look at the figure. Circle the mineral with fracture.

## What other physical properties are used to identify minerals?

Some minerals have interesting physical properties. Magnetite is a mineral that acts like a magnet. Another mineral, calcite, affects light. When light passes through calcite, two separate light rays form. Calcite also fizzes when hydrochloric acid is placed on it.

Remember, a mineral's appearance and color may not be enough information to identify it. Minerals have other physical properties to help identify them. You may need to test hardness, luster, specific gravity, streak, and cleavage or fracture.

## After You Read

### Mini Glossary

**cleavage:** physical property of some minerals that causes them to break along smooth, flat surfaces

**fracture:** physical property of some minerals that causes them to break with uneven, rough, or jagged surfaces

**hardness:** measure of how easily a mineral can be scratched

**luster:** how a mineral reflects light from its surface

**specific gravity:** ratio of a mineral's weight compared with the weight of an equal volume of water

**streak:** color of a mineral when it is in powdered form

**1.** Review the terms and definitions in the Mini Glossary. Think about a time you saw a glass break. Does glass break with cleavage or with fracture? Explain your answer.

______________________________________________

______________________________________________

**2.** Use the table below to identify three "mystery" minerals that have the following physical properties:

**Properties of Minerals**

| Mineral | Hardness | Streak |
|---|---|---|
| Copper | 2.5–3 | copper-red |
| Galena | 2.5 | dark gray |
| Gold | 2.5–3 | yellow |
| Hematite | 5.5–6.5 | red to brown |
| Magnetite | 6–6.5 | black |
| Silver | 2.5–3 | silver-white |

**1.**
- mineral has a hardness of 2.5
- mineral has dark gray streak

Mineral is ____________________.

**2.**
- mineral has hardness of 6
- mineral has reddish brown streak

Mineral is ____________________.

**3.**
- mineral has hardness of 3
- mineral has silver streak

Mineral is ____________________.

**3.** This section contained several new vocabulary words. You highlighted the key terms and their meanings. Describe another strategy to remember definitions of new words.

______________________________________________

______________________________________________

Visit **earth.msscience.com** to access your textbook, interactive games, and projects to help you learn more about the physical properties of minerals.

# Minerals

## section 3 Uses of Minerals

### Before You Read

What do you think of when you hear the word *metal?* On the lines below describe things made from metal.

______________________________________________

______________________________________________

______________________________________________

**What You'll Learn**

- how gems are used
- what minerals contain useful elements

### Read to Learn

#### Gems

Gems are often used in jewelry. A **gem** is a valuable mineral that is both rare and beautiful. When it is cut and polished, a gem has a beautiful color and shine. Most gems are special types of common minerals. They are clearer, brighter, or more colorful than common samples of that mineral.

**Study Coach**

**Think-Pair-Share** Work with a partner. As you read the text, discuss what you already know about the topic and what you learn from the text.

#### Why are some gems important?

All gems are valuable. Sometimes gems are valuable because of their size. A carat is a unit used to measure the weight of gems. In 1905, a huge diamond was found in South Africa. The diamond, called the Cullinan diamond, was the largest ever discovered. It weighed 3,106.75 carats. That is equal to 621 grams.

Sometimes gems are valuable because of their unusual color. Most diamonds are colorless. The Hope diamond, the world's most famous diamond, has a slight blue color. In 1830, Henry Hope bought the diamond, which was then named after him. Today, the 45.52 carat (9 gram) Hope diamond is on display at the Smithsonian Institution in Washington, D.C.

**FOLDABLES™**

**C Organize Information** Use two quarter sheets of notebook paper. On one write about gems. On the other write about useful elements in minerals.

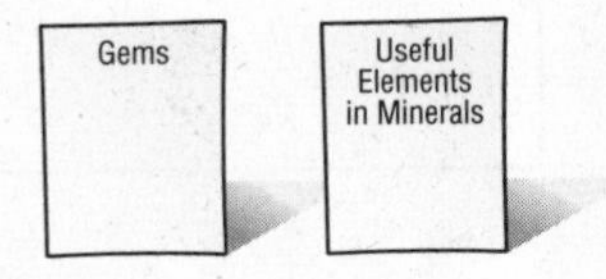

1. **Infer** What physical property of diamonds makes them useful for cutting stone and metal?

___________________

### Which gems are useful?

Some gems are as useful as they are beautiful. Diamonds are the hardest minerals which makes them very useful. In industry, diamonds are used on blades to cut through stone or metal. Today, most diamonds used in industry are artificial, or synthetic. They are made by humans. Scientists have studied the structure of natural diamonds and copied it to make synthetic diamonds that are just as hard.

Rubies are used to produce laser light. Laser light helps with many different tasks. It sends television signals, guides machinery, and even helps doctors in the operating room.

Quartz crystals are used in electronics and in clocks and watches. When an electric field comes in contact with quartz crystals, the quartz vibrates steadily. This vibration helps control electronic devices and keeps timepieces accurate.

## Useful Elements in Minerals

Gemstones are perhaps the best-known use of minerals, but they are not the most important. Look around your home. How many things made from minerals can you name? Do you see anything made from iron?

**Reading Check**

2. **Identify** Iron comes from what ore?

___________________

### What is an ore?

You probably have many things that are made out of iron. Frying pans, nails, and other objects may contain iron. Iron is obtained from its ore, hematite. An **ore** is a mineral or rock that contains a useful substance that is mined at a profit. ☑

**Picture This**

3. **Apply** another item made from aluminum.

Aluminum comes from the ore bauxite. Before it can be used, the aluminum in bauxite must be refined. In the refining process, aluminum oxide powder is separated from unwanted materials. The aluminum oxide powder is changed to aluminum by a process called smelting. Aluminum can be made into soft drink cans, bikes, cars, airplanes, and many other useful things. Some items made from aluminum are shown in the figure above.

## What are vein minerals?

Under some conditions, metallic elements can dissolve in liquids. The liquid flows through weaknesses in rocks and forms mineral deposits. Weaknesses in rocks include natural fractures or cracks, faults, and surfaces between layered rock formations. Mineral deposits left behind that fill in the open spaces created by the weaknesses are called vein mineral deposits. ☑

Sometimes vein mineral deposits fill in the empty spaces after rocks collapse. Sphalerite is a mineral that sometimes fills in spaces in collapsed limestone. Sphalerite is a source of the element zinc, which is used in batteries.

**Reading Check**

4. **Describe** Where do vein mineral deposits form?

______

______

______

## Which minerals contain titanium?

Titanium is a metallic element. It comes from minerals that contain this metal in their crystal structures. Two minerals, ilmenite (IHL muh nite) and rutile (rew TEEL), are sources of titanium. Ilmenite and rutile are found in rocks that form when magma cools and solidifies.

## How is titanium used?

Titanium is used in many products, like the ones shown below, because it is lightweight, strong, and long lasting. It is used to make some parts of automobiles and aircraft, frames for eyeglasses, wheelchair parts, bicycles, golf clubs, and tennis rackets. It is used in hip or knee replacements. Titanium is a useful element that improves the lives of many humans.

**Think it Over**

5. **Infer** Why might a tennis player use a racket made of titanium instead of one made of wood?

______

______

______

**Products Made from Titanium**

## After You Read

### Mini Glossary

**gem:** valuable mineral that is both rare and beautiful, often used in jewelry

**ore:** mineral or rock that contains a useful substance that can be mined for profit

1. Review the terms and their definitions in the Mini Glossary. Why do people value gemstones?

_______________________________________________

_______________________________________________

2. Use the Venn diagram below to show how gems and ores are the same and how they are different. In the outside circles, write how each is different. In the center, write how they are both the same.

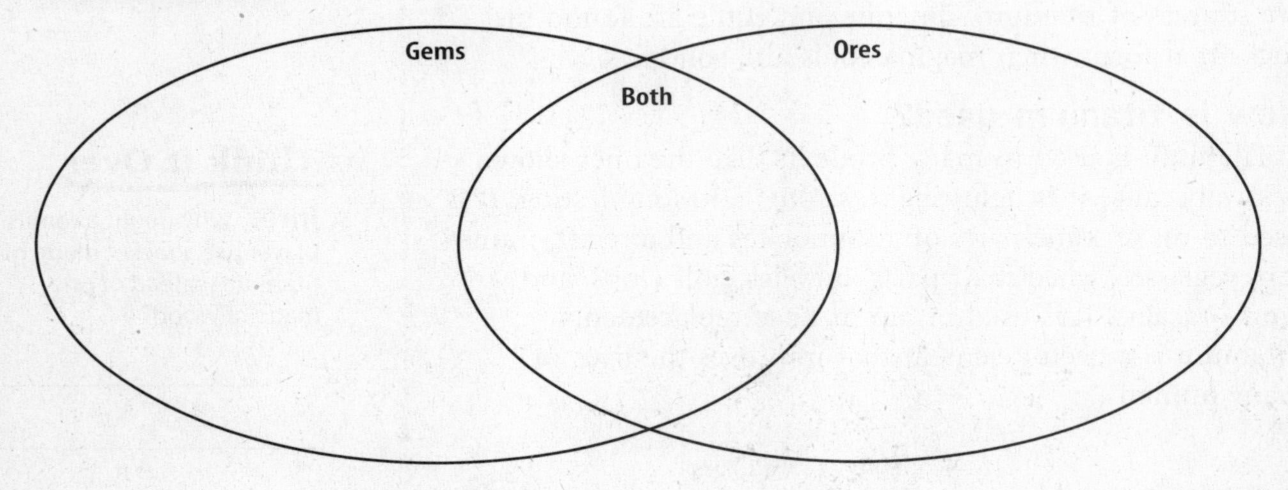


3. You and your partner talked about gems and ores. How did this help you understand what you read?

_______________________________________________

_______________________________________________

_______________________________________________

Science Online Visit **earth.msscience.com** to access your textbook, interactive games, and projects to help you learn more about uses of minerals.

# Rocks

## section 1 The Rock Cycle

## Before You Read

Think about all the different rocks you have seen. Some may have been shiny, others dull. Describe how the rocks you have seen are different.

______________________________________________

______________________________________________

______________________________________________

**What You'll Learn**
- what the rock cycle is
- how rocks change

## Read to Learn

### What is a rock?

Different rocks have different characteristics. Some are smooth, some are rough, some are striped, and some are spotted.

### What are common rocks?

Most buildings and public monuments are made from common rock. Rock used for building stone often contains one or more common materials, called rock-forming minerals. Two rock-forming minerals are quartz and calcite. A **rock** is a mixture of rock-forming minerals and other materials such as volcanic glass, organic material, or other natural materials.

### The Rock Cycle

Scientists have created a model to show how rocks slowly change over time. The **rock cycle** shows the processes that create and change rocks. The three types of rocks shown in the rock cycle are igneous, metamorphic, and sedimentary. The rock cycle shows how rocks can change from one type of rock to another.

Mark the Text

**Underline** As you read, underline key words and definitions. Underline ideas and explanations that help you understand the text.

FOLDABLES™

**A Draw and Label** Make a Foldable from two half sheets of notebook paper to list facts about rocks and to describe the rock cycle.

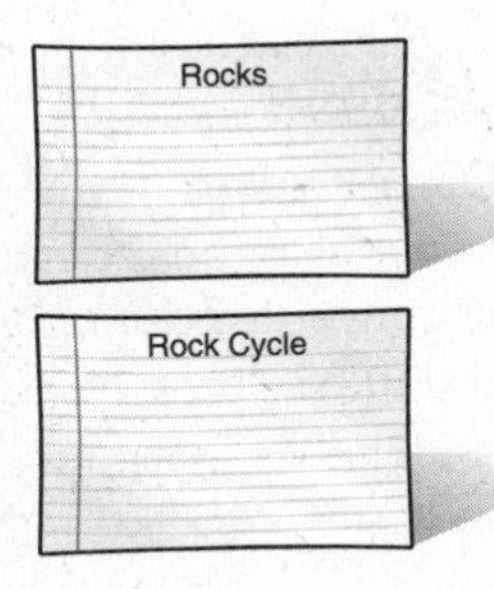

## Picture This

**1. Identify** Use colored pencils to add information to the rock cycle. Color arrows that involve heat red, weathering and erosion green, cooling blue, and compaction brown.

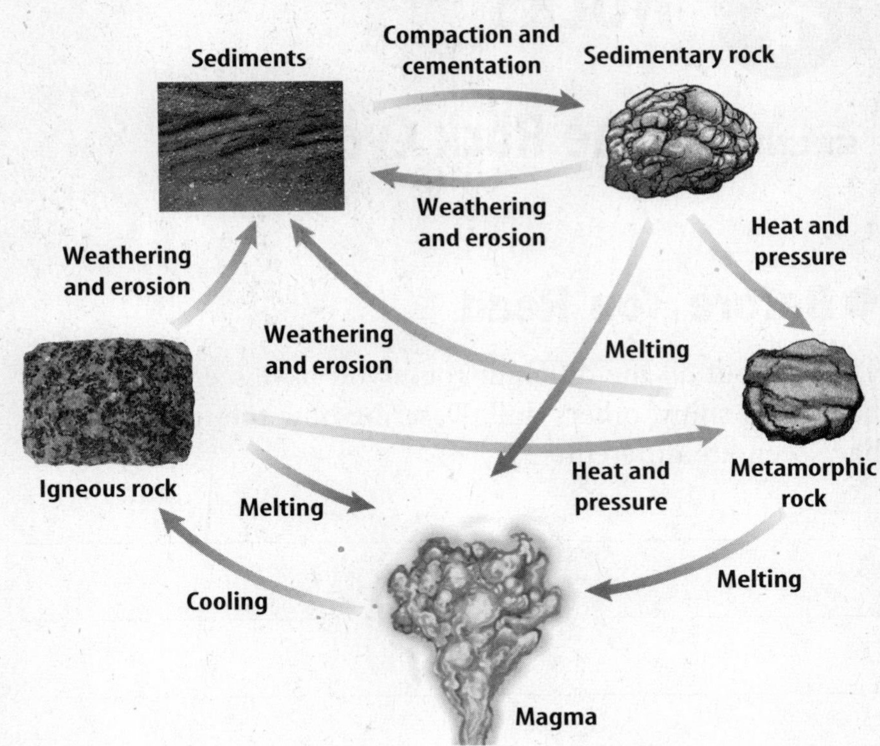


## How do rocks change?

The rock cycle in the figure above shows there are several processes that change rocks. Weathering breaks down rocks into tiny mineral grains, or sediments. Erosion moves the sediments by wind or water. Layers of sediments pile up. They are compacted, or packed down, by more layers of sediment piling on top of them. Over time, the pressure of compaction turns the sediment into sedimentary rock.

Heat and pressure deep inside Earth may change sedimentary rock into metamorphic rock. The metamorphic rock can then melt and later cool to form igneous rock. The igneous rock may then be weathered into mineral grains. The grains eventually form new sedimentary rock. Any rock can change into any of the three major types of rock. A rock can even change into another rock of the same type. No matter what happens, the mineral material is never lost or destroyed. It is conserved, or used in other forms. ☑

**2. Determine** Is mineral material in rocks destroyed or conserved during the changes?

_______________

## Who discovered the rock cycle?

Scottish scientist James Hutton noticed that some rocks have straight layers, while others are tilted. He saw that some rocks are weathered, while others are not. Hutton observed that rocks change constantly over time.

## After You Read

### Mini Glossary

**rock:** mixture of rock-forming minerals and other materials

**rock cycle:** model that shows how rocks slowly change over time

1. Review the terms and their definitions in the Mini Glossary. Then write a sentence about the rock cycle that explains how rocks change.

_______________________________________________

_______________________________________________

_______________________________________________

2. Fill in the blanks in the boxes below.

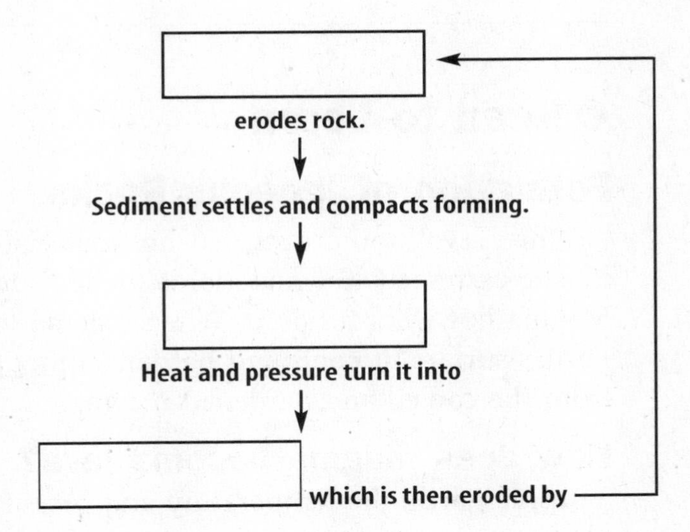


3. You underlined the main words, facts, and ideas in this section. How did underlining help you learn about and remember the different types of rocks?

_______________________________________________

_______________________________________________

**Science Online** Visit **earth.msscience.com** to access your textbook, interactive games, and projects to help you learn more about the rock cycle.

**End of Section**

# Rocks

## section 2 Igneous Rocks

### What You'll Learn

- how igneous rocks form
- how igneous rocks are grouped

### Before You Read

Think about an erupting volcano you may have seen on TV or in the movies. On the lines below, describe what comes out of an erupting volcano.

_______________________________________________

_______________________________________________

_______________________________________________

Study Coach

**Map Complete Definitions** Create a definition map. Write each vocabulary word, its definition, and ideas that tell about each word. Your map should answer the questions: "What is it?" "What is it like?" and "What are some examples?"

FOLDABLES™

**B Identification** Make a layered book with two sheets of notebook paper. Write about igneous rock as you read this section. You can complete your Foldable as you read Sections 3 and 4.

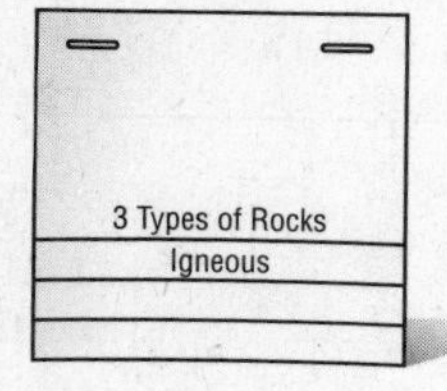


### Read to Learn

## Formation of Igneous Rocks

When a volcano erupts, red-hot material may flow out of it. The extremely hot material is melted rock, called magma. Magma flows like a liquid. When magma flows near or onto Earth's surface, it cools and hardens. **Igneous rock** forms from the cooled and hardened magma.

### How does magma become lava?

Inside Earth, the temperature and pressure in certain places are just right to melt rocks. As a result, magma forms. Magma can be found at depths ranging from near Earth's surface to about 150 km below the surface. The temperature of magmas range from about 650°C to 1,200°C.

Where does the heat come from that melts rock inside Earth? Some heat comes from the decay of radioactive elements in rocks. Some heat is left from when Earth was formed. At first, Earth was very hot, molten material.

Magma is less dense than the solid rock around it. Because it is less dense, it is forced up toward Earth's surface. When magma reaches Earth's surface and flows from volcanoes, it is called **lava.**

## What are intrusive igneous rocks?

Magma is melted rock made up of common elements and liquids. Magma cools as it rises toward Earth's surface. As magma cools, the atoms and compounds in the liquid rearrange themselves into new crystals called mineral grains. As cooling continues, mineral grains grow together to form rocks. Sometimes this process takes place beneath the surface. **Intrusive** igneous rocks form from cooling magma beneath Earth's surface, as shown in the figure below. It takes a long time for magma beneath Earth's surface to cool. Cooling is so slow, mineral grains grow quite large. Intrusive igneous rock has large mineral grains.

Intrusive igneous rocks can be found on Earth's surface. After many years, the layers of rock and soil that once covered them are removed by erosion. Erosion occurs when the rocks are pushed up by forces inside Earth.

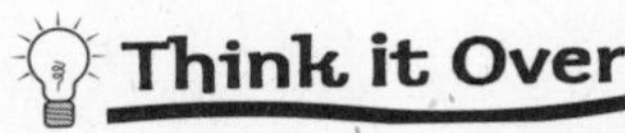


**1. Make Connections** The prefix "in" means "inside." Where do intrusive rocks form?

______________________________

______________________________

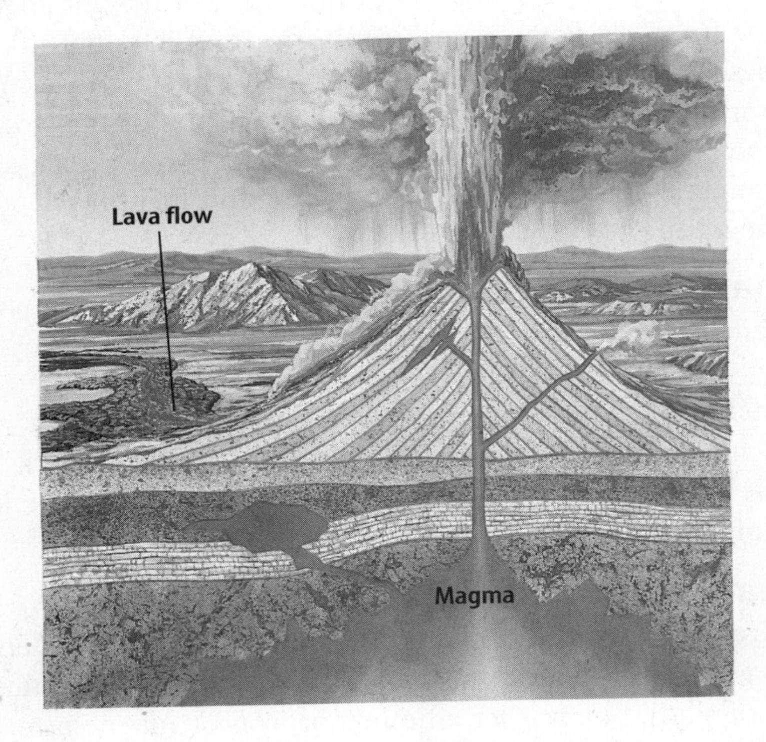


**Picture This**

**2. Label** Write *intrusive rock* where this rock forms. Write *extrusive rock* where this rock forms.

## What are extrusive igneous rocks?

**Extrusive** igneous rocks form as lava cools on the surface of Earth, as shown in the figure above. When lava reaches the surface, it is exposed to air and water, which cools it quickly. The atoms in the liquid do not have time to arrange into large crystals. Therefore, the mineral grains in extrusive igneous rock are quite small.

**Think it Over**

**3. Make Connections** The prefix "ex" means "outside." Where do extrusive rocks form?

______________________________

______________________________

### What is volcanic glass?

Sometimes, lava that comes out of a volcano cools so quickly that few or no mineral grains form. A rock that forms from this quickly cooling lava is called volcanic glass. Volcanic glass has few or no crystals because the atoms are not arranged in an orderly pattern.

Obsidian is a volcanic glass that looks like shiny black glass. Pumice and scoria are also volcanic glasses, but they do not look like glass. They have lots of holes, or pores. These materials form from a gooey liquid that contains pockets of gases. Some of these gases escape and holes are left where the rock formed around the gas pocket.

## Classifying Igneous Rocks

Igneous rocks can be grouped as either intrusive or extrusive depending on how they are formed. Igneous rocks can also be grouped according to the type of magma they come from. An igneous rock can form from basaltic, andesitic, or granitic magma. The type of magma that cools to form an igneous rock affects the properties of that rock. Some of the chemical and physical properties of a rock are its mineral composition, density, color, and melting temperature.

**Think it Over**

4. **Infer** The word "igneous" comes from the Latin word *ignis*, which means "fire." Why is igneous a good word for this type of rock?

___

___

### What are basaltic rocks?

Igneous rocks that are dense and dark-colored are **basaltic** (buh SAWL tihk). They form from magma containing a lot of iron and magnesium, but little silica, which is made of silicon and oxygen. Basalt gets its dark color from the iron and magnesium it contains. Basaltic lava is fluid and flows freely.

### What are granitic rocks?

**Granitic** igneous rocks are light-colored and not as dense as basalt. They form from thick, stiff magma that contains lots of silica, but smaller amounts of iron or magnesium. Stiff granitic magma can build up lots of gas pressure. This pressure is released in violent volcanic eruptions. ☑

**Reading Check**

5. **Determine** What is one way basaltic and granitic igneous rocks differ?

___

___

### What are andesitic rocks?

Andesitic igneous rocks have mineral compositions between those of basalt and granite. Like granitic magma, andesitic magma can produce violent volcanic eruptions.

# After You Read

## Mini Glossary

**basaltic:** dense, dark igneous rock that forms from magma containing a lot of iron and magnesium, but little silica

**extrusive:** igneous rocks that form when lava cools on the surface of Earth

**granitic:** light-colored igneous rock that forms from thick magma containing lots of silica, but little iron or magnesium

**igneous rock:** rock that forms as magma or lava cools and hardens

**intrusive:** igneous rocks that form beneath the surface of Earth

**lava:** molten rock that flows from volcanoes onto Earth's surface

1. Review the terms and their definitions in the Mini Glossary. Then write one sentence that explains how igneous rock forms.

_______________________________________________

_______________________________________________

2. Fill in the blank boxes with words from this section.

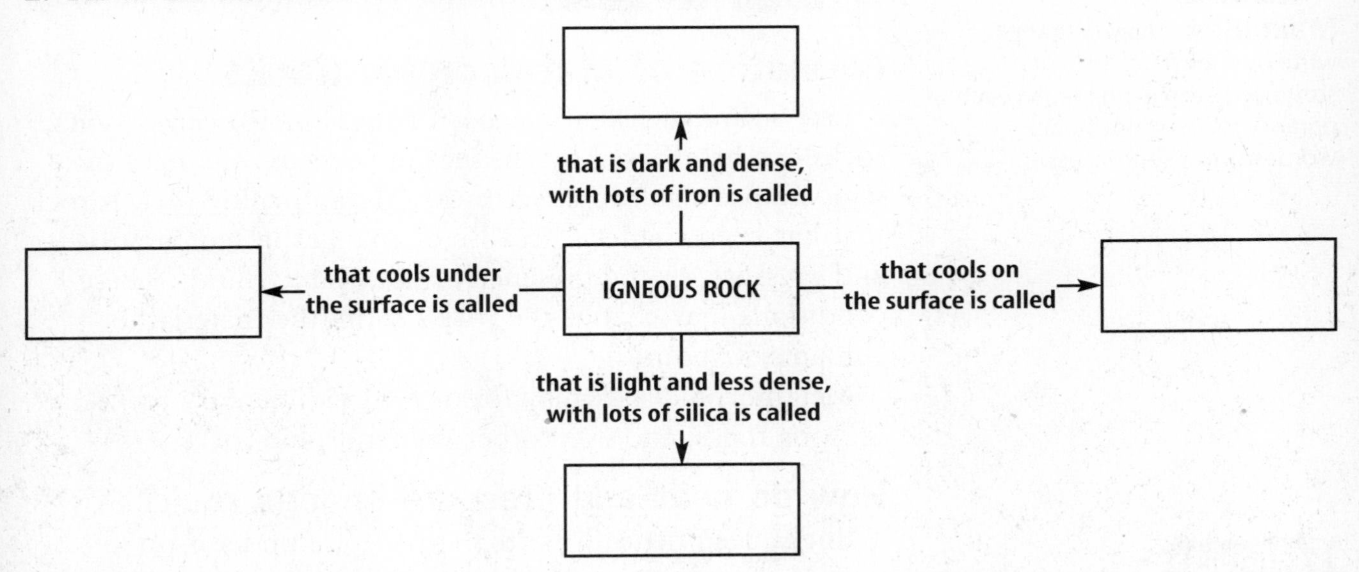


3. How did using your definition map help you learn and remember the vocabulary words in this section?

_______________________________________________

_______________________________________________

**ScienceOnline** Visit **earth.msscience.com** to access your textbook, interactive games, and projects to help you learn more about igneous rocks.

End of Section

# Rocks

## section 3 Metamorphic Rocks

**What You'll Learn**
- how metamorphic rock forms
- how metamorphic rock is classified

### Before You Read

Think about a time you packed a sandwich for lunch and placed a can of juice or soda on top of it. Describe how the sandwich looked at lunchtime and why it looked that way.

______________________________________________

______________________________________________

Study Coach

**Main Ideas** As you read, write one sentence to summarize the main idea in each paragraph. Use vocabulary words in your sentences.

### Read to Learn

#### Formation of Metamorphic Rocks

Like a sandwich that has been flattened by a can of soda, rocks can be affected by changes in pressure. Changes in temperature also can affect rocks. **Metamorphic rock** is rock that has been changed because of changes in temperature and pressure, or the presence of hot, watery fluids. These conditions may change the rock's form, the material it contains, or both.

Metamorphic rock may form from sedimentary rocks, igneous rocks, and even other metamorphic rocks.

#### How do heat and pressure change rock?

Rocks deep beneath Earth's surface are under great pressure from the layers of rock above them. Temperature also increases with depth. In some areas, the pressure and the temperature are just right to melt rock. The melted rock forms magma. Different types of metamorphic rock may form from the magma.

In other places deep inside Earth where there is a lot of liquid, rocks do not melt. Instead, some mineral grains dissolve in the liquid and then form new crystals. Under these conditions, minerals sometimes exchange atoms with surrounding minerals and new minerals form.

FOLDABLES™

**B List and Identify** As you read this section, use your layered Foldable to list types of metamorphic rock and to describe each type.

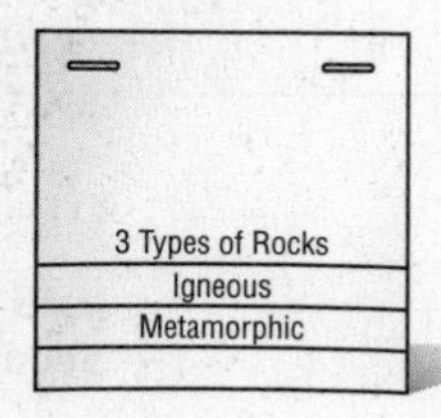

## How does shale change to gneiss?

Depending on the amount of pressure and the temperature under Earth, one type of rock can change into several different types of metamorphic rock. For example, shale, a sedimentary rock, will change into slate, a metamorphic rock. As the temperature and pressure on it increase, the slate can change into phyllite, then into schist, and finally into gneiss (NISE).

## How do hot fluids change rock?

Hot fluids from magma flow through spaces in and between underground rocks. The hot fluids are mostly water, but they also contain dissolved elements and compounds. These fluids can react with the rock they flow through and change its composition. As shown in the figure below, the hot fluid flows into the rock and chemically changes it into a type of metamorphic rock.

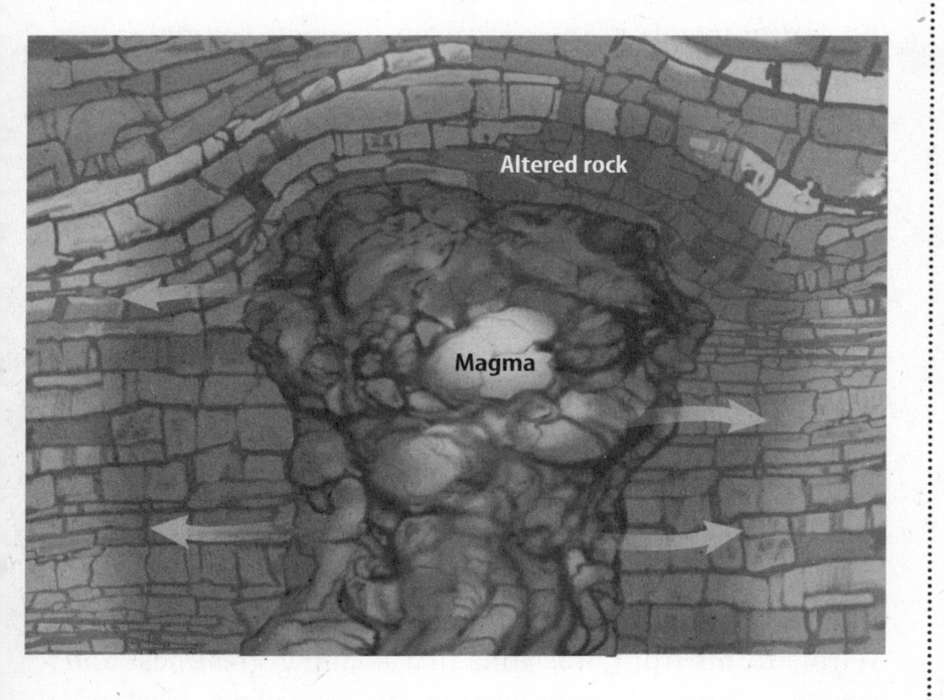


**Picture This**

1. **Highlight** Use a colored marker or pencil to highlight the rocks in the figure that are being changed by the hot fluid from magma.

# Classifying Metamorphic Rocks

Metamorphic rocks form from igneous, sedimentary, or other metamorphic rocks. Heat, pressure, and hot fluids cause these rocks to change. The types of metamorphic rocks that form can be classified based on their composition and texture. ☑

**Reading Check**

2. **Identify** Name three things that cause rocks to change into metamorphic rock.

_______________

_______________

_______________

## What are foliated rocks?

As some metamorphic rocks form, their mineral grains line up in parallel layers. Metamorphic rocks with a **foliated** texture have parallel layers of mineral grains. For example, slate is a metamorphic rock that forms from shale, a sedimentary rock. When shale is exposed to heat and pressure, it changes. Its mineral grains line up in parallel layers to form slate, a foliated metamorphic rock. ☑

**Reading Check**

3. **Describe** How do the layers look in a foliated metamorphic rock?

______________________

**Slate** Slate's parallel layers of mineral grains are pressed so tightly together that water cannot pass between them easily. Slate also breaks into smooth, flat pieces. Because it sheds water and splits smoothly, slate is often used for paving stones and roofing tiles.

**Gneiss** Gneiss is a foliated rock that forms when granite and other rocks are changed by heat and pressure. The foliated texture of gneiss is easily seen in its light and dark bands. As gneiss forms, the movement of atoms separates the dark minerals in the rock from the light-colored minerals in the rock.

## What are nonfoliated rocks?

Some metamorphic rocks are formed without layers. In these rocks, the mineral grains grow and rearrange, but do not form layers. **Nonfoliated** rocks are metamorphic rocks that form without a layered texture. ☑

**Reading Check**

4. **Determine** What is the name of metamorphic rocks that form without layers?

______________________

**Sandstone** Sandstone is a sedimentary rock made mostly of quartz grains. When it is heated under a lot of pressure, sandstone is changed into quartzite. Heat and pressure cause the sandstone's quartz grains to grow larger and lock together like pieces of a jigsaw puzzle. The quartz grains in quartzite are not in layers, so quartzite is a nonfoliated rock.

**Marble** Another nonfoliated metamorphic rock is marble. Marble forms from limestone that is under great heat and pressure. Limestone contains the mineral calcite. Heat and pressure change the calcite into marble, which does not have a layered texture. In fact, marble's fine, smooth texture makes it the perfect material for sculptures and buildings.

## After You Read

### Mini Glossary

**foliated:** metamorphic rock whose mineral grains are lined up in parallel layers

**metamorphic rock:** rock that has been changed by heat, pressure, or the presence of hot fluids

**nonfoliated:** metamorphic rock whose minerals grow and rearrange themselves in a nonlayered texture

1. Review the terms and their definitions in the Mini Glossary. Write a sentence that describes the difference between foliated metamorphic rock and nonfoliated metamorphic rock.

_______________________________________________

_______________________________________________

2. Fill in the blanks in the boxes below to organize the information from this section.

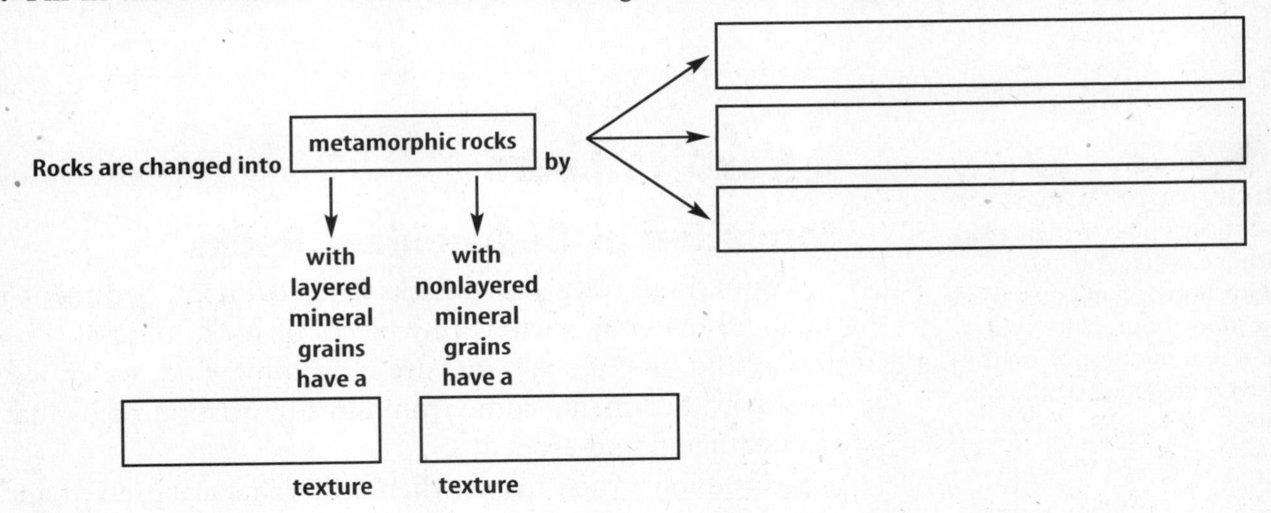


3. In this section you were asked to write sentences using the vocabulary words to summarize the main ideas. How did this help you understand what you read?

_______________________________________________

_______________________________________________

_______________________________________________

ScienceOnline Visit **earth.msscience.com** to access your textbook, interactive games, and projects to help you learn more about metamorphic rocks.

# Rocks

## section 4 Sedimentary Rocks

**What You'll Learn**
- how sedimentary rocks form
- how sedimentary rocks are classified

## Before You Read

Imagine you are stacking slices of bread, one on top of the other. Then you put a heavy book on top of the stack and leave it there overnight. Describe how the slices of bread might look the next day.

______________________________________________

______________________________________________

______________________________________________

Study Coach

**Sticky-note Discussion** Use sticky notes to mark places in the text that you find interesting or that you have a question about. Write your comment or question on the sticky note and stick it to the page.

## Read to Learn

### Formation of Sedimentary Rocks

Weathering breaks down rocks into sediment. **Sediment** is the loose material, such as tiny pieces of rock, mineral grains, and bits of shell, that are moved by wind, water, ice, or gravity. Sediments come from already-existing rocks that are weathered and eroded.

**Sedimentary rock** forms when sediments are pressed and cemented together, or when minerals form from solutions. About 75 percent of the rocks you see on Earth's surface are sedimentary rocks.

FOLDABLES™

**B Organize** Use your Foldable to help you organize sedimentary rocks into groups based on their characteristics.

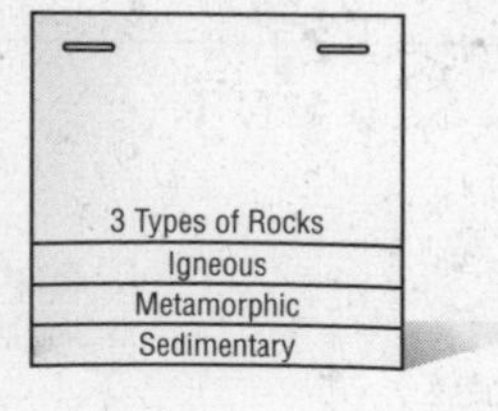


### What do sedimentary rocks look like?

Sedimentary rocks often form as layers, like a stack of papers. The older layers are at the bottom because they were deposited first. The newer layers are at the top because they were deposited later. If sedimentary rock is not disturbed, the layers will remain in place, with the oldest at the bottom and youngest at the top.

Sometimes, though, forces within Earth overturn layers of sedimentary rock. Then, the oldest layers are no longer on the bottom. The order of the layers is changed.

## Classifying Sedimentary Rocks

Sedimentary rocks can be made of just about any material in nature. Sediments come from weathered and eroded sedimentary, metamorphic, and igneous rock. Sediments also can come from the remains of some organisms. The composition of a sedimentary rock depends on what types of sediments formed it.

Sedimentary rocks are classified by what they are made of. They are also classified by the way in which they formed. Sedimentary rocks are classified as detrital, chemical, or organic. ☑

**Reading Check**

1. **Identify** the three classes of sedimentary rock.

______________________

______________________

______________________

## Detrital Sedimentary Rocks

The word *detrital* (dih TRI tul) comes from the Latin word *detritus,* which means "to wear away." Detrital sedimentary rocks are made from the broken pieces of other rocks. The tiny pieces are compacted and cemented together to form solid sedimentary rock.

### What are weathering and erosion?

Weathering is the process in which air, water, or ice breaks down rocks into smaller and smaller pieces. The movement of weathered material is called erosion.

### What is compaction?

Erosion moves sediments to a new place, where they are deposited in a thin layer. Over time, layer upon layer of sediment builds up. The weight of the top layers pushes down on the lower layers. Downward pressure causes small sediments to stick together and form solid rock. The process in which layers of sediments are pressed together to form rock is called **compaction.** The figure below shows how rock pieces are compacted to form sedimentary rock.

**Picture This**

2. **Describe** Use a colored pencil to color in the spaces between the sediments in each figure. What happens to the spaces as the sediments form rock?

______________________

**Compaction of Sediments**

## Picture This

**3. Outline** with your pencil the spaces between sediments where water and dissolved minerals move.

**Cementation**

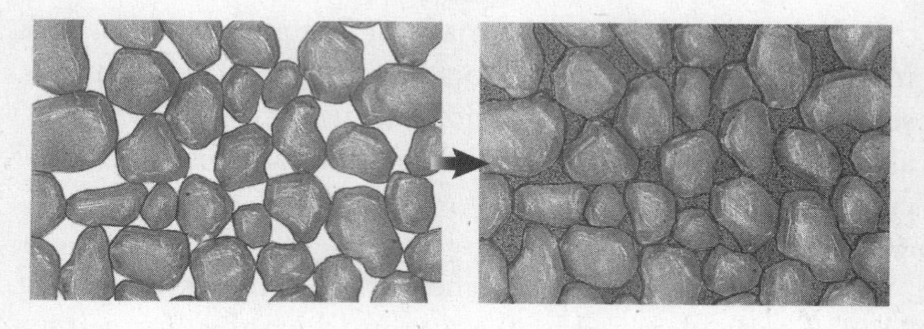

## What is cementation?

Large sediments, like sand grains and pebbles, cannot form rock from pressure alone. These large sediments form rock only if something helps them stick together.

As water moves through rock and soil, it picks up materials released by the weathering of minerals. The resulting solution of water and dissolved minerals moves through open spaces between larger sediments. The solution acts as a kind of glue that holds the large sediments together. **Cementation** is the process in which sediments are held together by dissolved minerals produced when water moves through rock. Minerals such as quartz, calcite, and hematite make the best cement for holding large sediments together. ☑

**Reading Check**

**4. Identify** What is necessary for the process of cementation to take place—wind, water, or air?

______________________

## What are the shapes and sizes of sediments?

Detrital rocks have a grainy texture, like grains of sugar. They are named according to the shapes and sizes of the sediments that form them. For example, conglomerate and breccia (BRECH uh) are detrital rocks that form from large sediments. If the sediments are rounded, the rock is called conglomerate. If the sediments have sharp angles, the rock is called breccia. The farther sediments are carried by wind, water, or ice, the more rounded they become.

**Think it Over**

**5. Infer** Which has probably been carried farther by wind and water—the sediment in conglomerate rocks or breccia rocks?

______________________

## What materials are found in sedimentary rocks?

Conglomerate and breccia are formed from gravel-sized sediments that are cemented together by quartz or calcite. These sediments may come from the minerals quartz or feldspar, or may contain chunks of other rocks, such as gneiss, granite, or limestone.

Sandstone forms from small sediments. The sand-sized sediments in sandstone can come from almost any mineral, though they usually come from quartz and feldspar. Shale is a detrital sedimentary rock that is made from the smallest clay sediments.

## Chemical Sedimentary Rocks

When water evaporates from a salt solution, salt grains remain. In a similar way, when the water in a lake evaporates, its minerals remain. The remaining mineral deposits form sediments which, in turn, form rocks. Chemical sedimentary rocks form when dissolved minerals come out of solution and form sediments that become rocks. ☑

### How does limestone form?

Calcium carbonate is found dissolved in ocean water. Calcium carbonate comes out of solution as the mineral calcite. Calcite forms crystals, which bond to form limestone, usually on the bottom of lakes and shallow seas. Long ago, the central United States was covered with a shallow sea. Over time, the water evaporated. As a result, much of the central United States has limestone bedrock.

### How does rock salt form?

Some bodies of water contain a lot of dissolved salts. When the water evaporates, it deposits the mineral halite, or rock salt. Rock salt is mined. It is used in manufacturing glass, paper, and soap. It is also made into table salt.

**Reading Check**

6. **Explain** What do chemical sedimentary rocks form from?

______________________

## Organic Sedimentary Rocks

Rocks made of materials that were once living things are called organic sedimentary rocks. One of the most common organic sedimentary rocks is fossil-rich limestone. It is made of the remains of once-living ocean organisms. Ocean animals, such as clams and snails, make their shells out of calcium carbonate, which eventually becomes calcite. When the animals die, their shells pile up and become cemented together to form fossil-rich limestone. ☑

### What are other organic sedimentary rocks?

**Chalk** Chalk is an organic sedimentary rock that is made up of extremely tiny bits of animal shells. When you write with chalk, you are crushing and smearing the calcite shell remains of once-living ocean animals.

**Coal** Coal is a useful organic sedimentary rock that forms when pieces of dead plants are buried under other sediments in swamps. The plant material is chemically changed. The resulting sediments are compacted to form coal. Today, coal is a fuel used in power plants to make electricity.

**Reading Check**

7. **Determine** What do all organic sedimentary rocks contain?

______________________

______________________

# After You Read

## Mini Glossary

**cementation:** process in which sediments are held together by dissolved minerals produced when water moves through rock

**compaction:** process in which layers of sediments are pressed together to form rock

**sediment:** loose material, such as tiny pieces of rock, mineral grains, and bits of shell, that are weathered from rocks and carried by wind or water

**sedimentary rock:** rock that forms when sediments are pressed and cemented together, or when minerals form from solutions

1. Review the terms and their definitions in the Mini Glossary. Then write one or two sentences that describe how sediments form sedimentary rocks. Use the terms in your answer.

2.

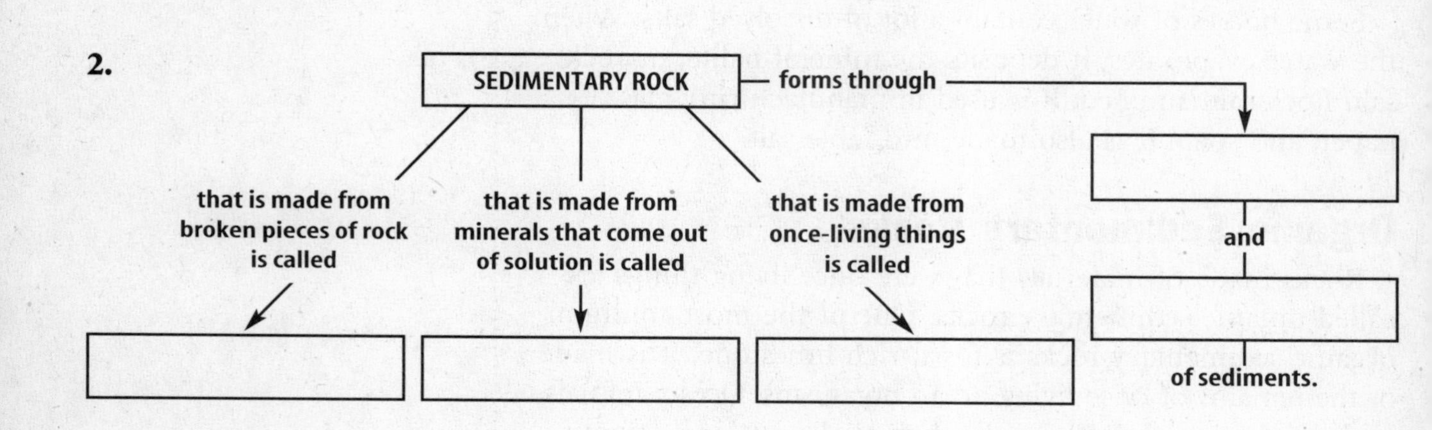


3. You used sticky notes to write comments or questions about this section. How did using sticky notes help you understand sedimentary rocks?

ScienceOnline Visit **earth.msscience.com** to access your textbook, interactive games, and projects to help you learn more about sedimentary rocks.

# Earth's Energy and Mineral Resources

## section 1 Nonrenewable Energy Resources

### Before You Read

What do you think of when you hear the word *fuel*? Write your ideas on the lines below.

_______________________________________________

_______________________________________________

_______________________________________________

**What You'll Learn**

- about formation and uses of nonrenewable resources
- about advantages and disadvantages of using fossil fuels and nuclear energy

### Read to Learn

#### Energy

The world depends on energy. Energy is the ability to cause change. People use many different energy resources. Some of the energy resources on Earth are being used faster than natural Earth processes can replace them. Energy resources that cannot be replaced, or renewed, are called nonrenewable resources.

**Mark the Text**

**Highlight** Identify the key terms and their meanings as you read this section.

#### Fossil Fuels

Fossil fuels are nonrenewable energy resources that formed over millions of years from the remains of dead plants and other organisms. Fossil fuels include coal, oil, and natural gas. Coal is a type of sedimentary rock formed from layers of ancient plant matter. Oil, or petroleum, is a liquid hydrocarbon. Hydrocarbons are compounds that contain both hydrogen and carbon atoms. Oil and natural gas formed from tiny organisms that lived millions of years ago. Fossil fuels are used to make gasoline for cars, to heat homes, to generate electricity, and for many other uses.

**FOLDABLES™**

**A Compare and Contrast** Use quarter-sheets of notebook paper to compare coal, oil, natural gas, and nuclear energy.

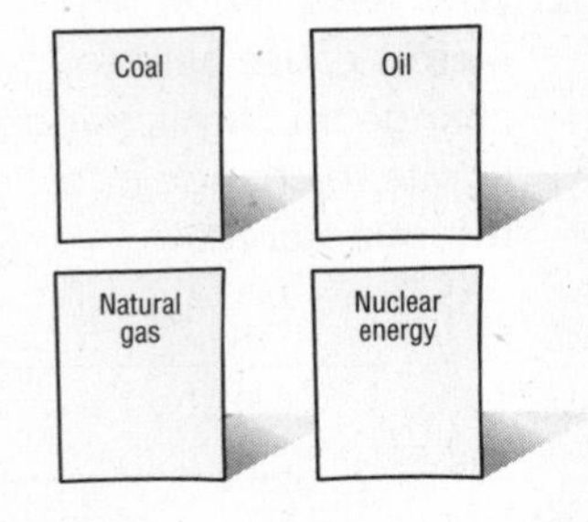

## Picture This

**1. Locate and Identify** Mark where you live on the map. Is it an area where coal is found?

_______________

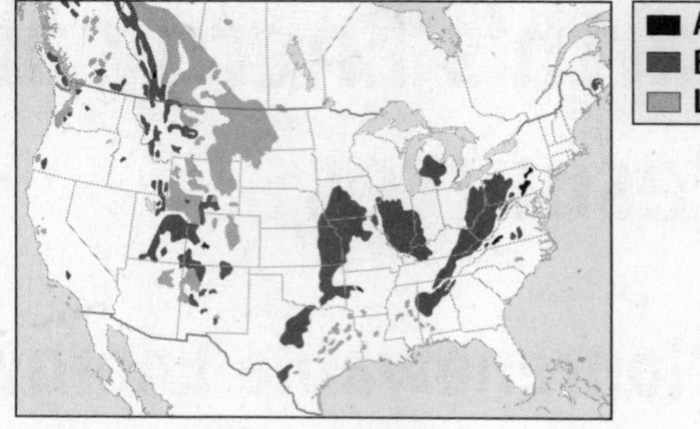


## What is coal?

Coal is the most abundant fossil fuel in the world. The United States has many coal reserves, as shown on the map above. The world's supply of coal should last 250 more years at the present rate of use. **Coal** is sedimentary rock that contains at least 50 percent decayed plant material. Coal starts to form when plants die in a swampy area and are covered by more plants, water, and sediment. Over millions of years, this material becomes coal.

## What are synthetic fuels?

Synthetic fuels are made from solid organic material, such as coal, and may be liquids or gases. Liquid synthetic fuels are used to make gasoline and fuel oil. Gaseous synthetic fuels are used to produce electricity.

## How is coal formed?

Coal forms in several stages. Each stage yields a different fuel. In the first step, peat forms.

**Peat** Dead plant material builds up in swamps, forming a layer of organic sediment, or peat. As decaying plant matter loses gas and moisture, the concentration of carbon increases. When peat burns, it gives off large amounts of smoke because it contains water and impurities.

**Lignite** If peat is buried under more sediment, it becomes lignite, a soft, brown coal with much less moisture than peat. Lignite forms as heat and pressure force water from the peat, further increasing the concentration of carbon. When lignite is burned, it releases more energy and less smoke than peat.

**Bituminous Coal** Over time, if layers of lignite are buried deeper, bituminous coal, or soft coal, forms. Bituminous coal is compact, black, and brittle. It provides lots of heat energy when it is burned. Bituminous coal contains some sulfur which can pollute the air.

**2. Identify** What two factors cause peat to change into lignite and lignite to change into bituminous coal?

_______________

_______________

**Anthracite Coal** If enough heat and pressure are applied to buried layers of bituminous coal, anthracite coal forms. Anthracite is also called hard coal. It has the highest amount of carbon of all kinds of coal. As a result, it is the cleanest burning of all coals. ☑

**Reading Check**

**3. Explain** Why is anthracite the cleanest burning form of coal?

______________________

______________________

## What other fossil fuels are used for energy?

Coal isn't the only fossil fuel that is used for energy. Two other fossil fuels, oil and natural gas, provide large amounts of energy. **Oil** is a thick, black liquid formed from the buried remains of organisms that once lived in the oceans. **Natural gas** is also formed from the buried remains of ancient ocean organisms. Natural gas is gaseous rather than liquid. Although oil and natural gas both come from ocean organisms, the compounds found in natural gas are lighter than the compounds found in oil.

People in the United States use large amounts of oil and natural gas each day. The circle graph below shows how much of the energy we use comes from these two fuels. Natural gas is used mostly for heating and cooking. Oil is used in many ways, including making heating oil, gasoline, and plastics.

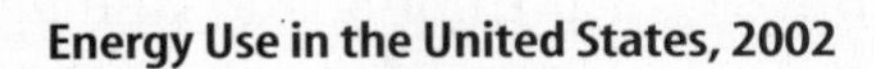


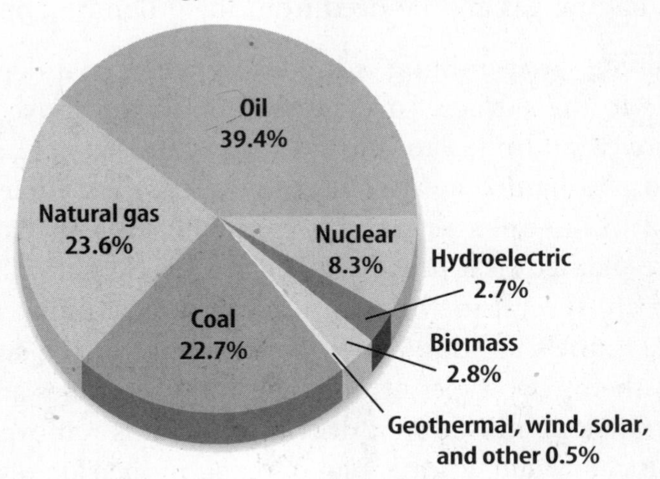


**Applying Math**

**4. Evaluate** Add together the percentages of coal, oil, and natural gas shown in the graph. How much of the graph is *not* one of these three fuels? Show your work.

______________________

______________________

______________________

## How do oil and natural gas form?

It takes millions of years for oil to form from the remains of tiny marine organisms. The process begins when tiny organisms called plankton die and fall to the seafloor.

Sediment covers the dead plankton. Over time, more plankton and sediment pile up. As the piles grow thicker, pressure builds up and temperatures rise. The increased heat causes the dead plankton to slowly become oil and natural gas.

## What is a reservoir rock?

Oil and natural gas often are found in layers of rock that have been tilted or folded. Fossil fuels are not as dense as water, so they tend to move upward. Some rock layers, such as shale, stop this upward movement. A folded layer of shale can trap the oil and natural gas as shown in the figure below. The rock layer underneath the trapped oil and natural gas is called a reservoir rock.

**Picture This**

**5. Identify** Where can oil and natural gas be found?

______________________

______________________

## Removing Fossil Fuels from the Ground

Coal is removed from the ground by many methods. Two of the most common are strip mining and underground mining. Oil and natural gas are removed from the ground by pumping.

**Coal Mining** Strip mining is used only when coal deposits are close to the surface. Layers of soil and rock above coal are removed and piled to one side. The exposed coal is removed and hauled away. The soil and rock are then returned to the open pit and covered with topsoil. Trees and grass are planted in a process called land reclamation.

There are several methods of underground mining. In one method, tunnels are dug and coal is brought to the surface through them. Two types of underground coal mines are drift mines and slope mines. In a drift mine, coal is removed from the ground through a horizontal opening in the side of a hill or mountain. In slope mining, an angled opening and air shaft are dug in the side of a mountain to remove coal. ☑

**6. Identify** Name two types of underground coal mines.

______________________

______________________

**Oil and Gas Pumping** Oil and natural gas are pumped from underground deposits. First, a narrow hole, or a well, is drilled down through rock to the oil deposit. Next, equipment is put into the well to control the oil flow. Finally, the rock that surrounds the oil and gas is broken to allow the fuels to flow into the well. The oil and gas are pumped to the surface.

## Fossil Fuel Reserves

The terms *reserve* and *resource* are often used when fossil fuels are discussed. The amount of a fossil fuel that can be taken from the ground for a profit, using current technology, is called a **reserve.** A fossil fuel resource is a deposit that contains enough fuel to be taken from the ground in useful amounts, whether or not a profit can be made. A fossil fuel resource is called a reserve if the fuel in it can be mined or pumped at a profit. ☑

### What are methane hydrates?

Current reserves of natural gas in the United States are expected to last about 60 years. The main substance in natural gas is methane. Recent studies show a new source of methane may be located under the seafloor. This new source of methane is an icelike substance called methane hydrate. Methane hydrates are found in sediments on the ocean floor. They form where temperatures are cold and pressure is high. Scientists estimate that methane hydrates contain more fuel than all of today's fossil fuel deposits. As a fuel, methane burns cleanly. If methane could be removed from the sea, the world's supply of clean energy would be much greater.

### How can fossil fuels be conserved?

You can avoid wasting fossil fuels by turning off lights when you leave a room. Make sure doors and windows are shut tightly during cold weather so heat won't leak out of your home. If you have air conditioning, run it as little as possible. Check with an adult in your home to see if more insulation could be added to your home to help save energy. See if an insulated jacket could be put on the water heater.

## Energy from Atoms

In the United States, most electricity comes from power plants that burn fossil fuels. But, there are other ways to get energy. **Nuclear energy** is an alternate source of energy produced from atoms. ☑

Atoms give off energy during atomic reactions. The center of the atom is called the nucleus. The nucleus is made up of particles called protons and neutrons. When the nucleus of a heavy element is split, lighter elements form and energy is released. This energy, produced by nuclear fission, can be used to make electricity or power a submarine.

**✔ Reading Check**

7. **Explain** What is the difference between a reserve and a resource?

______________________

______________________

______________________

**✔ Reading Check**

8. **Identify** Name a source of energy that comes from atoms.

______________________

## Picture This

**9. Identify** Look at the figure. Circle the two kinds of rods in the nuclear reactor. Name them.

_______________

_______________

**Nuclear Power Plant**

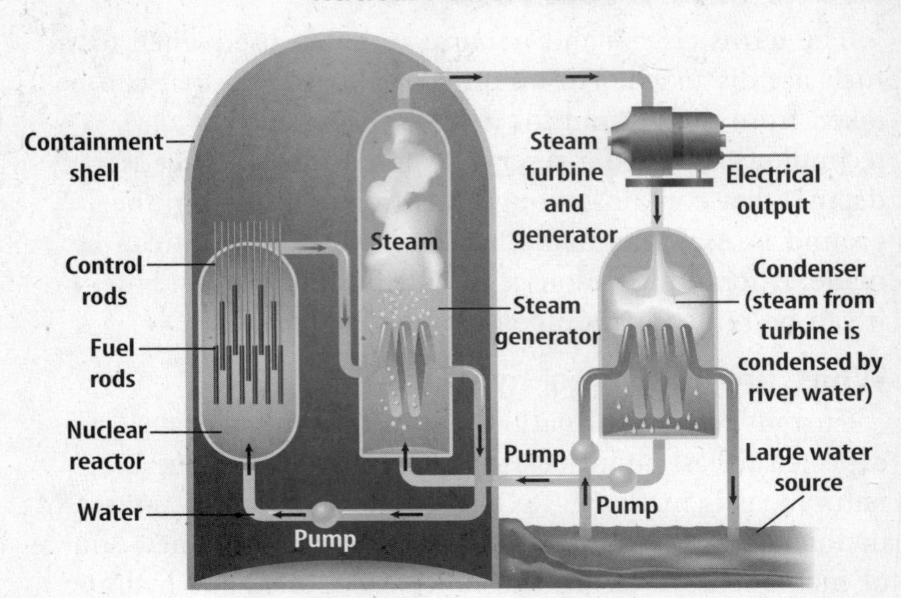


## How does nuclear energy make electricity?

A nuclear power plant has a large chamber called a nuclear reactor, which houses fuel rods containing uranium-235. The fuel rods sit in a pool of cooling water. Neutrons are fired into the fuel rods. When the neutrons hit uranium-235 atoms, the atoms split apart and fire out more neutrons that hit other atoms. Those atoms also split, starting a chain reaction.

As each uranium atom splits, it does two things: it fires neutrons and it releases heat. In a nuclear power plant, that heat can be used to boil water to make steam. The steam drives a turbine which turns a generator to produce electricity.

Nuclear energy from fission presents problems. Uranium-235, the fuel, is a nonrenewable energy source. Nuclear energy also produces waste. Nuclear waste is highly radioactive and must be stored away from people and the environment for at least 10,000 years. ☑

**Reading Check**

**10. Explain** What is one problem with nuclear energy?

_______________

_______________

## How does fusion provide energy?

Fusion is a nuclear reaction that occurs when two atoms are joined, or fused, to form one atom. During this process, large amounts of energy are released. The Sun is a natural fusion power plant that provides energy for Earth and the solar system. Scientists have not yet developed controlled technology to produce energy by nuclear fusion. If the technology is developed, nuclear energy would no longer be considered a nonrenewable fuel resource.

## After You Read

### Mini Glossary

**coal:** sedimentary rock formed from decayed plant material

**fossil fuels:** nonrenewable energy resources, such as coal and oil, that formed over millions of years from the remains of dead plants and other organisms

**natural gas:** fossil fuel that forms in a gaseous state from the buried remains of marine organisms

**nuclear energy:** alternate source of energy produced from atoms

**oil:** thick, black liquid fossil fuel formed from the buried remains of organisms that lived in the seas; known as petroleum

**reserve:** amount of fossil fuel that can be taken from Earth at a profit using current technology

1. Review the terms and their definitions in the Mini Glossary. Explain how a fossil fuel resource is classified as a reserve.

_______________________________________________

_______________________________________________

_______________________________________________

2. Circle the correct answer and fill in any missing information in the following table.

| Fuel | Fossil Fuel? | Renewable? | How Long Will It Last? | Uses? |
|---|---|---|---|---|
| Coal | yes / no | yes / no | | |
| Oil | yes / no | yes / no | 100 years | Heating, travel, industry, electric power |
| Natural gas | yes / no | yes / no | 60 years | |

3. How did highlighting the text help you learn and understand the key terms and their meanings?

_______________________________________________

_______________________________________________

Visit **earth.msscience.com** to access your textbook, interactive games, and projects to help you learn more about nonrenewable energy resources.

# Earth's Energy and Mineral Resources

## section 2 Renewable Energy Resources

### What You'll Learn

- what inexhaustible and renewable energy resources are
- why these energy resources are used less than nonrenewable resources

### Before You Read

Think about the last time you were really tired. What is the best thing to do when you feel exhausted?

___

___

___

Study Coach

**Authentic Questions** As you read this section, write down any questions you may have about each topic. Discuss your questions with your teacher or another student.

### Read to Learn

#### Inexhaustible Energy Resources

How soon will the world run out of fossil fuels? That depends on how fast they are used. Fortunately, there are some sources of energy that are inexhaustible. Inexhaustible resources will never run out. They include the Sun, wind, water, and geothermal energy.

#### How can people use energy from the Sun?

Solar energy is energy from the Sun. Solar energy is clean, inexhaustible, and can be used to produce electricity. You already know that the Sun's energy heats Earth. It also causes winds in the atmosphere and currents in oceans.

People can use solar energy in a passive way or an active way. Windows are passive solar collectors. Windows on the south side of a building trap sunlight. The sunlight warms the inside of the building. Solar cells actively collect energy from the Sun and change it into electricity. Solar cells are used to power calculators, street lights, and experimental cars. In sunny regions, people put solar cells on their roofs to produce electricity for their homes.

FOLDABLES™

**B Compare** Make a layered Foldable using three sheets of paper. Use it to compare energy resources.

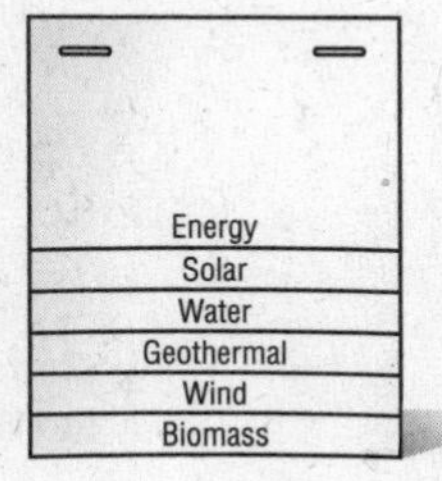

## What are the disadvantages of solar energy?

Solar energy does have some drawbacks. Solar cells don't work as well on cloudy days. They cannot work at all at night. Batteries can be used to store solar energy for use at night or on cloudy days. However, it is difficult to store large amounts of energy in batteries. Also, old batteries must be disposed of in a way that does not pollute the environment.

## How can wind be used for energy?

Wind is a source of energy. Wind powers sailing ships. Windmills have been built that used wind energy to grind corn or pump water. Today, windmills are used to produce electricity. A large number of windmills placed in one area to generate electricity is called a **wind farm**. ☑

Wind energy has advantages and disadvantages. Wind energy does not cause pollution and it's free. It does little harm to the land and produces no waste. But, only a few places in the world have winds that are strong enough to generate electricity using windmills. Also, wind isn't steady. At times it blows too hard. At other times, it is too weak or even stops completely. For reliable power, an area must have steady winds that blow at the right speed.

## How is flowing water used for power?

Running water also can be used to generate electricity. **Hydroelectric energy** is electricity produced by waterpower. To produce hydroelectric energy, a large dam is built across a river. The dam holds back the water, causing a lake to form behind the dam. As the figure below shows, when water is released from the lake, it turns turbines at the base of the dam. The turbines then turn generators that make electricity.

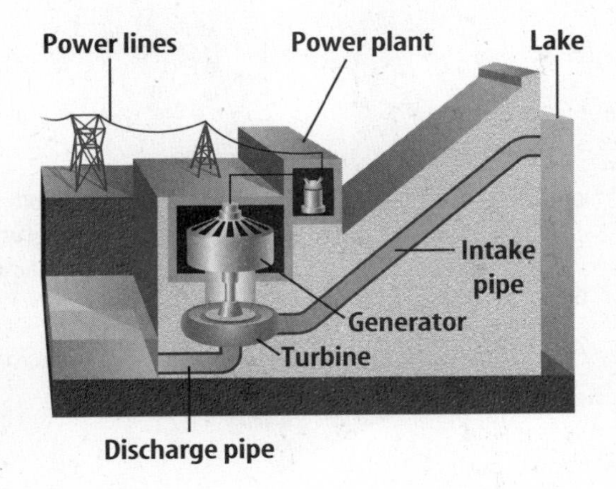


**Reading Check**

**1. Explain** How are windmills used today?

______________________

______________________

______________________

**Picture This**

**2. Interpret Diagrams** Place arrows on the figure to show the direction of water flow through the hydroelectric power plant.

## What are the disadvantages of water power?

Hydroelectric energy does not create air pollution. But dams do cause some harm to the environment. When dams are built, land is flooded and wildlife habitats are damaged. Lakes created by dams can slowly fill with silt. Silt destroys the habitats of organisms that live in the water. In the stream below the dam, erosion can become a problem.

## How can Earth's heat supply power?

Earth's heat can be used to generate electricity. **Geothermal energy** is an inexhaustible energy resource that uses hot magma or hot, dry rocks from below Earth's surface to produce steam to generate electricity. ☑

If magma rises fairly close to the surface, it can heat large pockets of water in the ground. The heat turns the water to steam. Geothermal power plants, such as the one in the figure below, can use this steam to turn turbines and generators that produce electricity. There are a few places on Earth where steam is naturally produced. In another type of geothermal power plant, water is pumped deep underground where it can flow through hot, dry rocks. The water becomes steam that can be pumped back to the surface and used to produce electricity. One advantage of this type of geothermal plant is that hot, dry underground rocks are found just about everywhere on Earth.

**Reading Check**

3. **Determine** What supplies the heat to make steam in geothermal energy?

______________________

______________________

**Picture This**

4. **Evaluate** Look at the figure. How deep into Earth do the pipes from the geothermal plant reach? How many meters is that?

______________________

______________________

**Geothermal Power Plant**

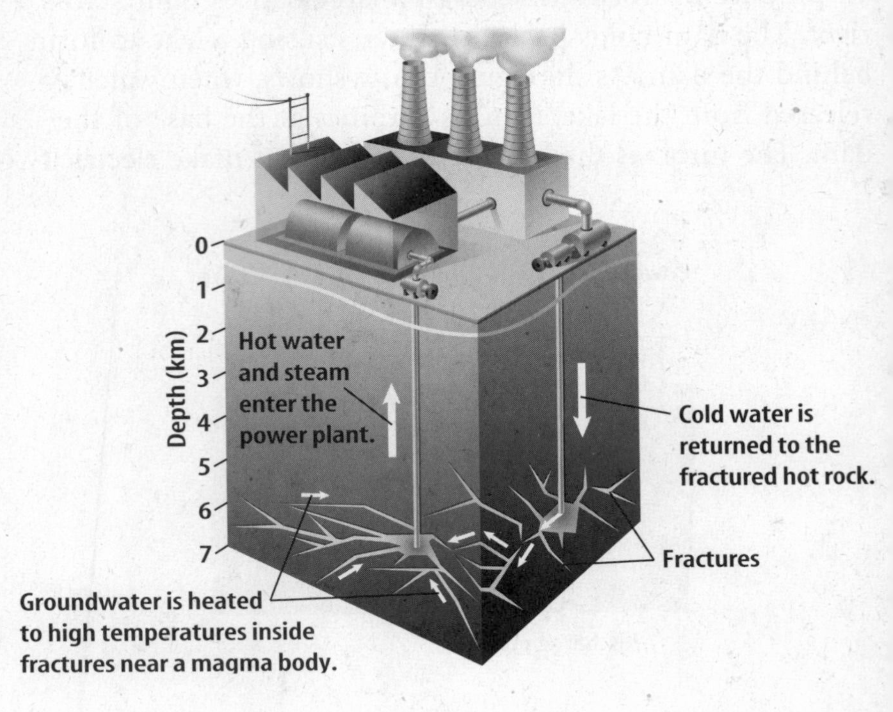

## Renewable Energy Resources

When you are exhausted, a good night's sleep can renew you. In the same way, time and rest can renew some energy resources. Renewable energy resources can be replaced by nature or by people within a person's normal lifetime. For example, trees can be a renewable resource. As one tree is cut down, another tree can be planted in its place. ☑

**Reading Check**

5. **Explain** What makes an energy resource renewable?

______________________

______________________

______________________

## Biomass Energy

Biomass materials are an important renewable energy resource. **Biomass energy** is renewable energy that comes from burning organic materials such as wood, alcohol, or garbage. The word *biomass* comes from combining the words *biological* and *mass.*

**Wood** If you ever sat near a campfire or a fireplace to keep warm, then you have used renewable energy from wood. As a tree grows, it stores energy from the Sun in its wood. When the wood burns, that stored solar energy is released as heat energy. People have long used wood as a source of energy. Much of the world still cooks with wood. Using wood has its problems. Gases and ashes that can pollute the air are released when wood is burned. When trees are cut down, natural habitats may be destroyed.

**Alcohol** Some biomass materials can be changed into cleaner-burning fuels. Corn is a biomass fuel that can be used to produce a kind of alcohol called ethanol. Ethanol mixed with gasoline is called gasohol. Gasohol can be used as fuel for cars and trucks. Using gasohol can cut down on the amount of fossil fuel needed to produce gasoline. But there is a problem with the process that produces ethanol. Growing the corn and making the ethanol often uses more fossil fuel energy than is saved by using gasohol.

**Think it Over**

6. **Compare** What is a problem that occurs with burning garbage and wood?

______________________

______________________

**Garbage** Every day humans throw away mountains of garbage. Two-thirds of it could be burned as a fuel. Burning garbage as a fuel has three benefits. It's a cheap source of energy, it cuts down on the need for fossil fuels, and it reduces the amount of material dumped into landfills. Unfortunately, burning garbage can pollute the air. Toxic ash residue remains after certain wastes are burned.

## After You Read

### Mini Glossary

**biomass energy:** renewable energy that comes from burning organic materials such as wood and alcohol

**geothermal energy:** inexhaustible energy resource that uses the energy from hot magma or hot, dry rocks deep below Earth's surface to generate electricity

**hydroelectric energy:** electricity produced by waterpower using large dams in a river

**solar energy:** power from the Sun that is clean, inexhaustible, and can be used to produce electricity

**wind farm:** area where many windmills use wind to generate electricity

1. Review the terms and their definitions in the Mini Glossary. Then choose one of the four kinds of energy listed above. Explain why it is a renewable resource.

_______________________________________________

_______________________________________________

2. Make a data record to compare the four kinds of renewable energy you learned about in this section. You may have to infer some answers.

**Biomass energy source** _______________________

**Advantages:** _______________________

**Disadvantages:** _______________________

**Geothermal energy source** _______________________

**Advantages:** _______________________

**Disadvantages:** _______________________

**Hydroelectric energy source** _______________________

**Advantages:** _______________________

**Disadvantages:** _______________________

**Solar energy source** _______________________

**Advantages:** _______________________

**Disadvantages:** _______________________

End of Section

# Earth's Energy and Mineral Resources

## section 3 Mineral Resources

### Before You Read

Do you recycle wastes at your home or your school? Write the items you recycle on the lines below. If you don't recycle, list items that could be recycled.

______________________________

______________________________

______________________________

**What You'll Learn**

- how minerals are classified as ores
- what metallic and nonmetallic mineral resources are

### Read to Learn

#### Metallic Mineral Resources

Look carefully around your home. You probably will find many metal items. The frame of your bed, soft-drink cans, and spoons are all made from metal. Metals are obtained from Earth materials called metallic mineral resources. A **mineral resource** is a resource from which metal is obtained.

#### What are ores?

Deposits of minerals that are large enough to be mined at a profit are **ores.** Most often, the word *ore* is used to describe deposits of metals. Hematite is an iron ore. Bauxite is an aluminum ore. Both are metallic ores.

When is a mineral deposit considered an ore? First, people must have a need for the mineral. Second, there must be enough of the mineral present in the deposit to make it worth removing. Third, it must be fairly easy to separate the mineral from the material in which it is found. Economic factors largely determine what an ore is. If any one of these three conditions is not met, the deposit might not be considered an ore.

Study Coach

**Make Flash Cards** As you read this section, make flash cards for each vocabulary term or unknown word. On one side of the card, write the term or word. On the other side of the card, write the definition.

FOLDABLES™

**C Contrast** Make a two-tab Foldable to explain the differences between metallic and nonmetallic mineral resources.

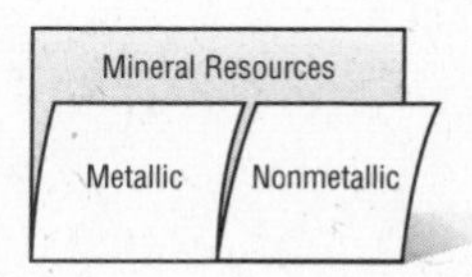

**Refining Oil** There are two steps for separating a useful mineral from its ore—concentrating and refining. Concentrating is the first step. After a metallic ore is mined from Earth's crust, it is crushed and the waste rock is removed. The waste rock that must be removed before a mineral can be used is called gangue (GANG). ☑

**Reading Check**

1. **Explain** What is gangue?

______________________

______________________

______________________

Refining is the second step in separating a mineral from its ore, producing a pure or nearly pure mineral. One way to refine metal is by smelting, a chemical process that removes unwanted elements. For example, iron ore, or hematite, contains iron oxide. During smelting the concentrated iron ore is combined with a certain chemical and heated. The chemical mixes with the oxygen in iron oxide, leaving behind pure iron. Smelting uses a fossil fuel resource to produce the heat needed to obtain another resource, in this case, iron.

## Nonmetallic Mineral Resources

Any mineral resources not used as fuels or as sources of metals are called nonmetallic mineral resources. These resources are mined for the nonmetallic elements they contain or for their physical or chemical properties. In general, nonmetallic mineral resources fall into two groups—industrial minerals or building materials. Some materials, such as limestone, belong to both groups. ☑

**Reading Check**

2. **Classify** What are the two groups of nonmetallic mineral resources?

______________________

______________________

**Industrial Minerals** Industrial minerals are sources of many useful chemicals. For example, sandstone is a source of silica, which is used to make glass. Sylvite, a mineral that forms when seawater evaporates, is used to make fertilizers for farms and gardens. Table salt comes from halite, a nonmetallic mineral resource. Halite also is used to melt snow and ice on roads and sidewalks and to help soften water.

Some industrial minerals are useful because of their physical properties. Garnet is a hard mineral that can scratch most other materials. Tiny pieces of garnet are glued on heavy paper to make sandpaper.

**Building Materials** Building materials are used to construct roads and buildings. An important nonmetallic mineral resource is aggregate, which is crushed stone or a mix of gravel and sand. It has many uses as a building material. Concrete is made by mixing gravel and sand with cement and water. Concrete is used for sidewalks, roads, driveways, basements, and foundations.

**Gypsum and Rock** When seawater evaporates, gypsum forms. It is a building material used to make plaster and wallboard.

Rock also is used as building stone. Some buildings in your area are probably made from granite, limestone, or sandstone. Some rock also is used to make statues and other artwork.

## Recycling Mineral Resources

Mineral resources are nonrenewable. They cannot be replaced by natural Earth processes within an average person's lifetime. Most mineral resources take millions of years to form. Have you ever thrown away an empty soft-drink can? These cans end up as solid waste. It would be better if cans and other items made from mineral resources were recycled into new items.

**Recycling** is using old materials to make new items. Recycling has many benefits. It reduces the need for new mineral resources. Also, recycling often costs less than making the same item from new materials. Minerals are nonrenewable resources. The graph below shows the percentage of mineral resources that are imported to and produced in the United States. In the future, supplies of some minerals might become limited. Recycling might be the only way to meet the need for them.

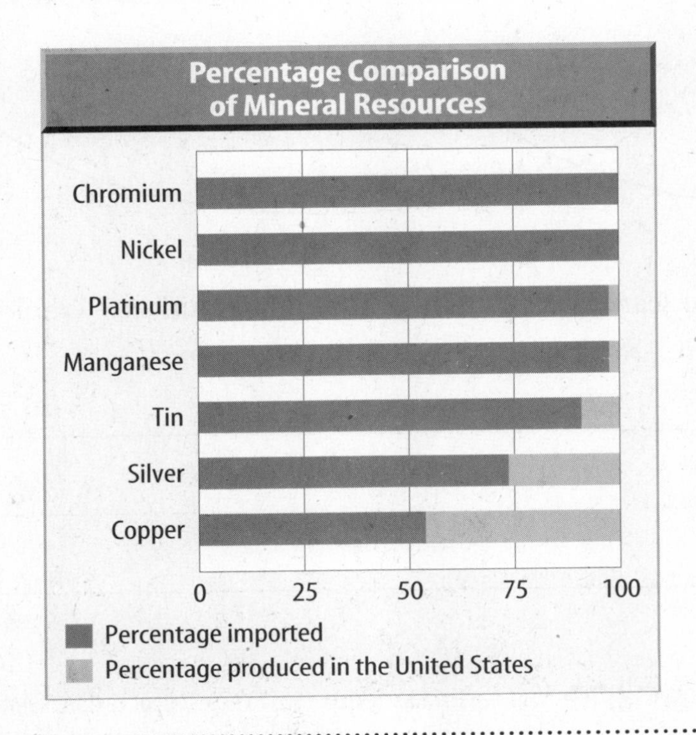


### Think it Over

**3. Explain** What is one benefit of recycling?

______________________

______________________

### Picture This

**4. Interpret** According to the graph, what is the most widely produced mineral resource in the U.S.?

______________________

## After You Read

### Mini Glossary

**mineral resource:** resource from which metal is obtained

**ore:** deposits of minerals that are large enough be mined at a profit

**recycling:** processing old materials to make new items

1. Review the terms and their definitions in the Mini Glossary. Then explain how a mineral resource and an ore are similar and different.

_______________________________________________

_______________________________________________

_______________________________________________

2. Use the following diagram to describe the similarities and differences of metallic and nonmetallic mineral resources.

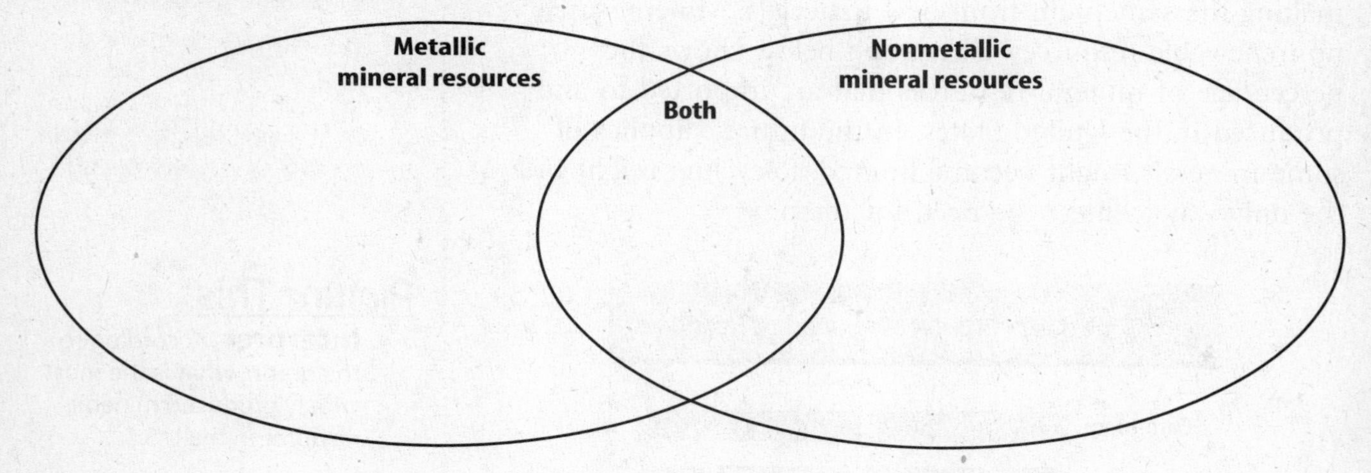


3. How did making flash cards help to learn the new terms and the definitions? Would you use this strategy again?

_______________________________________________

_______________________________________________

_______________________________________________

Visit **earth.msscience.com** to access your textbook, interactive games, and projects to help you learn more about mineral resources.

# chapter 6 Views of Earth

## section 1 Landforms

### Before You Read

If you could look at the area where you live from a hot air balloon, what would you notice about the land? Describe what you would see. Include any important landforms.

______________________________________________

______________________________________________

**What You'll Learn**

- differences between plains and plateaus
- about folded, upwarped, fault-block, and volcanic mountains

### Read to Learn

## Plains

Whether viewed from the ground, the sea, or the air, Earth's landforms show great variety. There are vast plains, rocky plateaus, tall mountains, and deep valleys. Plains are one of the most common landforms on Earth.

**Plains** are large, flat areas often found in the interior, or middle, regions of continents. Plains are good places to grow crops. They are flat and often have rich soil. Thick grass grows well in this type of soil. This makes plains ideal places for cattle and other grass-eating animals to live. Foxes, snakes, and ground squirrels also live on plains.

Coastal plains are the flat areas along seacoasts. Interior plains and coastal plains make up half of all the land in the United States.

### What are lowlands?

Another name for coastal plains is lowlands. A lowland is lower than the land surrounding it. Coastal plains are part of the continental shelf. The continental shelf is the part of the continent that extends into the ocean. The Atlantic Coastal Plain is a good example. It stretches along the east coast of the United States from New Jersey to Florida. This area has low rolling hills, swamps, and marshes. A marsh is a grassy wetland that usually is flooded with water.

Study Coach

**Two-Column Notes** As you read, organize your notes into two columns. In the left column, write the main idea. Next to it, in the right column, write details about it.

FOLDABLES™

**A Record Information** Make a layered book from two sheets of paper to record differences among major landforms discussed in this section.

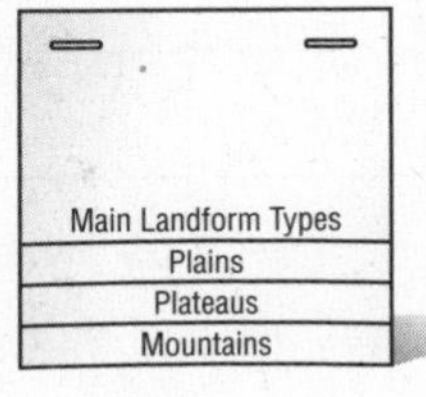

## Picture This

**1. Classify** List the terms found on the map that are used to name landforms.

______________________

______________________

______________________

______________________

**Major U.S. Landforms**



## How do coastal plains form?

The Atlantic Coastal Plain began forming about 70 million years ago. Sediment began to build up on the ocean floor. When the sea level dropped, the sea floor was exposed. This sea floor became the coastal plain.

The size of coastal plains changes over long periods of time. When the level of the sea rises, water covers part of the plain. When the level of the sea falls, more of the coastal plain is exposed. Coastal plains were larger during the last ice age because much of Earth's water was frozen in glaciers.

The Gulf Coastal Plain includes the lowlands in the southern United States that border the Gulf of Mexico. Much of this plain was formed from sediment deposited by the many rivers that flow into the gulf. Both coastal plains are shown in the map above.

## What are interior plains?

Interior plains cover most of the central part of the United States. They are located between the Rocky Mountains, the Appalachian Mountains, and the Gulf Coastal Plain.

Within the interior plains is a large area called the Great Plains. The Great Plains are located between the Mississippi River and the Rocky Mountains. This area is flat, dry, and covered with grass and few trees. The Great Plains also are referred to as the high plains. They rise from 350 m to 1500 m above sea level. The Great Plains consist of layers of sedimentary rocks. ☑

**Reading Check**

**2. Explain** Why are the Great Plains called the high plains?

______________________

______________________

______________________

## Plateaus

**Plateaus** (pla TOHZ) are flat, raised areas of land made up of horizontal rocks that have been uplifted by forces inside Earth. Plateaus are different from plains because they have steep edges. These edges rise sharply away from the land around them.

Plateaus often have deep river valleys or canyons. The Grand Canyon was formed by the Colorado River cutting into the rock layers of the Colorado Plateau, which is shown on the map on the previous page. The Colorado Plateau is located mostly in what is now a dry region. As a result, only a few rivers have developed on its surface. If you hiked around on this plateau, you would be in a high, rugged area.

## Mountains

Mount Everest is the world's highest mountain. Mount Everest is more than 8,800 m above sea level. It is located in the Himalaya in southeast Asia. The highest mountain in the United States is just over 6,000 m. Mountains vary in how they are formed. The four main types of mountains are folded, upwarped, fault-block, and volcanic. ☑

### How do folded mountains form?

**Folded mountains** form when strong forces within Earth squeeze rock layers. These forces cause the rock layers to fold. The figure below shows how the rock layers fold. The folds look like a rug that has been pushed up against a wall.

The Appalachian mountains are folded mountains. They formed between 480 million and 250 million years ago and are one of the oldest and longest mountain chains in North America.

**Folded Mountains**

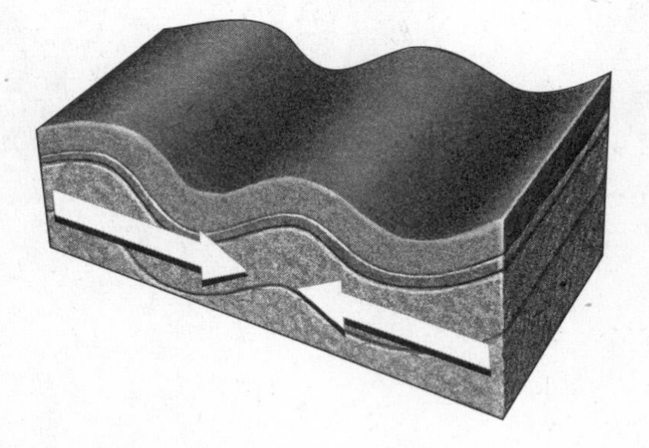

**Reading Check**

3. **Identify** List the four main types of mountains.

_______________

_______________

_______________

_______________

**Picture This**

4. **Describe** Which of these words best describes a folded mountain: pointed, flat, craggy, or wavy?

_______________

## How do upwarped mountains form?

**Upwarped mountains** form when blocks of Earth's crust are pushed up by forces inside Earth. Their shape is shown in the figure below. Over time, wind and weather wear down, or erode, the mountain's top layer of soil and soft rock. The blocky shapes of upwarped mountains slowly erode into sharp peaks and ridges. This exposes the hard rock beneath. The Adirondack Mountains in New York are upwarped mountains.

### Think it Over

**5. Infer** What do the arrows represent in the figure of fault-block mountains?

______________________

______________________

## Where can fault-block mountains be found?

The Grand Tetons in Wyoming and the Sierra Nevada in California are fault-block mountains. **Fault-block mountains** are made of huge tilted blocks of rocks as shown below. Faults separate the blocks of rocks from the rocks around them. Faults are large cracks in Earth's crust formed when layers of rock shift up and down. Sometimes, when one block of rock shifts up, another shifts down. This creates mountains with sharp peaks and steep slopes.

## How do volcanic mountains form?

**Volcanic mountains** are formed when hot, melted material reaches the surface through a weak area of crust. Hot rocks, ashes, smoke, and gases are thrown out, or ejected. The materials build up, layer by layer. Over time, a cone-shaped mountain forms, as shown in the figure below.

Mount Shasta in California and Mount St. Helens in Washington are volcanic mountains. The Hawaiian Islands are volcanic mountains. Their bases rest on the ocean floor. Their peaks form the islands that are seen in the water.

### Picture This

**6. Describe** Which best describes upwarped mountains: block-shaped and wavy or block-shaped and peaked?

______________________

Upwarped

Fault-Block

Volcanic

# After You Read

## Mini Glossary

**fault-block mountains:** mountains formed from huge, tilted blocks of rock; separated by faults from surrounding rock

**folded mountains:** mountains formed when horizontal rock layers are squeezed from opposite sides, causing them to buckle and fold

**plain:** large, flat landform that often has thick, fertile soil and is usually found in the interior region of a continent

**plateau (pla TOH):** flat, raised landform made up of nearly horizontal rocks that have been uplifted

**upwarped mountains:** mountains formed when blocks of Earth's crust are pushed up by forces inside Earth

**volcanic mountains:** mountains formed when molten material reaches Earth's surface through a weak area in Earth's crust and piles up into a cone-shaped structure

1. Review the terms and their definitions in the Mini Glossary. Write two sentences explaining the differences between two different types of mountains.

_______________________________________________

_______________________________________________

_______________________________________________

2. Write these terms in the correct area in the Venn diagram below: *flat, raised, interior region, uplifted rocks, landform.*

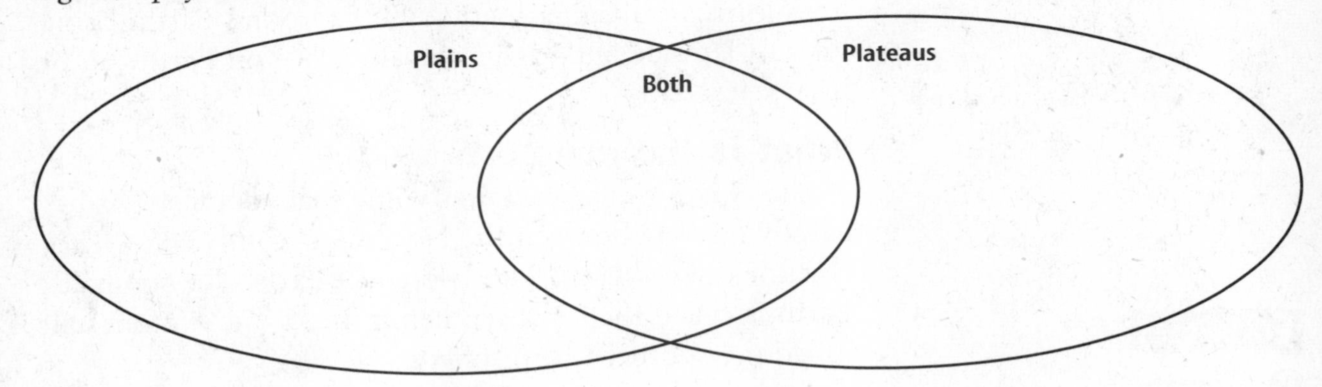


3. You identified and wrote the main ideas and details as you read this section. How did this strategy help you learn the material?

_______________________________________________

_______________________________________________

Visit **earth.msscience.com** to access your textbook, interactive games, and projects to help you learn more about landforms.

# Views of Earth

## section 2 Viewpoints

### What You'll Learn

- about latitude and longitude
- about the equator
- about the prime meridian

## Before You Read

If your body were Earth, where would the North Pole and the South Pole be located?

_______________________________________________

_______________________________________________

Mark the Text

**Identify** Highlight the main idea in each paragraph in this section.

## Read to Learn

### Latitude and Longitude

During hurricane season, scientists track storms as they form in the Atlantic Ocean. The exact location of the storm can be identified using latitude and longitude lines. These lines form an imaginary grid system around Earth. Using lines of latitude and longitude, any place on Earth can be located exactly.

### What is the equator?

The **equator** is an imaginary line that encircles Earth exactly halfway between the North and South Poles. The equator divides Earth into two equal parts. The top half of Earth is called the northern hemisphere. The bottom half is called the southern hemisphere.

FOLDABLES™

**B Identify** Make a ten-tab vocabulary Foldable to help you learn some of the terms in this section.

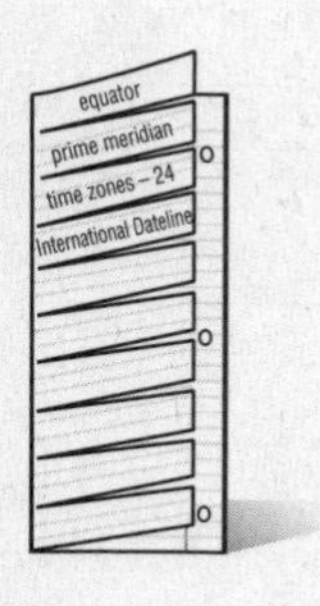


### What are lines of latitude?

Imaginary lines that run parallel to the equator are called lines of **latitude,** or parallels. Latitude is the distance, measured in degrees, north or south of the equator. Because these imaginary lines are parallel to the equator, they do not cross one another.

The equator is at 0° latitude. The North Pole is at 90° north latitude. The South Pole is at 90° south latitude. Each degree is divided into 60 minutes. Each minute is further divided into 60 seconds.

## Latitude and Longitude

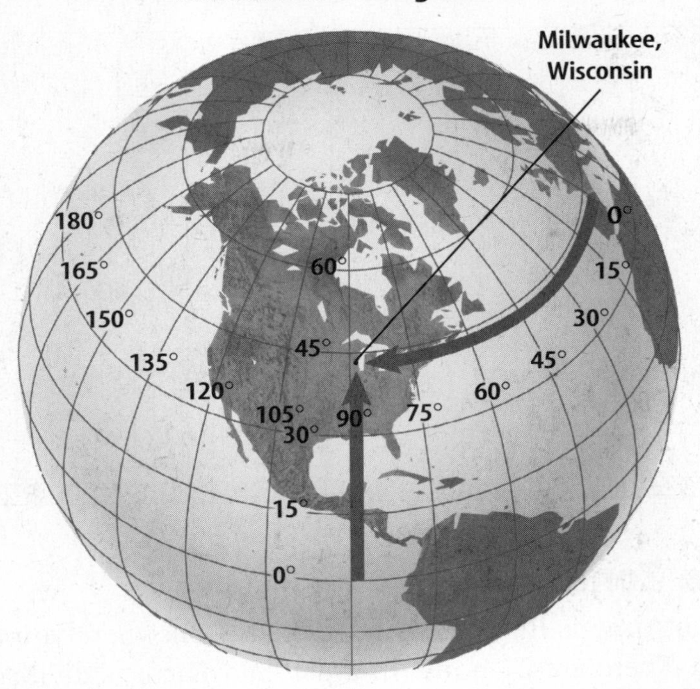


## What is longitude?

Imaginary vertical lines that run north and south are called meridians, or lines of longitude (LAHN juh tewd). Lines of longitude use the **prime meridian** as a reference point. The prime meridian runs from the North Pole through the city of Greenwich (GREN ihtch), England, to the South Pole. The prime meridian represents 0° longitude. **Longitude** is the distance in degrees east or west of the prime meridian. ☑

Look at the figure at the top of the page. Find the lines of longitude running from north to south. Points west of the prime meridian have west longitude measured from 0° to 180°. Points east of the prime meridian have east longitude from 0° to 180°. The city of Milwaukee, Wisconsin is located at 43° north latitude and 88° west longitude. Latitude is always listed before longitude.

## Does the prime meridian circle Earth?

The prime meridian and the equator are both reference points. They are both imaginary lines. However, the prime meridian does not circle Earth as the equator does. It runs from the North Pole to the South Pole through Greenwich, England. The line of longitude on the opposite side of Earth from the prime meridian is the 180° meridian. Any point on Earth can be located by knowing both its latitude and its longitude.

### Picture This

1. **Locate** Highlight or trace over the equator in red.

### Reading Check

2. **Determine** Longitude is a measure of degrees east or west of what reference point?

______________________

### Think it Over

3. **Explain** one difference between longitude and latitude.

______________________

______________________

**Picture This**

4. **Identify** Into how many time zones is Earth divided?

______________________

**World Map with Time Zones**

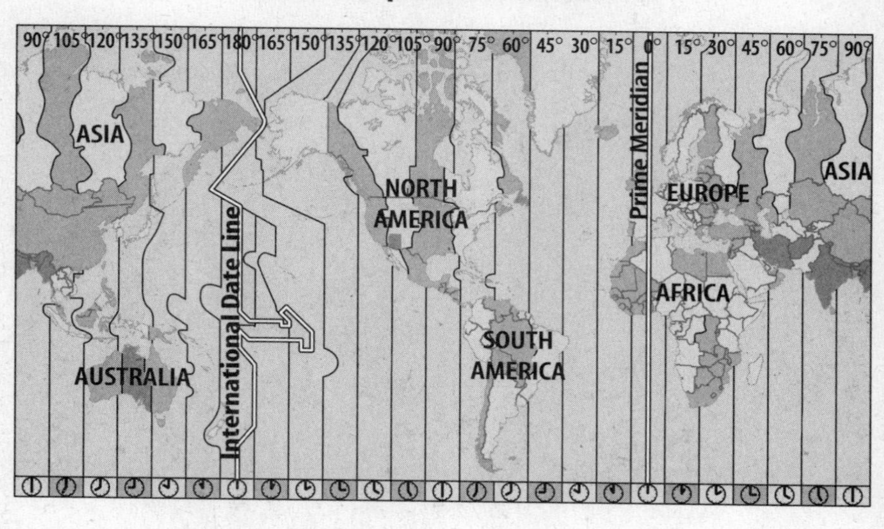


# Time Zones

What time is it? The answer depends on where you are on Earth. There are 24 hours in a day and Earth is divided into 24 time zones. Time is measured by tracking Earth's movement in relation to the Sun. As Earth rotates, the Sun's rays strike different parts of the planet at different times.

## How are time zone lines established?

Each time zone is about 15° of longitude wide. Each time zone is about one hour different from the time zones on each side of it. Look at the figure above. Not all time zones follow lines of longitude exactly. Some time zone boundaries are adjusted in local areas. It would be confusing to have cities split between two different time zones.

# Calendar Dates

In each time zone, one day ends and the next day begins at midnight.

## What is the International Date Line?

You gain time or lose time when you move from one time zone to another. If you travel far, you can gain or lose a whole day. On the map above, find the International Date Line. It is located near the 180° meridian. This is the boundary line for calendar days. If you travel west and cross the International Date Line, you would move your calendar forward one day. If you travel east and cross it, you would move your calendar back one day. ☑

**Reading Check**

5. **Describe** At about what degree longitude is the date line located?

______________________

# ● After You Read

## Mini Glossary

**equator:** imaginary line that wraps around Earth at 0° latitude, halfway between the north and south poles

**latitude:** distance in degrees north or south of the equator; lines of latitude are also known as parallels

**longitude:** distance in degrees east or west of the prime meridian; lines of longitude are also called meridians

**prime meridian:** imaginary line that represents 0° longitude and runs from the north pole through Greenwich, England, to the south pole

1. Review the terms in the Mini Glossary. Then write a sentence that explains the difference between the equator and the prime meridian.

_______________________________________________

_______________________________________________

2. Locate the *equator* and the *prime meridian* and write these terms in the correct boxes. In the other two boxes write *longitude* or *latitude* to identify these lines.

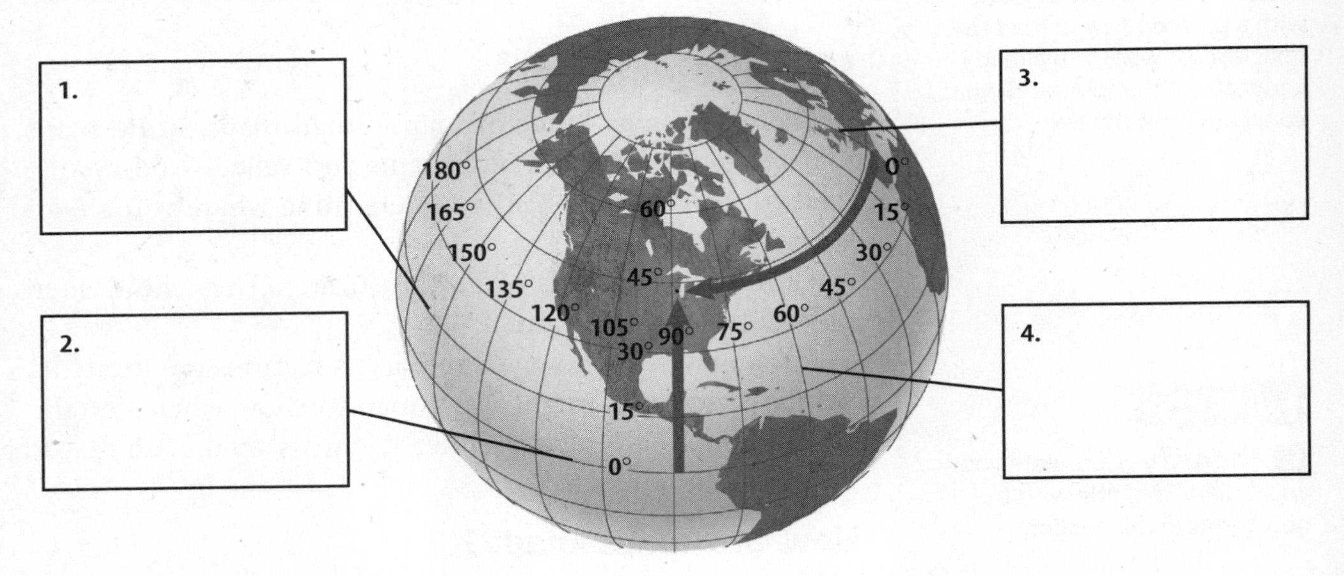


3. You highlighted the main idea in each paragraph as you read the section. How would this strategy help you study for a test?

_______________________________________________

_______________________________________________

Visit **earth.msscience.com** to access your textbook, interactive games, and projects to help you learn more about latitude and longitude.

# chapter 6 Views of Earth

## section 3 Maps

**What You'll Learn**
- about different map projections
- how to use information on maps

## Before You Read

Suppose someone hands you a city map and asks you to find city hall. What do you need to know in order to locate city hall?

_______________________________________________

_______________________________________________

**Study Coach**

**Think-Pair-Share** Work with a partner. As you read the text, discuss what you already know about the topic and what you learn from the text.

## Read to Learn

### Map Projections

There are many kinds of maps—road maps, world maps, maps that show Earth's mountains and valleys, and even treasure maps. Maps help you determine where you are and where you are going.

All maps are models of Earth's surface. They show where different places are located.

Maps may also show where Earth's features are located. An Earth scientist might use a map to show where certain types of rocks are found. Other scientists might use maps to locate ocean currents.

**FOLDABLES**

**C Identify** Complete your vocabulary Foldable with the new terms in this section.

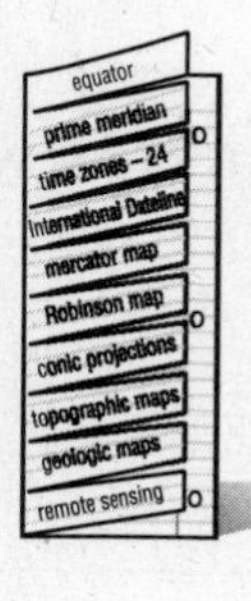


#### How are maps made?

Many maps are made as projections. A map projection is made when points and lines on the surface of a globe are transferred to paper. There are different ways to make a map projection. However, all types of map projections distort the shapes and sizes of land masses. As a result, the land masses may look larger or smaller or may even be shaped a little differently.

## Mercator Projection

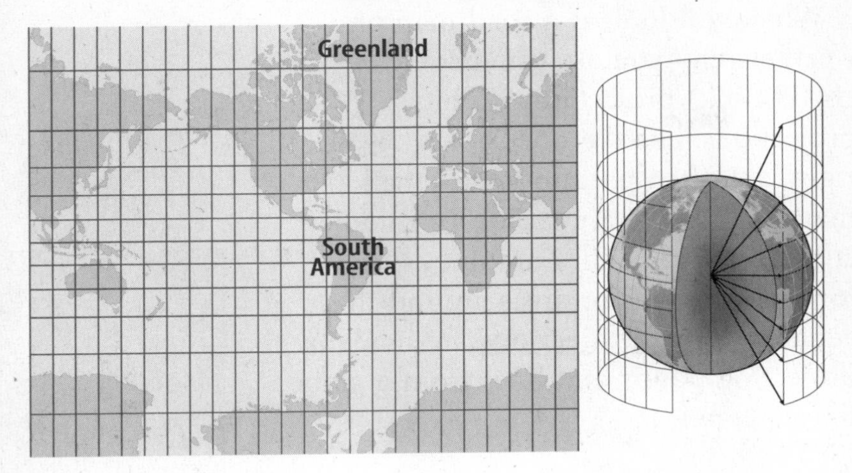


## What is a Mercator projection?

Mercator (mer KAY ter) projections project the correct shapes of continents but the areas are distorted. Lines of longitude are projected onto the map parallel to each other. As you learned earlier, only latitude lines are parallel. Longitude lines meet at the poles. When longitude lines are projected as parallel, areas near the poles appear bigger than they are. In the figure above, Greenland appears to be larger than South America. However, Greenland is smaller. Mercator projections are used mainly on ships.

## Robinson Projection

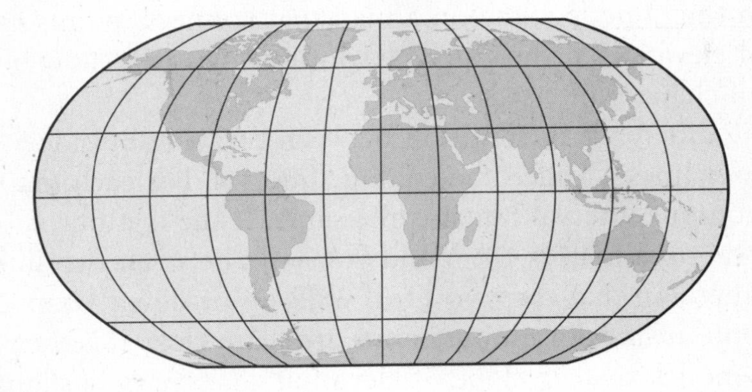

## What is a Robinson projection?

Like a Mercator projection, a Robinson projection shows accurate shapes of continents. It also shows land areas accurately. As the figure above shows, the lines of latitude in a Robinson projection are parallel. The lines of longitude curve as they do on a globe. As a result, land areas near the poles do not look as distorted.

### Picture This

1. **Explain** Trace around Greenland and South America. Why does Greenland appear larger on a Mercator projection?

______________________

______________________

______________________

**FOLDABLES™**

**D Organize** Use two half sheets of paper to take notes on map projections and topographic maps.

Map Projections

Topographic Maps

### Picture This

2. **Interpret** Look at the Robinson projection. Why aren't the shapes of continents distorted?

______________________

______________________

## Picture This

**3. Differentiate** Does the conic projection display a shape like a box or a shape like a cone?

_______________

### What is a conic projection?

When you look at a road map or a weather map you are using conic (KAH nihk) projections. Conic projections, like the one shown on the right, are often used to make small area maps. The maps are useful for middle latitude regions. They are not as useful for polar regions or areas around the equator. **Conic projections** are made by projecting points and lines from a globe onto a cone.

Conic Projection

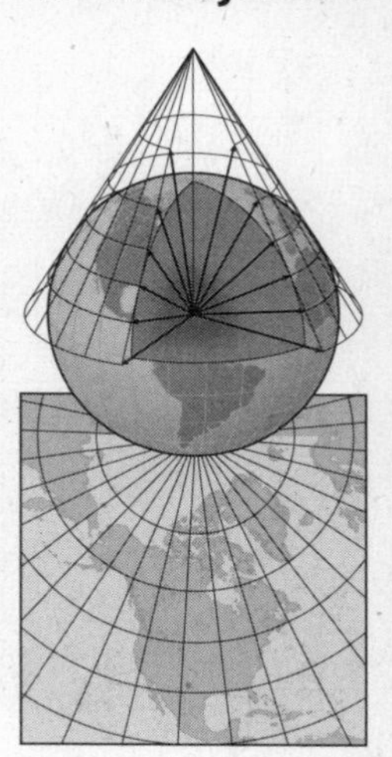

## Topographic Maps

If you were going to hike in a new area, you would look at a conic map projection first. It would show which roads to follow to get to the area. Once you arrived, a topographic map would show you the terrain. **Topographic** (tah puh GRA fihk) **maps** model the changes in the elevation of Earth's surface. They show the hills, valleys, and other landforms. They also show roads, cities, and other structures built by humans.

## Think it Over

**4. Infer** Which kind of hiker would be likely to climb a mountain whose map showed contour lines close together—a beginning hiker or an experienced hiker?

_______________

### What are contour lines?

Before your hike, you study the contour lines on your topographic map to see the change in elevation in the trail. A **contour line** is a line on a map that connects points of equal elevation, or height. The contour lines will show how a trail changes in elevation.

The difference in elevation between two side-by-side contour lines is called the contour interval. For each map, the contour interval remains the same. If the contour interval on a map is 10 m and you walk between two lines anywhere on that map, you will walk up or down 10 m.

Contour lines in mountainous areas are close together. This models a steep slope. If elevation changes only slightly, the contour lines will be farther apart.

### What are index contours?

If the contour interval is 5 m, you can determine the elevation of other lines around the index contour by adding or subtracting 5 m from the elevation shown on the index contour. Some contour lines are marked with their elevation. They are called index contours.

**Map Scales and Legends** The map scale shows the relationship between the distances on the map and the distances on the ground. A map's scale is expressed as a fraction or a ratio. If one inch on the scale equals a distance of 100 miles, it would be shown on the map as 1:100. A map scale may appear as a small bar divided into sections. ☑

A map legend explains what each symbol used on the map means. The legend is usually at the bottom of the map.

**Map Series** Topographic maps cover different amounts of Earth's surface. A map series includes maps that have the same dimensions of latitude and longitude. A map series may include maps that focus on small areas of Earth's surface, such as 7.5 minutes of latitude by 7.5 minutes of longitude, or larger areas.

## Geologic Maps

Geologic maps show the arrangements and types of rocks on Earth's surface. Earth scientists use geologic maps to determine how rock layers may look beneath Earth's surface.

Topographic maps and geological maps are two-dimensional maps used to study Earth's features. Computers can create maps that show a three-dimensional view of Earth's features. These maps are useful because they show not only how long and wide an area is, but also what is beneath the surface.

The block diagram below is a 3-D model that shows a solid section of Earth. The top surface is the geologic map. Side views are cross sections. Geologic maps and cross sections provide information that helps locate natural resources.

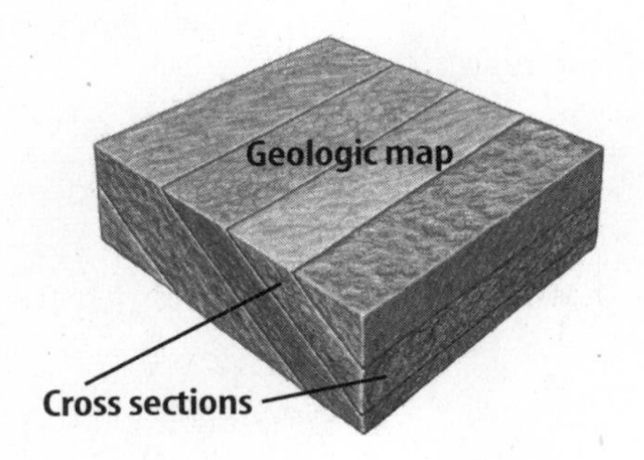


**Map Uses** Maps are tools that give information about Earth's surface. A road map is the best tool for finding a route. A topographic map can be useful when hiking. A Mercator projection is useful when you don't have a globe.

**Reading Check**

5. **Determine** What shows the relationship between distance on a map and distance on the ground?

______________________

**Picture This**

6. **Identify** What dimensions are shown on 3-D maps and models?

______________________

______________________

______________________

# After You Read

## Mini Glossary

**conic projection:** map made by projecting points and lines from a globe onto a cone

**contour line:** line on a map that connects points of equal elevation

**map legend:** explains the meaning of symbols used on a map

**map scale:** relationship between distances on a map and distances on Earth's surface that can be represented as a ratio or as a small bar divided into sections

**topographic map:** map that shows the changes in elevation of Earth's surface and indicates such features as roads and cities

**1.** Review the terms and their definitions in the Mini Glossary. Then write a sentence that explains how map legends and map scales are used.

______________________________________________

______________________________________________

______________________________________________

**2.** Fill in each blank with the name of the correct map projection.

**Map Projections**

| ____________ | ____________ | ____________ |
|---|---|---|
| • shows correct shapes of continents<br>• shows correct areas of continents<br>• shows lines of longitude curving as they are on a globe | • shows correct shapes of continents<br>• shows distorted areas of continents<br>• projects lines of longitude parallel | • made by projecting points and lines from a globe onto a cone<br>• used for smaller areas<br>• road maps and weather maps |

**3.** If you were planning a hike, what type of map would you consult to know what kind of terrain you'll be traveling?

______________________________________________

______________________________________________

**Science Online** Visit **earth.msscience.com** to access your textbook, interactive games, and projects to help you learn more about maps.

# Weathering and Soil

## section 1 Weathering

### Before You Read

Think of a time when you had soil on your hands. That soil might once have been at the top of a mountain. Describe how soil might make its way down a mountain.

______________________________________________

______________________________________________

______________________________________________

**What You'll Learn**
- how weathering changes Earth's surface
- how climate affects weathering
- the difference between mechanical and chemical weathering

### Read to Learn

#### Weathering and Its Effects

Tiny moss plants, earthworms, and even oxygen weaken and break apart rocks at Earth's surface. The surface processes that break down rock are called **weathering**. Weathering breaks rock into smaller and smaller pieces called sediment. Sand, silt, and clay are three different sizes of sediment. Sand grains are larger than silt. Silt is larger than clay.

Over millions of years, weathering has changed Earth's surface. The process continues today. There are two different types of weathering—mechanical weathering and chemical weathering. Both types work together to change Earth's surface.

Mark the Text

**Main Ideas** As you read, underline or highlight the main idea in each paragraph.

FOLDABLES™

**A Compare and Contrast** Make a two-tab Foldable to help you organize notes about mechanical weathering and chemical weathering.

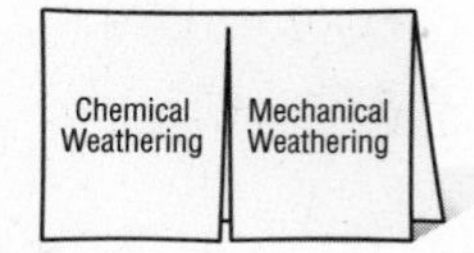


#### Mechanical Weathering

**Mechanical weathering** occurs when rocks are broken apart by physical processes. Mechanical weathering breaks rocks into small pieces, changing only the size and shape of the rock. The chemical makeup of the small pieces is the same as the chemical makeup of the original rock.

## How do plants and animals cause weathering?

Water and nutrients collect in the cracks of rocks. Seeds that land in the cracks are able to grow. As a plant grows, its roots grow larger and move deeper into the crack in the rock. As the roots get bigger, they make the crack larger. You may have seen how the roots of a tree can lift and crack a piece of sidewalk. This is one way plants cause mechanical weathering.

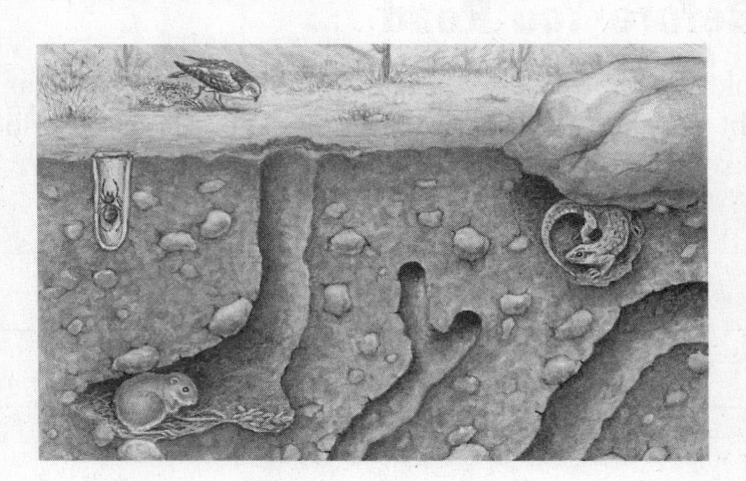

### Picture This

**1. Explain** what happens to soil when animals burrow in the ground.

______________________

______________________

______________________

Animals also cause mechanical weathering. Look at the figure above. Small burrowing animals, such as voles, dig tunnels in the ground. Burrowing loosens small rocks and sediment in soil. The animal pushes these small pieces of rock to the surface. Once these small rocks and sediment are out of the ground, other weathering processes act on them.

## What is ice wedging?

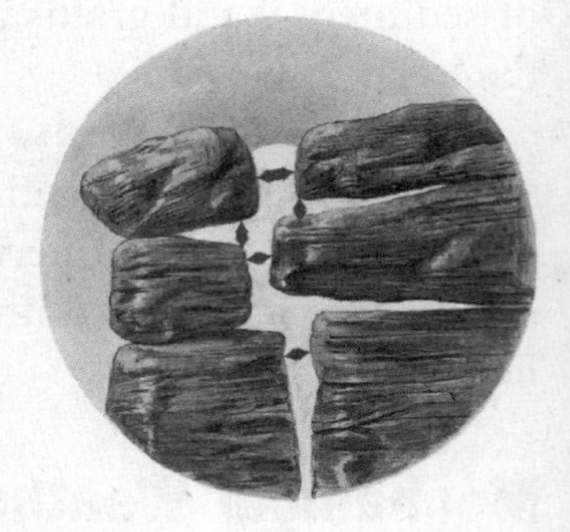

### Picture This

**2. Determine** Use a marker to highlight the place in the figure where ice wedging is most likely to occur.

**Ice wedging** is the mechanical weathering process that occurs when water freezes in the cracks of rocks. Water may seep into a crack in a rock. As the water turns to ice, it expands and pushes against the sides of the crack. The crack gets wider and deeper, as shown in the figure. The pressure of the ice in the crack is so great it can break the rock apart.

When temperatures rise, the ice melts. Because the crack is larger now, more water can enter the crack. When the water freezes again, the ice will again put pressure on the crack. After many years of this freezing and melting cycle, the rock will break up completely.

### Where does ice wedging occur?

Ice wedging is often seen in mountains, where warm days and cold night are common. Ice wedging is a process that wears down mountain peaks. The cycle of freezing and thawing also breaks up roads. When water seeps into cracks in the pavement and freezes, it forces the pavement apart. Ice wedging in roads is one cause of potholes.

### How does mechanical weathering affect surface area?

Mechanical weathering by plants, animals, and ice wedging breaks rocks into smaller pieces. These small pieces have more surface area than the original rock had. As the surface area increases, more rock is exposed to water and oxygen. This speeds up a different kind of weathering, called chemical weathering.

## Chemical Weathering

Chemical weathering occurs when chemical reactions dissolve the minerals in rocks or change them into different minerals. Like mechanical weathering, chemical weathering changes the size and shape of rocks. But it also changes the chemical makeup of rock. These chemical changes weaken the rock.

### How do natural acids weather rock?

Naturally formed acids can weather rock. Carbonic acid is a natural acid formed when water reacts with carbon dioxide gas in the air or soil. Even though carbonic acid is a weak acid, it causes chemical weathering in rocks.

Over thousands of years, carbonic acid can form caves in limestone. Calcite is the main mineral in limestone. When carbonic acid reacts with calcite, it causes the calcite to dissolve. Over time, enough calcite in the limestone may dissolve to form a cave. ☑

Other naturally occurring acids weather other types of rock. Granite, some types of sandstone, and other rocks all contain the mineral feldspar. Over many years, feldspar is broken down into a clay mineral called kaolinite (KAY oh luh nite). Kaolinite clay is found in some soils. Clay is an end product of weathering.

**Think it Over**

**3. Compare and Contrast** How are mechanical and chemical weathering the same?

______________________

______________________

______________________

**Reading Check**

**4. Identify** What natural acid dissolves calcite?

______________________

## How do plant acids cause chemical weathering?

Some plant roots give off acids. Rotting or decaying plants also give off acids. These natural acids can dissolve minerals in rock. When the minerals dissolve, the rock is weakened. Over time, the rock cracks and breaks into smaller pieces. As the rock weathers, nutrients become available to plants. ☑

**Reading Check**

**5. Identify** How do plant roots and rotting plants affect rock?

______________________

______________________

## How does oxygen cause chemical weathering?

Oxygen causes chemical weathering. You have probably seen rusty swing sets, nails, and cars. Rust is caused by oxidation (ahk sih DAY shun). **Oxidation** occurs when some materials are exposed to oxygen and water. When minerals containing iron are exposed to water and oxygen, the iron in the mineral reacts to form a new material. This new material looks like rust.

Oxidation occurs in the mineral magnetite. When the iron in magnetite is exposed to water and oxygen, it forms limonite, a rustlike material. Oxidation of minerals gives some rock layers a red or rust color.

# Effects of Climate

Climate affects how quickly weathering occurs. **Climate** is the pattern of weather in a region over many years. In cold climates, with frequent freezing and thawing, mechanical weathering happens rapidly. Freezing and thawing cause ice wedging that breaks down rock.

Chemical weathering is more rapid in warm, wet climates. So, chemical weathering occurs quickly in tropical areas, such as rain forests. Chemical weathering is slower in deserts where there is little water. The constant low temperatures in polar regions also slow down chemical weathering.

**Think it Over**

**6. Draw Conclusions** Why does chemical weathering occur slowly in polar regions?

______________________

______________________

## Do all rocks weather at the same rate?

Different types of rock weather at different rates. In wet climates, marble weathers rapidly and discolors. Granite weathers more slowly in humid climates.

Weathering affects rocks, caves, mountains, streets, and even statues. Weathering is an important part of the rock cycle. Weathering also begins the process of forming soil from rock and sediment.

# After You Read

## Mini Glossary

**chemical weathering:** occurs when chemical reactions dissolve minerals or change them into different minerals

**climate:** the pattern of weather in a region over many years

**ice wedging:** mechanical weathering process caused when water freezes inside the cracks of rocks

**mechanical weathering:** occurs when rocks are broken apart by physical processes

**oxidation:** chemical weathering process that occurs when some materials are exposed to oxygen and water

**weathering:** surface processes that weaken and break rock into smaller and smaller pieces

1. Review the terms and their definitions in the Mini Glossary. Then write a sentence explaining how ice wedging breaks down rocks. Use at least two terms in your answer.

______________________________________________________________________

______________________________________________________________________

______________________________________________________________________

2. In each box below, write two examples of weathering.

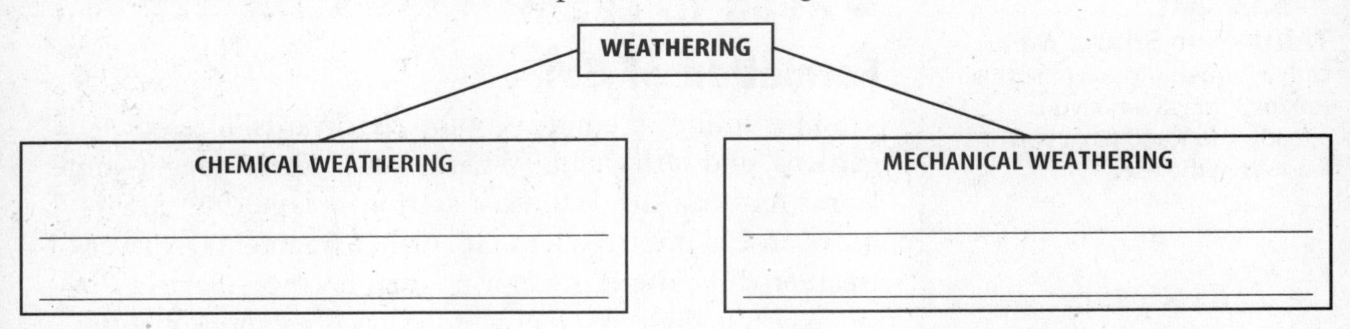


3. As you read this lesson, you highlighted the main idea in each paragraph. How did highlighting the main idea help you understand the different types of mechanical and chemical weathering?

______________________________________________________________________

______________________________________________________________________

______________________________________________________________________

Visit **earth.msscience.com** to access your textbook, interactive games, and projects to help you learn more about weathering and its effects.

# Weathering and Soil

## section 2 The Nature of Soil

**What You'll Learn**
- how soil forms
- what factors affect soil development
- what the characteristics of soil are

### Before You Read

Think of the plants that grow where you live. Do you think your area would be a good place to grow vegetables? Explain why or why not on the lines below.

______________________________

______________________________

______________________________

**Study Coach**

**Think-Pair-Share** Work with a partner. As you read this section, discuss what you already know and what you learn from the text.

### Read to Learn

#### Formation of Soil

Soil is found in empty city lots, backyards, forests, gardens, and farm fields. What is soil? Where does it come from? As you read in the last section, weathering slowly breaks rocks into smaller and smaller fragments. A layer of weathered rock and mineral fragments covers Earth's surface. But these fragments don't become good quality soil until plants and animals live in them. Plants and animals add organic matter, the remains of once living organisms. Organic matter can include leaves, twigs, roots, and dead worms and insects. **Soil** is a mixture of weathered rock, decayed organic matter, mineral fragments, water, and air.

#### What factors affect soil formation?

Soil can take thousands of years to form. In some places soil is 60 m thick, but in other places it is only a few centimeters thick. Five factors—climate, slope of the land, types of rock, types of plants, and the amount of time that rock has been weathering—affect soil formation. For example, different types of soil develop in tropical areas than in polar regions. Soils that form on steep slopes are different from soils that develop on flat land.

## How does soil form?

Over time, soil can form from rock. Natural acids in rainwater begin to weather the surface of rock. Water seeps into cracks in the rock and freezes, causing the rock to break apart. Then, plants start to root in the cracks. As the roots of the plants grow, they continue breaking down the rock. A thin layer of soil begins to form. Organisms like grubs and worms live among the plant roots, adding organic matter to the soil. As organic matter increases, the layer of soil gets thicker. Over time, a rich layer of soil forms that can support trees and plants with larger roots.

# Composition of Soil

As you have learned, soil is made up of rock and mineral fragments, organic matter, air, and water. The rock and mineral fragments found in soils come from rocks that have been weathered. Most of the pieces of rock and mineral are small particles of sediment such as sand, silt, and clay. However, some larger pieces of rock also might be present.

## Where does organic matter come from?

Most of the organic matter in soil comes from plants. Animals add organic matter to soil when they die. After plant and animal material gets into the soil, bacteria and fungi help it decay. The decayed organic matter turns into a rich, dark-colored material in soil called **humus** (HYEW mus). Humus is a source of nutrients for plants.

Animals are important in making good soil. As worms, insects, and rodents, like mice, burrow in the soil, they mix the humus with the fragments of rock. Good soil is made up of equal amounts of humus and weathered rock material.

There are many small spaces between the sediment and humus particles in soil that are filled with water or air. The spaces in moist soil hold water that plants need to grow. During a drought, the spaces in the soil are filled with air.

# Soil Profile

If you've ever seen a deep hole, you've probably seen different layers of soil. Most plant roots grow in the top layer of soil. The top layer is usually darker than the layers below it. The different layers of soil are called **horizons.** All the horizons of a soil form a **soil profile.** Most soils have three horizons that are referred to as A, B, and C horizons.

### Think it Over

1. **Explain** What are some things you would look for to find out if the soil in an area was rich?

_______________

_______________

_______________

**FOLDABLES™**

**B Organize Information**
Make a three-tab Foldable to organize information about the three horizons in a soil profile.

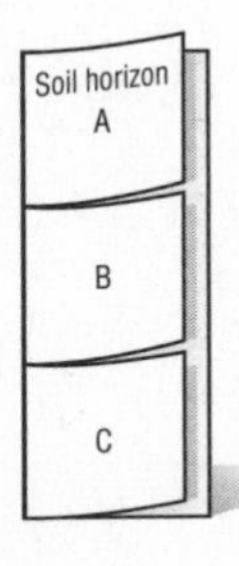

## What is found in the A horizon?

The top layer of soil is called the A horizon. In a forest, the A horizon might be covered with organic plant litter. **Litter** consists of twigs, leaves, and other organic matter that will become humus. Litter traps water and keeps the A horizon moist. Litter also prevents erosion. When litter decays and turns into humus, it provides nutrients for plant growth. ☑

Another name for the A horizon is topsoil. Topsoil has more humus and fewer rock and mineral fragments than other layers. The A horizon is usually dark-colored and rich, or fertile.

**Reading Check**

2. **Explain** one way that plant litter improves the soil?

______________________

______________________

______________________

## What is found in the B horizon?

The layer below the A horizon is the B horizon. The B horizon is lighter in color than the A horizon because it contains less humus. As a result, the B horizon is not as fertile. Sometimes material from the A horizon moves down into the B horizon through a process called leaching.

**Leaching** is the carrying away of minerals that have been dissolved in water. Water seeps through the A horizon. The water reacts with humus and carbon dioxide to form acid. This acid dissolves some of the minerals in the A horizon. The acid seeps deeper into the soil, carrying the minerals into the B horizon. The figure below shows how leaching occurring in forest soil.

**Picture This**

3. **Determine** Trace over the arrows that show leaching. What occurs at the surface to begin the process of leaching?

______________________

______________________

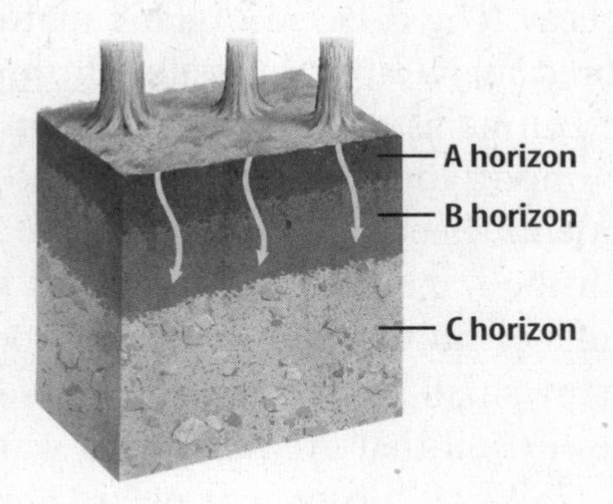


## What is found in the C horizon?

The C horizon is the bottom horizon in a soil profile. It is often the thickest soil horizon. The C horizon contains partly weathered rock and little organic matter from plants and animals. Leaching generally does not affect the C horizon.

At the bottom of the C horizon lies rock—the rock that formed the soil above it. This rock is called the parent material of the soil. Because the C horizon is nearest to the parent material, it is most like the parent material.

## How do glacial deposits affect soil?

Many places on Earth are covered by a thick layer of sediment that was deposited by melting glaciers. This material is a mixture of clay, silt, sand, and boulders. It covers much of the northern United States. Rich soil developed from this sediment. Much of the farmland in the Midwest is covered with this rich soil. ☑

A soil profile of the land in the Midwest would probably show that the sediment in the A and B horizons does not match the parent material below the C horizon. The topsoil came from sediment carried by the glaciers.

**Reading Check**

4. **Explain** Why is the Midwest soil so rich?

______________________

______________________

______________________

## Soil Types

If you travel around the country and look at soils, you will notice that they are not all the same. Some soils are thick and red. Others soils are brown and rocky. The figure below shows some of the different soil types in various areas in the United States.

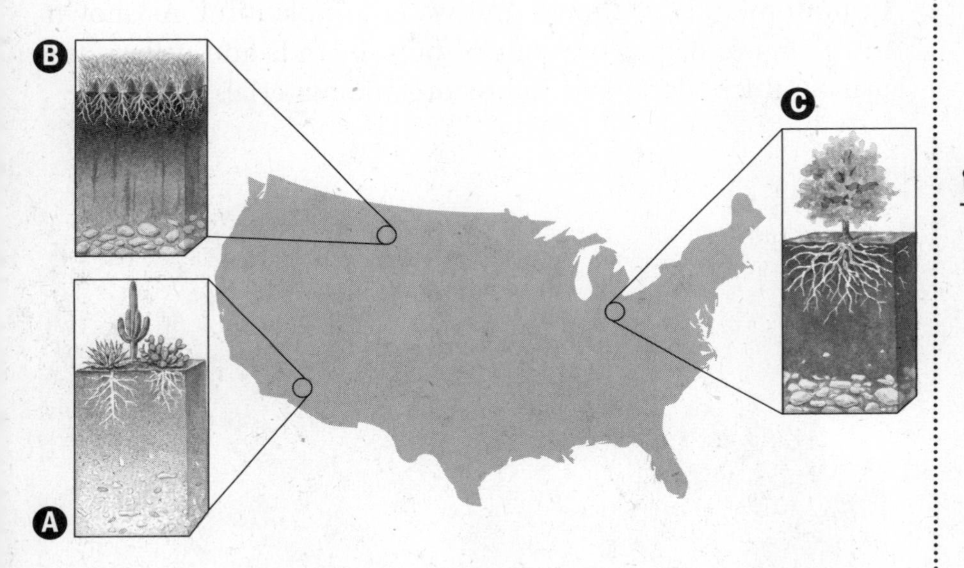


**Picture This**

5. **Analyze** Look carefully at the figure. Which soil appears to have the thinnest A horizon?

______________________

## How does climate affect soil?

Different regions of Earth have different climates. Different climates produce different types of soil. Deserts are dry and have few plants. Because desert soil contains little organic matter, desert soil is thin and light. On the grasslands of the prairie, the soil contains plenty of organic matter from grasses. Lots of organic matter gives prairie soils thick, dark A horizons. Other regions, such as temperate forests and tropical areas, have their own particular types of soil.

## What other factors affect soil type?

Parent material affects the kind of soil that develops from it. Clay soils often form on parent rocks like basalt. The minerals in basalt weather to form clay. If the parent material is sandstone, it weathers into sand, producing sandy soil.

**Plants** Rock type also affects the types of plants that grow in an area. Different rocks provide different nutrients that plants need to grow. The type of plant that grows then affects the formation of soil.

**Time** Soil development is affected by time. If rock has been weathering for a short time, the sediment will be similar to the parent material. If rock has been weathering for a long time, the soil is less like the parent material.

**Slope** The slope of the land is a factor in soil profiles. On steep slopes, soils are poorly developed because sediment does not have time to build up before it is carried downhill. In bottomlands, sediment and water are plentiful. As shown in the figure below, the soils of bottomlands and valleys are usually thick, dark, and full of organic material.

### Think it Over

6. **Explain** What is one way rock type affects soil type?

______________________

______________________

______________________

### Picture This

7. **Analyze** Label the A horizon in the figure. Where is the A horizon thickest?

______________________

______________________

## After You Read

### Mini Glossary

**horizon:** layer in the soil

**humus:** dark-colored material formed by decaying organic matter

**leaching:** process in which dissolved minerals are carried through soil by water

**litter:** twigs, leaves, and other organic matter that covers the forest floor and will become humus

**soil:** mixture of weathered rock, decayed organic matter, mineral fragments, water, and air

**soil profile:** all the horizons of a soil

1. Review the terms and their definitions in the Mini Glossary. Write one sentence explaining what makes A horizons richer than B or C horizons. Use at least two terms.

_______________________________________________

_______________________________________________

2. Fill in the graphic below to identify five things that make up soil.

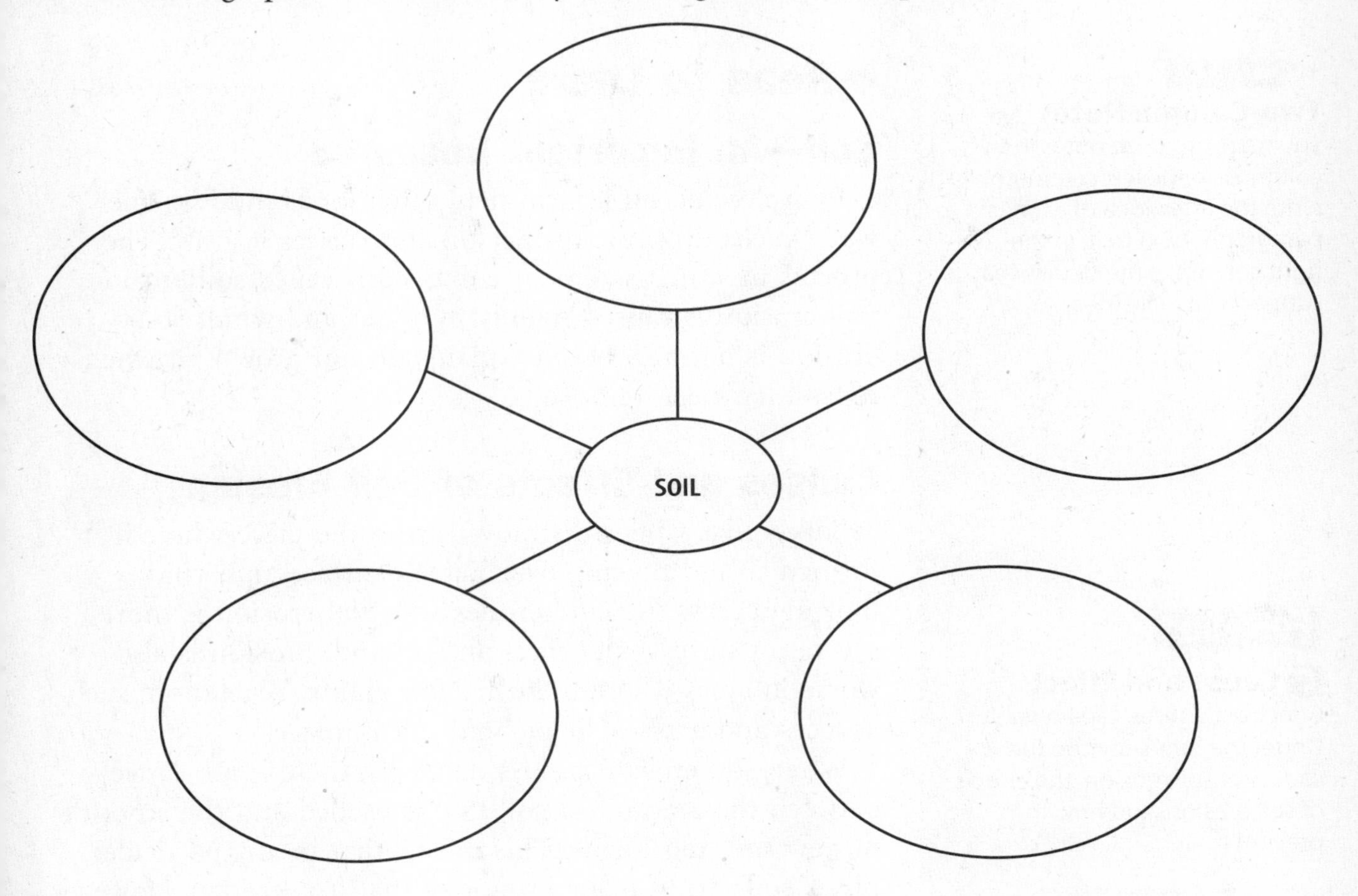


**Science Online** Visit **earth.msscience.com** to access your textbook, interactive games, and projects to help you learn more about the nature of soil.

**End of Section**

# Weathering and Soil

## section 3 Soil Erosion

### What You'll Learn

- why soil is important
- how human activity has affected Earth's soil
- how soil erosion can be reduced

## Before You Read

Imagine that you planted flower seeds in your backyard garden. What might heavy rainfall do to the seeds and soil?

______________________________________________

______________________________________________

______________________________________________

Study Coach

**Two-Column Notes** As you read, organize notes in two columns. In the left column, write the main idea of each paragraph. Next to it, in the right column, write details that support the main idea.

## Read to Learn

### Soil—An Important Resource

Rain flowing off a farm field often looks muddy. The water picks up some of the soil and carries it away. The process in which sediment is moved is called soil erosion. Soil erosion is caused mainly by water and wind. Soil erosion is harmful because plants do not grow well when topsoil has been removed.

### Causes and Effects of Soil Erosion

Soil erodes when it is moved from the place where it formed. Water flowing over Earth's surface and wind blowing across the land erodes soil. Soil erosion is more severe on steep slopes than on flat land. Erosion is also worse in areas where there are few plants. Vegetation, such as trees and grasses, helps hold soil in place.

When soil erosion occurs naturally, there is a balance between the amount of soil that is eroded and the amount of new soil that forms. This means that in any particular place, soil forms at the same rate that it is eroded. However, humans sometimes upset this balance causing soil erosion to occur faster than new soil can form.

FOLDABLES™

**C Cause and Effect** Construct a three-tab Foldable. Under the tabs describe the causes of soil erosion, the effects of soil erosion, and how to prevent it.

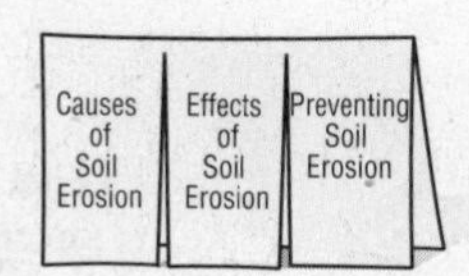

## How does soil erosion affect farming?

Soil erosion is a serious problem for agriculture. Topsoil holds water and contains nutrients that plants need. Topsoil also has pores, or open spaces, that are good for plant growth. But when topsoil erodes, the quality of the soil decreases. The eroded soil has fewer nutrients that plants need and may not hold enough water for plants.

In nature, decaying plant parts add nutrients to the soil, while plant roots take some nutrients out of the soil. This is called nutrient balance. Soil erosion upsets this balance. If topsoil erodes rapidly, there are not enough nutrients for plants to grow. Farmers may have to add fertilizers to the soil to make up for the lost soil nutrients. It is more difficult for plants to take root and grow well in poor-quality, eroded soil. ☑

## How does removing forests affect erosion?

Trees protect the soil from erosion. In large forests, soil is protected from erosion by trees and roots that hold soil in place. But, when trees are cut down, soil is exposed and erosion increases. Erosion of forest land is a problem in many parts of the world, but it is especially severe in tropical rain forests.

Each year, huge areas of rain forest are cut down for lumber, farming, or grazing. In rain forests, most of the nutrients are in only the top few centimeters of soil. Farmers can grow crops in rain forest soil for only a few years before the topsoil is gone. Then they must clear more rain forest land somewhere else to grow their crops. The cycle continues. More trees are cut down, more topsoil erodes, and then more soil needs to be cleared.

## How does overgrazing affect erosion?

Sheep or cattle feed on grasses. In many places, sheep and cattle can graze on the land without causing much damage to soil. In areas that get little rain, grasses do not grow quickly. If grazing animals eat too much of the vegetation, there is no ground cover to protect the soil. The topsoil is exposed and may blow away in the wind before new grasses have a chance to grow. The exposed soil also loses moisture. In time, an overgrazed grassland can become like a desert where the poor, dry soil cannot support plants.

**Reading Check**

1. **Explain** What is nutrient balance?

______________________

______________________

______________________

**Think it Over**

2. **Draw Conclusions** Why is rain forest land a poor choice for growing crops?

______________________

______________________

______________________

## How does sediment affect the environment?

In places where soil erosion is severe, sediment can damage the environment. Severe erosion may occur when land is exposed at large construction sites or strip mines. The soil removed from these sites is moved to a new location where it is deposited. Huge mounds of soil deposits are easily eroded by rain and wind. The soil may be carried into streams, or people may deposit it directly in a stream. Large amounts of sediment might fill the stream channel, damaging or destroying the stream environment and the fish and wildlife that depend on it. ✔

**Reading Check**

3. **Determine** How does sediment from soil erosion damage streams?

______________________

______________________

# Preventing Soil Erosion

Every year more than 1.5 billion metric tons of soil are eroded in the United States. Soil is a natural resource that must be protected and managed. There are several things people can do to protect soil.

## How can managing crops protect the soil?

Farmers use several methods to slow down soil erosion. They plant rows of trees, or shelter belts, to break the force of the wind that would otherwise erode topsoil. They plant crops to cover the ground and hold the soil after the main crops are harvested. In dry areas, farmers do not plow under their crops because this disturbs the soil and increases erosion. Instead, they graze animals on the leftover vegetation. If grazing is controlled, the animals leave enough vegetation on the land to reduce erosion.

**Think it Over**

4. **Infer** No-till farming has been a great benefit to birds and wildlife. Why do you think this is?

______________________

______________________

______________________

**No-till Farming** In the past, farmers would till or plow their fields one or more times each year. When the soil was plowed, loose soil was turned over. Wind and water eroded the loosened soil. In recent years, many farmers have begun to practice no-till farming to reduce erosion. **No-till farming** is a farming method in which the plant material left after harvesting remains in the field to decay over the winter. When it is time to plant again, the farmers plant their crop seed without plowing the land or clearing away last year's plants.

No-till farming leaves plant cover on the land all year, so topsoil erosion by wind and water is greatly reduced. One study showed that no-till farming leaves up to 80 percent of the soil covered and protected by plants or plant remains. Another benefit of no-till farming is that plant remains keep weeds from growing in farm fields.

**Contour Farming** Soil erosion occurs more quickly on slopes than on flat land. Farmers who plant crops on slopes take steps to limit soil erosion on their land. They plant crops using contour farming. **Contour farming** is planting crops along the land's natural contour, or slope shape. The figure below shows an orchard planted along the natural contours of a slope. The curving rows catch water that would otherwise flow downhill, carrying a lot of topsoil with it. Contour farming helps prevent water and sediment from flowing down the slope.

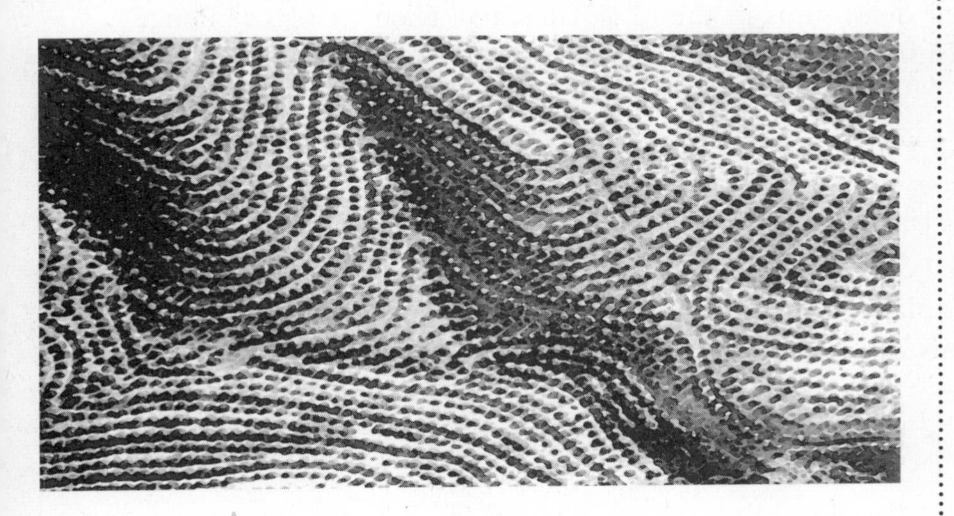

## Picture This

5. **Identify** Use a colored pencil to trace three rows of crops planted in contours on the hillside farm.

**Terracing** Where slopes are steep, farmers use terracing (TER uh sing) to conserve water and prevent soil erosion. **Terracing** is a method of farming in which steep-sided, flat-topped areas are built into the sides of steep hills so crops can be grown on the level areas. The flat-topped areas that are planted with crops are like terraces of land. The terraces reduce runoff of rainwater by creating flat areas and short sections of sloping land. The terraces also help prevent topsoil from eroding down the hill or mountainside.

## How can erosion be reduced in exposed soil?

There is a variety of ways to control erosion on soil that is exposed. During large construction projects, water may be sprayed on exposed soil to weigh it down and prevent erosion by wind. When the construction is complete, topsoil is added to the land and trees and other vegetation are planted to protect the soil. Soil removed at strip mines may also be protected so that it can be put back in place when mining stops. After mining is complete, vegetation may be planted to hold the soil and limit erosion on the reclaimed land. ☑

**Reading Check**

6. **Explain** Name one way to control erosion on bare soil.

______________________

______________________

# After You Read

## Mini Glossary

**contour farming:** planting crops along the land's natural contour, or slope shape

**no-till farming:** farming method in which the plant material left after harvest remains in the field to decay over the winter

**terracing (TER uh sing):** method of farming in which steep-sided, flat-topped areas are built into the sides of steep hills so crops can be grown on the level areas

1. Review the terms and their definitions in the Mini Glossary. Then write a sentence explaining how no-till farming helps prevent soil erosion.

______________________________________________

______________________________________________

2. Fill in the figure below. Explain how the first three activities can lead to soil erosion. Then explain what methods can be used to prevent soil erosion in each of the three areas.

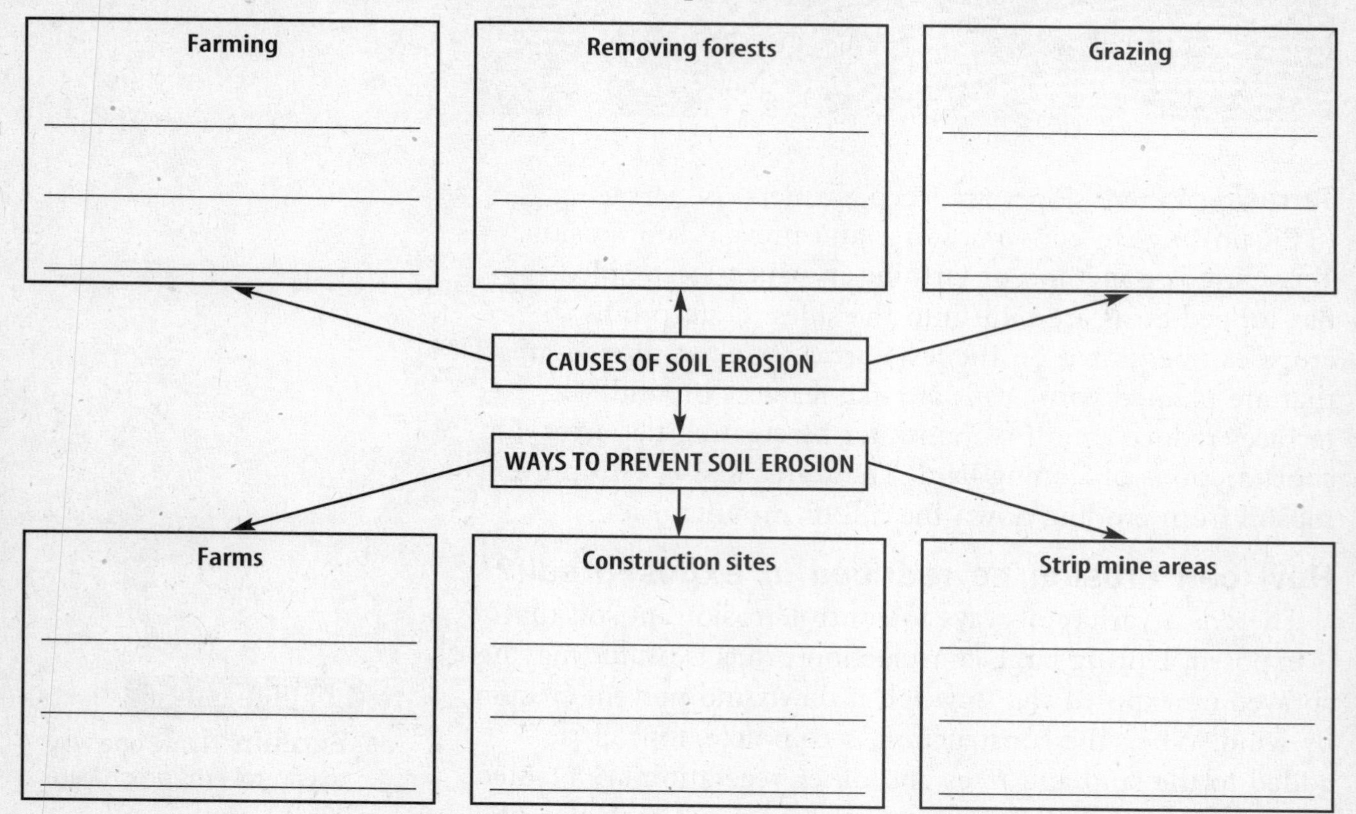

ScienceOnline Visit **earth.msscience.com** to access your textbook, interactive games, and projects to help you learn more about soil erosion.

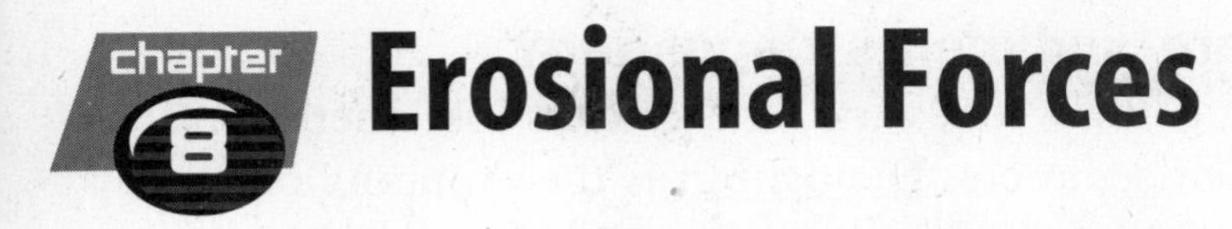

# chapter 8 Erosional Forces

## section 1 Erosion by Gravity

### Before You Read

How do things get worn out? List three things that wear out. Next to each, write what causes it to wear.

_______________________________________________

_______________________________________________

_______________________________________________

**What You'll Learn**
- the differences between erosion and deposition
- about slump, creep, rockfalls, rock slides, and mudflows
- why it is not wise to build on a steep slope

### Read to Learn

**Mark the Text**

**Identify the Main Point** After you read each paragraph, highlight the main point.

#### Erosion and Deposition

Did you ever see a landslide on the news? In a landslide, large piles of sediment and rock move downhill. Recall that sediments are loose materials such as small pieces of rocks, minerals, and once-living plants and animals. Heavy rains can lead to a landslide. The sediment and rock that move downhill are a result of erosion. **Erosion** is a process that wears away surface materials and moves them from one place to another.

**FOLDABLES™**

**A Build Vocabulary** Make a vocabulary book from a sheet of notebook paper. As you read, write each new term on the front and each definition inside.

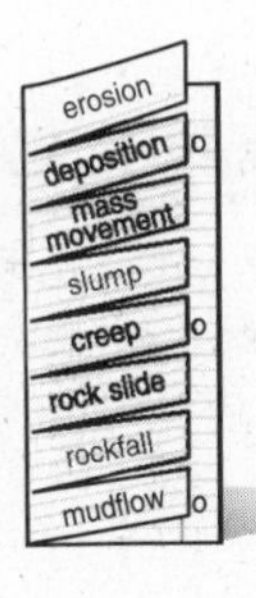


#### What moves sediments?

Suppose you put sand on a tray. If you tilt the tray far enough, the sand starts to move because of gravity. Gravity is the force that pulls all things toward Earth's center. Gravity is one cause of erosion. Other causes of erosion include water, wind, and glaciers. They all move sediments.

Water and wind can cause erosion if they are strong enough or have enough energy. On a calm day, air can't move much dust. But a strong wind can move dust and larger bits of soil. Glaciers also cause erosion. They move sediments that are trapped in solid ice. When the ice melts, the soil and rocks are dropped. Water from the melting ice can carry the sediment even farther.

## Why are sediments deposited?

In time, the water, wind, and glaciers lose energy and drop their sediment. **Deposition** is the dropping of sediments that occurs when water, wind, or glaciers lose their energy and drop the load. When sediments are eroded, they are not lost from Earth—they are just moved. ☑

**Reading Check**

1. **Determine** When deposition occurs, what is deposited?

______________________

## Mass Movement

The greater an object's mass is, the stronger its gravitational pull. Earth's mass is great, so the pull of Earth's gravity is a key force in erosion and deposition.

Sediments are pulled toward Earth's center by gravity. If the sediments are on a slope, they move downslope. A **mass movement** is any kind of erosion in which gravity moves sediments downslope. Some mass movements are slow. Others move quickly. Some of them cause disasters.

There are five common kinds of mass movement. They are slump, creep, rockfalls, rock slides, and mudflows. Landslides are mass movements that can be one of these types or a combination of these types of mass movement. ☑

**Reading Check**

2. **Identify** Name two kinds of mass movements.

______________________

______________________

### What is slump?

**Slump** is a mass movement that occurs when a mass of material slips down a curved surface. When the base materials cannot support the rock and sediment above, the rock and soil slip downslope. The material can break into parts or move as a large mass.

Sometimes a slump occurs when water moves to the base of a slipping mass of sediment. The water weakens the slipping mass. That can cause the mass to move downhill. Or, a strong rock layer may sit over a weaker layer, such as clay. The clay can weaken under the weight of the rock. When clay can't hold up the rock layer on the hillside, the layers start to move. The area left after slump occurs is usually a curved scar in the slope, as shown in the figure below.

**Think it Over**

3. **Infer** What force caused the slumped materials to move downslope?

______________________

## What is creep?

The next time you travel, watch for slopes where trees and fence posts lean downhill. Leaning trees and fence posts show mass movement called creep. **Creep** occurs as soils slowly move downhill. Creep is common where the soil freezes and thaws often. ☑

**Reading Check**

**4. Apply** What mass movement occurs as soils slowly move downhill?

_______________

## What causes rockfall?

Signs on the roads in many mountains warn of falling rocks. Rockfalls happen when blocks of rock break loose from a steep slope and fall. As they fall, the rocks crash into other rocks, knocking them loose. More and more rocks break loose and fall.

Rockfalls often happen in the spring. During the winter, ice freezes in the cracks in rocks. The freezing ice makes the cracks larger and larger. In the spring, the rocks break loose and fall down the mountainside.

## What is a rock slide?

Rock slides occur when layers of rock slip down a hill suddenly. Rock slides are common where there are steep rock layers, as in the figure. They can occur in mountains or on steep cliffs. Rock slides are fast-moving and can destroy property. Some rock slides start after heavy rains or earthquakes. However, they can start on any steep hill at any time without warning.

**Picture This**

**5. Interpret** Add these labels to the figure: *steep rock layer; rock slide.*

## What causes mudflows?

Mudflows occur where there are thick layers of loose sediment. They often happen after fire has burned off the plant material. When heavy rains fall, water mixes with sediment. The sediment becomes like thick paste. Gravity causes the thick mass to flow downhill.

When a mudflow reaches the bottom of a hill, it stops. It deposits the sediment and everything else it has carried. The mudflow often spreads out in a fan shape. A mudflow has more weight than an equal volume of water. Mudflows can cause more damage than floods.

## How are all mass movements alike?

All mass movements—slump, creep, rockfalls, rock slides, and mudflows—are alike in some ways. All are most likely to happen on steep slopes. Gravity causes all of them. They often occur after a heavy rain. The water adds mass to the soil causing more pressure between the grains and layers of sediment. This makes the sediment expand and weaken. ☑

**Reading Check**

**6. Identify** Where do all mass movements usually happen?

______________________

# Consequences of Erosion

Some people like to have the great view that hills and mountains provide. Others like to live in scenic areas away from noise and traffic. For these reasons, many homes are built on the sides of hills and mountains. Because gravity is one cause of erosion, steep slopes may not be the safest places to live.

## How does building affect steep slopes?

For people who build homes on steep hills, erosion will always be a problem. Sometimes builders make a slope even steeper or remove plant material. This speeds up the process of erosion. Some steep slopes have weak layers of sediment underneath which may cause slumps.

## How can steep slopes be made safer?

Growing plants on bare slopes is a good way to slow erosion. The plant roots help hold soil in place. Soil that is held together by plant roots isn't as likely to move. Plants also absorb water in the soil. Drainage pipes or tiles can be put into slopes to prevent water from building up, too. The pipes and tiles help make the slope more stable by allowing excess water to drain away more easily.

There are other ways to slow soil erosion. Concrete or stone walls can help hold soil in place.

It is not easy to stop mass movements of soil on a slope. Rain or earthquakes can weaken all types of soils. In time, weakened soils will move downhill.

People who live in areas with erosion problems spend a lot of time trying to save their land. Sometimes they can slow down erosion. Yet, they can't ever fully stop it. They can't get rid of the risk of mass movement. In time, gravity wins. Soils move downhill, changing the shape of the land.

**Think it Over**

**7. Explain** Name two ways to slow erosion on steep slopes.

______________________

______________________

## After You Read

### Mini Glossary

**creep:** occurs when soils move slowly downhill

**deposition:** dropping of sediments that occurs when water, wind, or glaciers lose their energy and can no longer carry the load

**erosion:** process that wears away surface materials and moves them from one place to another

**mass movement:** any kind of erosion in which gravity moves sediments downhill

**slump:** a mass movement that occurs when a mass of materials slips down a curved surface

1. Review the terms and their definitions in the Mini Glossary. Write two sentences that tell how erosion and deposition are alike and different.

_______________

_______________

_______________

2. Fill in the concept map below with the kinds of mass movements caused by gravity.

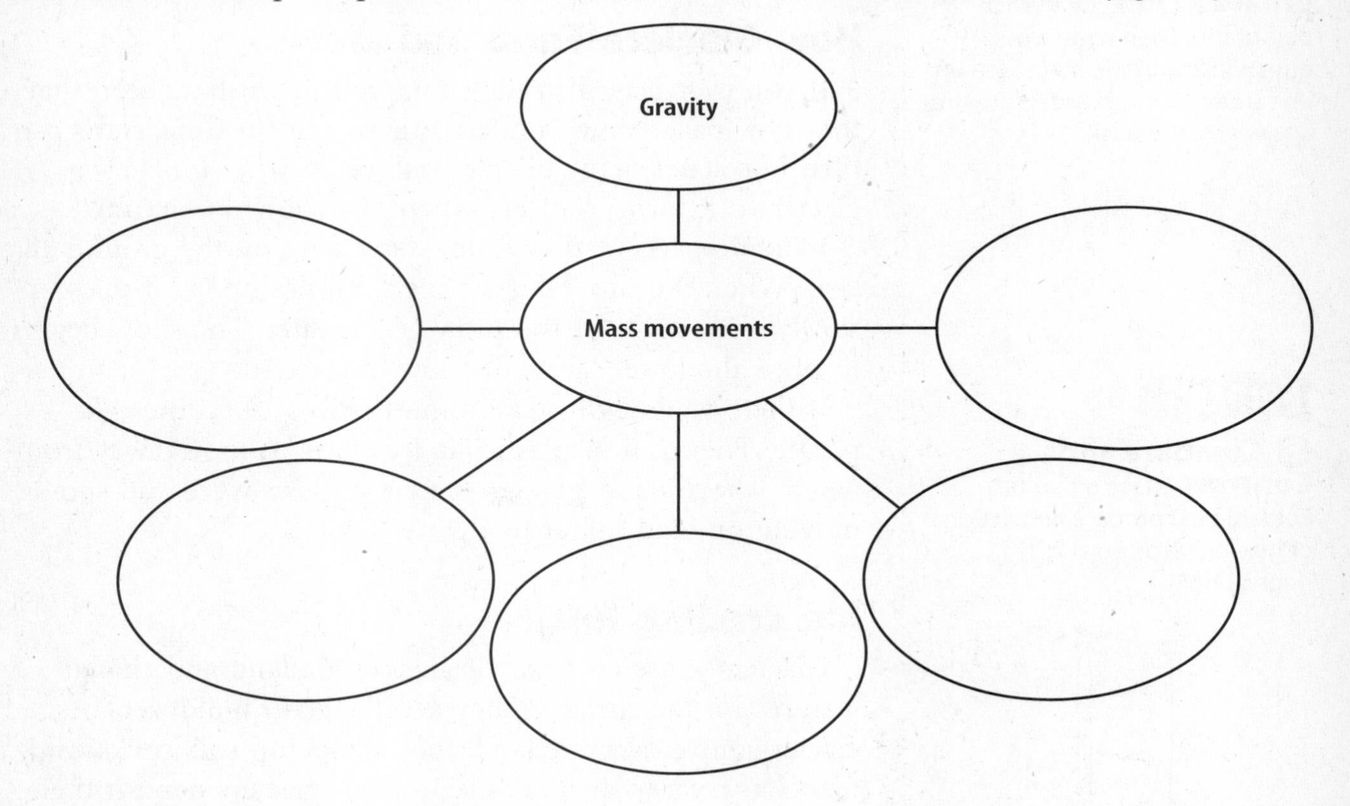


Science Online Visit **earth.msscience.com** to access your textbook, interactive games, and projects to help you learn more about the agents of erosion.

End of Section

# Erosional Forces

## section 2 Glaciers

**What You'll Learn**
- how glaciers move
- about glacial erosion and deposition
- how till and outwash compare

### Before You Read

Write down what you think of when you hear the word *glacier*. Describe what a glacier looks like and how it moves.

______________________________________________

______________________________________________

______________________________________________

Study Coach

**Create a Quiz** Read each paragraph. Then write a quiz question about the main idea on one side of a note card. Write the answer on the back.

FOLDABLES™

**B Compare and Contrast** Make a two-tab Foldable to compare two actions of glaciers: erosion and deposition.

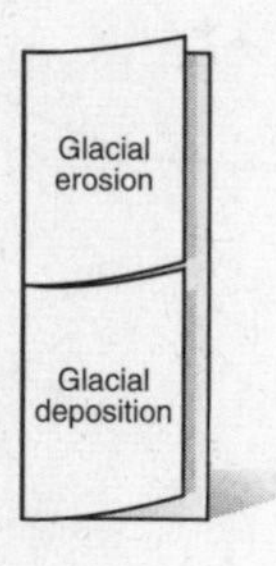


### Read to Learn

#### How Glaciers Form and Move

If you ever played in the snow, you might have seen that the snow packs down under your weight. In time, snow can turn into ice if many people walk on it. In a similar way, glaciers can form in places where the snow doesn't melt.

In some parts of the world, snow stays on the ground all year. When the snow doesn't melt, it piles up. As it piles up slowly, the weight of the snow gets greater. The snow begins to press the lower layers of snow into ice.

If there is enough pressure on the ice, it becomes like plastic. The mass of plasticlike ice starts to move away from where it formed. A **glacier** is a large mass of ice and snow moving on land under its own weight.

#### Ice Eroding Rock

Glaciers cause erosion. Glaciers erode land and change features on the surface. They act like giant bulldozers. As glaciers move over the land, they scrape up soil, rocks, and plants and carry them along. In time, glaciers deposit their load in other places. Glacial erosion and deposition change large areas of Earth's surface.

## What is plucking?

As it moves, a glacier wears down solid rock. When the ice in the glacier melts, the water flows into cracks in the rocks. Later, the water refreezes in these cracks. The water expands as it freezes, breaking the rocks. Pieces of rock are picked up by the moving glacier. **Plucking** is the action in which a moving glacier picks up loose pieces of rock. By plucking, a glacier can add rocks, gravel, and sand to its bottom and sides, as shown in the figure below.

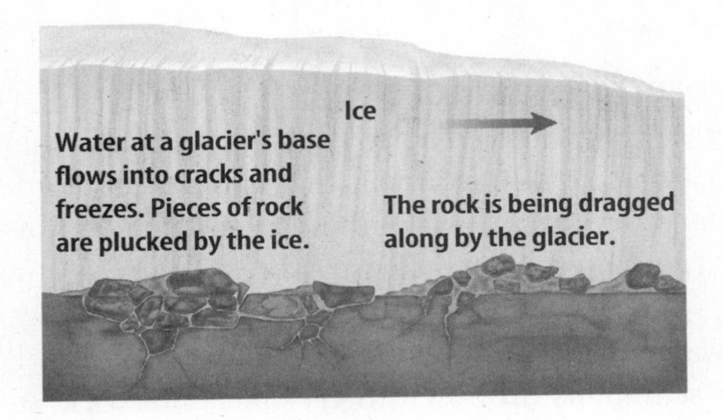


**Picture This**

1. **Infer** What happens to rock that is not plucked by the glacier?

______________________

______________________

## What is transporting and scouring?

A glacier carries huge amounts of sediment and rock as it moves. The action of carrying sediment and rock is called transporting. Plucked rocks and sand at the base of the glacier scrape the soil and bedrock like sandpaper against wood. This scraping action is called scouring.

When moving glaciers scour the bedrock, they leave marks on it. Glaciers leave two kinds of marks, grooves and striations. Grooves are long, deep, side-by-side scars on rocks. Shallow marks are called striations (stri AY shunz). Grooves and striations show the direction in which the glacier moved. ☑

**Reading Check**

2. **Identify** Name two kinds of marks that glaciers leave in rock.

______________________

______________________

# Ice Depositing Sediment

When a glacier starts to melt, it can't carry as much sediment. A melting glacier deposits the sediment on the land.

As a glacier melts and starts to shrink back, it is said to retreat. As a glacier retreats, it leaves behind a mix of rocks, clay, sand, and silt. This mix of different-sized sediments that are deposited from a retreating glacier is called **till**. Thousands of years ago, glaciers in the northern United States left behind large amounts of till. Today these till areas include farmland that runs north and west from Iowa to northern Montana.

## What are moraine deposits?

Till also is deposited at the end of a glacier when it is not moving forward. This type of till deposit does not cover as wide an area as the till deposited by a moving glacier.

Rocks and soil are moved to the end of the glacier. They move like items on a grocery store conveyor belt. As a result, a big ridge of soil and rocks piles up. It looks like something pushed along by a bulldozer. This ridge of rocks and soil deposited by a glacier when it stops moving is called a **moraine.** Moraines also are deposited along the sides of a glacier. ☑

**Reading Check**

**3. Describe** What does a moraine deposit look like?

______________________

______________________

## What are outwash deposits?

At times, the ice in a glacier starts to melt. The meltwater can carry and deposit sediment that is different from till. The water often deposits the material past the end of the glacier. The material deposited by the water from a melting glacier is called **outwash.**

Meltwater carries sediments and deposits them in layers. Heavier sediments drop first. Bigger pieces of rock drop closer to the glacier. Sometimes the outwash forms a fan-shaped layer of sand and gravel in front of a glacier.

## What are eskers?

Some kinds of outwash layers look like long, winding ridges. The ridges form from meltwater, or water from a melting glacier. The meltwater forms a stream inside a glacier's ice. Streams of meltwater also form on the glacier's surface, as shown in the diagram below. The streams carry sand and gravel and deposit them in the channels. A winding ridge of sand and gravel left behind after a glacier melts is called an esker (ES kur).

**Picture This**

**4. Identify** Put an X on the tunnel and meltwater stream where eskers will form.

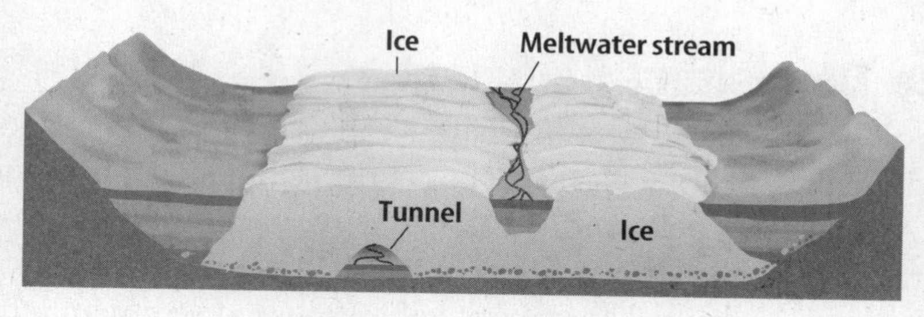


**Esker Formation**

## Continental Glaciers

The two types of glaciers—continental glaciers and valley glaciers—are found in different parts of the world.

Continental glaciers are wide, thick sheets of ice and snow. They cover huge areas of land, usually near the North and South Poles. About ten percent of Earth is now covered by continental glaciers. Some glaciers are thicker than some mountains. Glaciers make it impossible to see most of the land features in Antarctica and Greenland.

### How does climate affect continental glaciers?

In the past, continental glaciers covered 28 percent of Earth. The figure below shows ice over much of North America during the most recent ice age. An ice age is a time when ice covers large areas of land.

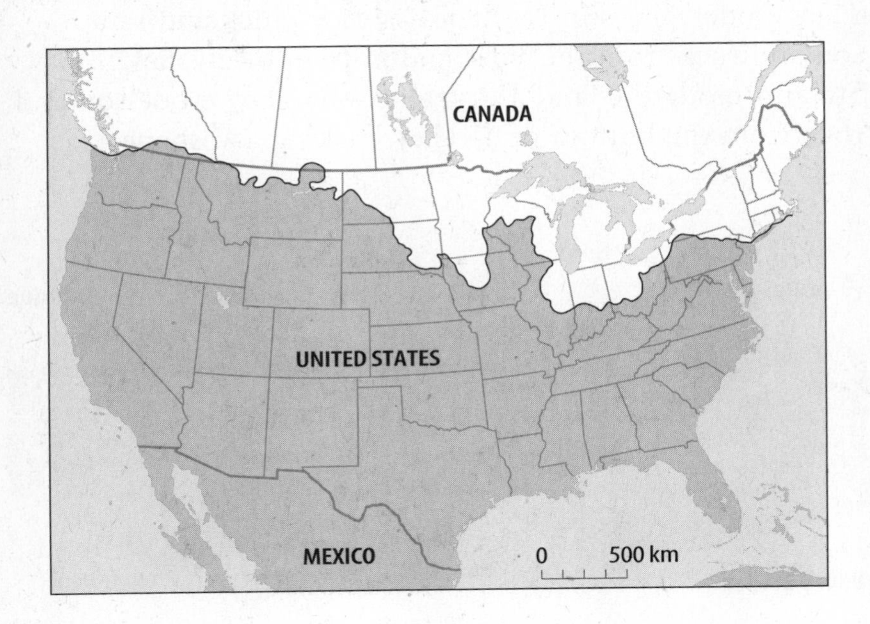


Glaciers have formed and melted many times as climates have changed. During the ice ages, Earth's air temperatures were about 5°C cooler than they are today.

The peak of the last ice age was about 18,000 years ago. Then the ends of ice sheets started to melt and shrink. Land that had been buried by ice was exposed.

## Valley Glaciers

Valley glaciers still occur today even though the world climate is warmer. They are found in some of the high mountains where the air stays cool in summer. Because the snow doesn't melt, valley glaciers grow and creep along.

**Applying Math**

5. **Calculate** Earth is about 150,000,000 square kilometers. If ten percent of Earth's surface is covered by glaciers, how many square kilometers are covered by glaciers? Show your work.

______________________

______________________

______________________

**Picture This**

6. **Observe** Put a star where you live. Was your area covered by ice during the last ice age?

______________________

## What are the signs that valley glaciers existed?

If you visit the mountains, you can tell if valley glaciers ever existed there. Look for grooves in the rocks. Look for signs of plucking. You might look for other signs as well. Those signs are caused by the way that glaciers erode mountains. ☑

The figures below show different ways that valley glaciers change mountains. Valley glaciers erode bowl-shaped basins, called cirques (SURKS). These round basins occur on the sides of mountains. If two side-by-side glaciers erode a mountain, a long ridge forms between them. This long ridge is called an arête (ah RAYT). Valley glaciers may erode a mountain from many sides. This action may form a sharp peak, called a horn.

Valley glaciers move down mountains and into valleys. They erode valleys in U-shapes as they pluck and scrape soil and rock from the sides and bottom of a valley. Streams erode the land a different way. They erode soil and rock from the bottom of a valley, making a V shape.

**Reading Check**

**7. Explain** What are two signs that valley glaciers existed?

______________________

______________________

**Picture This**

**8. Explain** How can you tell that the valleys in the figure on the right were formed by glaciers, not streams?

______________________

______________________

______________________

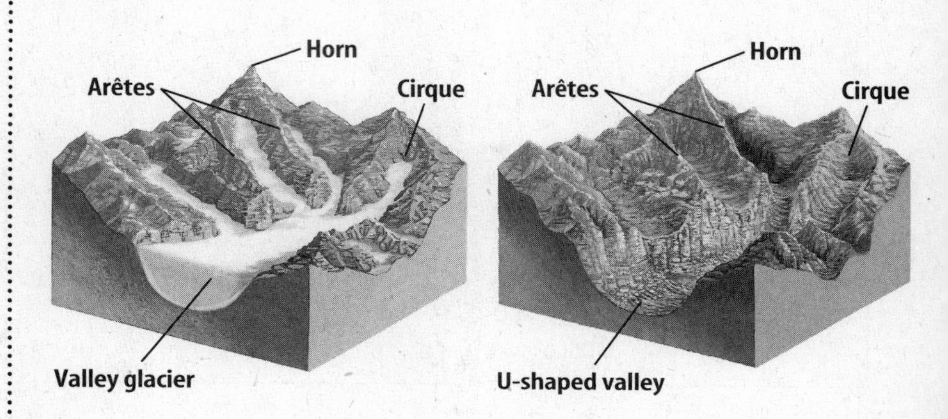


## Importance of Glaciers

Glaciers have changed Earth's surface. They have eroded mountains and changed valleys. Great ice sheets have deposited sediments on huge areas of land. Today, glaciers are still changing Earth's surface.

Glaciers do more than change the look of the land. When glaciers melt, they leave behind sediments. Sand and gravel deposits from glacial outwash and eskers are important resources. These deposits are used in the construction of roads and buildings.

## After You Read

### Mini Glossary

**glacier:** large mass of ice and snow moving on land under its own weight

**moraine:** ridge of rocks and soil deposited by a glacier when it stops moving

**outwash:** material deposited by the water from a melting glacier

**plucking:** action in which a moving glacier picks up loose pieces of rock

**till:** mix of different-sized sediments that is deposited from the base of a glacier as it retreats

**1.** Review the terms and their definitions in the Mini Glossary. Write two sentences explaining how glaciers change Earth's surface. Use at least three terms.

______________________________________________

______________________________________________

______________________________________________

**2.** Fill in the missing information from the following outline of this section.

A. Glaciers erode rock

1. By plucking
2. By ______________
3. By ______________

B. Glaciers deposit rocks and soil

1. As till
2. As ______________

C. Types of glaciers

1. ______________ glaciers
2. Valley glaciers

D. Glaciers are important because they

1. Change Earth's ______________
2. Deposit valuable ______________

**3.** At the beginning of this section, you were asked to create a quiz question for each paragraph. Explain how this helped you understand the information.

______________________________________________

______________________________________________

Visit **earth.msscience.com** to access your textbook, interactive games, and projects to help you learn more about glaciers.

## section 3 Wind

### What You'll Learn

- how wind causes deflation and abrasion
- how loess and dunes form

## Before You Read

Think about a windy day when dust or sand was blown in your face. Write about how it felt.

_______________

_______________

_______________

**Mark the Text**

**Identify the Main Point** After you read each paragraph, underline the main point.

## Read to Learn

### Wind Erosion

Erosion is the process that wears away surface materials and moves them from one place to another. You have read about gravity as a force of erosion. Gravity causes mass movements. Glaciers also are powerful forces of erosion that carve U-shaped valleys and cut grooves in Earth's surface. A third force of erosion is wind.

When air moves, it picks up tiny bits of loose material and moves them to other places. Air is different from other forces of erosion. Air cannot pick up heavy sediments.

Also, wind is different because it can carry and drop sediments over wide spaces. Rivers move sediment, but usually inside their own banks. Wind can pick up dust particles and ash from volcanoes and carry them high into the atmosphere. They can be deposited thousands of kilometers away.

**FOLDABLES™**

**C Summarize** Make a layered book from two sheets of paper. Write information about wind erosion as you read this section.

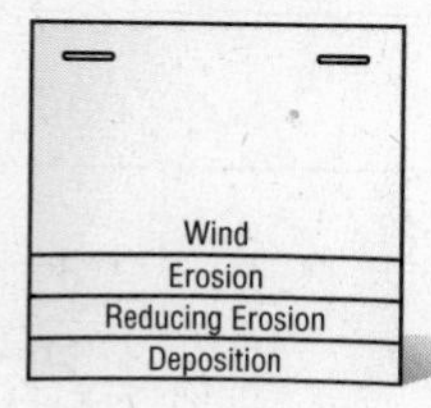


#### How does deflation erode land?

One way that wind erodes Earth's surface is by deflation (dih FLAY shun). **Deflation** is erosion that occurs when wind removes small pieces of sediment, such as silt and sand, and leaves heavier material, such as pebbles and boulders, behind.

## How does abrasion erode the land?

A sandblaster sprays a mix of sand and water. Workers use this material to clean buildings. The flying sand wears away dirt from stone, concrete, or brick walls. The friction polishes the walls and leaves them smooth.

Soil carried by the wind acts like a sandblaster. It scrapes and wears away rock. Erosion that occurs when wind scrapes and wears away rock is a process called **abrasion** (uh BRAY zhun). Abrasion leaves the rock pitted and worn down.

## Where do deflation and abrasion occur most often?

Deflation and abrasion erode all land surfaces, but occur most often in deserts, beaches, and plowed fields. There are fewer plants to hold the soil in these areas. The sediments can be eroded rapidly when winds blow over them. Grasslands have many plants that hold the soil in place, so there is little soil erosion caused by wind. ☑

**Reading Check**

1. **Explain** Why do deflation and abrasion occur often in deserts and beaches?

_______________

_______________

_______________

## What are sandstorms?

Even if the wind is strong, it seldom carries sand higher than 0.5 m from the ground. But sometimes sandstorms do occur. When strong winds blow in sandy deserts, the sand grains bounce into other sand grains. This causes more and more sand grains to rise into the air. In time, the blowing sand forms a low cloud just above the ground. Most sandstorms occur in deserts. But some happen in other dry regions.

## What causes dust storms?

When soil is moist, it stays packed on the ground. However, when soil dries out, it can be eroded by wind. Soil is made mostly of silt-sized and clay-sized particles. These small particles are even smaller than sand particles, so wind can move them high into the air.

Silt and clay particles are small and stick together. A faster wind is needed to lift these fine particles of soil than is needed to lift sand. However, after the tiny clumps of soil are in the air, wind can carry them long distances. Dust storms are clouds of fine soil blown by the wind. When the land is dry, dust storms can cover hundreds of kilometers. These storms blow topsoil from open fields, overgrazed areas, and places where plants have disappeared.

**Think it Over**

2. **Determine** What soil particles are smaller than sand?

_______________

_______________

## Reducing Wind Erosion

Wind erosion is most common in areas where there are no plants to protect the soil. One of the best ways to slow or stop wind erosion is to grow plants or vegetation. This helps save the soil and protects valuable farmland.

### How do windbreaks help stop erosion?

People in many countries plant vegetation to stop wind erosion. For hundreds of years, farmers have planted trees along their fields to act as windbreaks and prevent soil erosion. As wind hits the trees, it slows. It doesn't have enough energy to lift soil particles. ☑

**Reading Check**

**3. Explain** What is one way farmers worldwide slow the wind and reduce erosion?

___

___

### How do roots help stop erosion?

Often plants are used to stop wind erosion along seacoasts and deserts. Plants with fibrous roots are good for holding the soil. These fibrous roots grow over a wide area. The roots wrap tightly around many particles in the soil. They help hold the soil in place.

Planting vegetation is a good way to slow erosion. Even so, if the wind is strong and the soil is dry, nothing can stop erosion completely.

## Deposition by Wind

Sediments carried by the wind eventually are dropped, or deposited. Over time, some of the sediments develop into landforms such as dunes and loess (LES).

### What is loess?

**Loess** is wind deposits of fine-grained sediments. Strong winds blew across melting glaciers and picked up the fine sediment from outwash. Then the wind dropped the sediment on hills and valleys. Over time, the fine particles packed together, making a thick, yellowish-brown deposit. Loess is as fine as talcum powder. Many farms in the Midwest have rich soils that formed from loess deposits.

**Think it Over**

**4. Infer** Why might the loess deposits in the Midwest have come from melting glaciers that were great distances from where the loess was deposited?

___

___

___

### How do dunes form?

Have you ever noticed what happens when wind blows sediment against an object? The wind sweeps around or over the object. It drops the sediment behind the object. If the wind blows for a long time, the mound of sediment will become a dune. A **dune** is a mound of windblown sediment that piles up behind an object.

## Where are dunes found?

Dunes are common in deserts. You can also find dunes along the shores of oceans, seas, or lakes. If there is dry sediment where the winds blow daily, dunes build up. Sand or other sediment will keep building up to form a dune until the sand runs out or the object is removed.

## What are moving dunes?

Sand dunes have two sides. The side facing the wind has a gentler slope. The side away from the wind is steeper. You can tell which way the wind usually blows by looking at the two sides of a sand dune.

Most dunes will move, or migrate, in the same direction as the wind is blowing. This movement is shown in the figure below. Some dunes are called traveling dunes. They move quickly across desert areas. As they lose sand on one side, they build it up on the other.

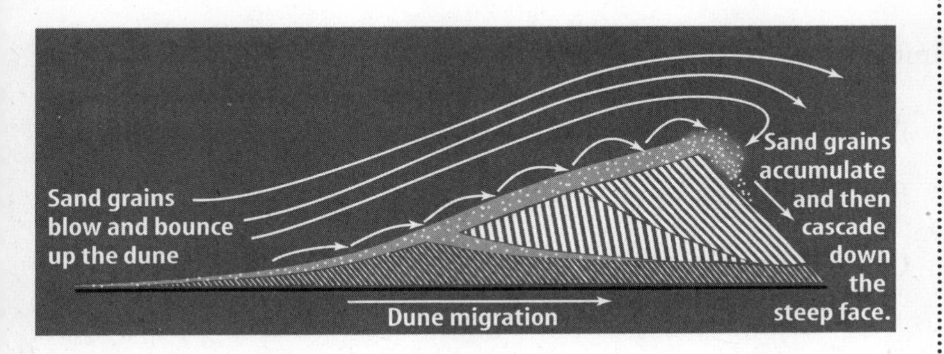


## Picture This

**5. Label** Draw an arrow to show the direction the wind is blowing to form this dune.

## Why do the shapes of dunes differ?

The shape of a dune depends on the amount of sediment available, the speed and direction of the wind, and the amount of vegetation present. One common dune with a curved shape is called a barchan (BAR kun) dune. The open side of a barchan dune faces the direction the wind is blowing. Transverse dunes have their long sides transverse, or crosswise, to the usual direction of the wind. Star dunes form where the direction of the wind changes often. A star dune has points in different directions.

## What happens when sediments shift?

When dunes and loess form, the land changes. Wind, like gravity, glaciers, and water, changes the shape of the land. Erosion and deposition are part of a cycle of change. They constantly shape and reshape the land.

## Think it Over

**6. Identify** Name three dune shapes that can form.

______________________

______________________

______________________

# After You Read

## Mini Glossary

**abrasion:** type of erosion that occurs when windblown sediment scrapes and wears away rock

**deflation:** type of erosion that occurs when wind removes small particles of sediment from the land and leaves heavy pieces of sediment behind

**dune:** mound of windblown sediment that piles up behind an object

**loess:** deposits of very fine-grained soil carried by the wind

**1.** Review the terms and their definitions in the Mini Glossary. Explain the difference between deflation and abrasion.

______________________________________________

______________________________________________

______________________________________________

**2.** Write the letter of the definition next to the correct term.

_____ windbreak

_____ dust storms

_____ moving dune

A. dune that travels in the same direction as the wind

B. trees planted to slow the wind

C. wind moves soil mostly made of silt-sized and clay-sized particles high into the air

**3.** You used three methods to identify the main idea of each paragraph as you read this chapter. The methods were highlighting, creating a quiz, and underlining. Which of these methods was most helpful? Why?

______________________________________________

______________________________________________

______________________________________________

Visit **earth.msscience.com** to access your textbook, interactive games, and projects to help you learn more about wind erosion.

# Water Erosion and Deposition

## section 1 Surface Water

### Before You Read

Is there a river near the place where you live? Write two sentences describing the biggest river you've ever seen or heard about.

_______________________________________________

_______________________________________________

_______________________________________________

**What You'll Learn**
- what causes runoff
- about rills, gullies, sheet, and stream erosion
- about the stages of stream development
- how alluvial fans and deltas form

### Read to Learn

Study Coach

**Authentic Questions** As you read this section, write down any questions you may have about each topic. Discuss your list with your teacher.

#### Runoff

Picture this. You pour a glass of milk and it overflows. Milk spills onto the table. Before you can grab a towel to clean up the mess, the milk has run through the crack of the table, over the edge, and onto the floor.

This is similar to what happens to rainwater when it falls to Earth. Some rainwater soaks into the ground. Some rainwater evaporates, turning into a gas. But some rainwater just runs over the ground, flowing into streams, lakes, or oceans. **Runoff** is any rainwater that does not soak into the ground or evaporate but flows over Earth's surface. ☑

**Reading Check**

1. **Identify** What is rainwater that flows over Earth's surface called?

_______________________

#### What are the main causes of runoff?

After it rains, most water either soaks into the ground or runs off. How much runoff there is depends on the amount of rain and the length of time it falls. Light rain that falls for a long time usually soaks into the ground. Heavy rain that falls in a short time often runs off. Rain will run off if it cannot soak into the ground fast enough. Rain also will run off if the ground cannot hold any more water.

## What other factors affect runoff?

Another factor that affects runoff is the slope or steepness of the land. Water flowing quickly down steep slopes has little time to soak into the ground. Water flowing more slowly down gentle slopes and across flat areas has more time to soak into the ground.

Vegetation like plants and trees also affects the amount of runoff. Just like milk running off a table, water will run off smooth surfaces that have little or no vegetation. But what would happen to the milk if the table had a tablecloth on it? The flow of the milk would slow down. In a similar way, runoff slows down when it flows around plants. By slowing runoff, plants and their roots help prevent soil erosion.

**Effects of Gravity** Gravity is the attractive force between all objects. The greater the mass of an object, the greater its force of gravity is. Earth has a greater mass than any objects on it. As a result, Earth's gravitational force pulls objects toward its center. Water flows down a slope because of the pull of gravity. As water starts to flow down a slope, it picks up speed. As water flows faster, it gains more energy. Fast-moving water carries more soil that slow-moving water.

**Think it Over**

**2. Identify** Name three factors that affect runoff.

_______________

_______________

_______________

# Water Erosion

Suppose you and several friends walk the same way to school each day through a field or empty lot. You always walk in the same footsteps as you did the day before. After a few weeks, you've worn a path through the field. When water runs down the same slope time after time, it also wears a path. The running water moves soil and rock from one place to another in a process called erosion.

## What is rill and gully erosion?

Sometimes running water cuts a groove or small ditch on the side of a hill. This is a sign of rill erosion. Rill erosion starts when a small stream forms during a heavy rain. As this stream flows along, it carries away soil.

Water flowing down the same path over and over wears a channel in the soil. A **channel** is a groove created by water moving down the same path. When water flows down the same channel many times, rill erosion may turn into gully erosion. Gullies are wider and deeper than rills. Gully erosion carries away large amounts of soil.

**FOLDABLES™**

**A Compare and Contrast** Make a three-tab Foldable to learn about three types of surface soil erosion caused by runoff.

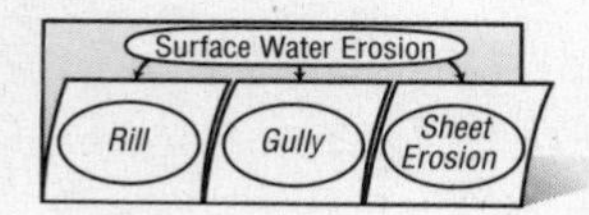

## What is sheet erosion?

Sometimes water erodes the soil without being in a channel. During a rainstorm, water often runs off first as thin, wide sheets, like water flowing off the hood of a car. Later, the runoff forms rills and streams.

Water also can flow as sheets if it runs out of its channel. Sometimes water spills out of a river and flows as sheets over the flat land beside the river. Streams flowing out of mountains may fan out, flowing as sheets away from the bottom of the mountain. **Sheet erosion** occurs when water flowing as sheets picks up sediments and carries them away. ☑

**Reading Check**

**3. Identify** What type of erosion occurs when water flowing as a sheet carries away sediment?

______________________

## How do streams erode?

Sometimes water keeps flowing along a low place it has formed. Water moves along in a stream, picking up sediment from the bottom and sides of its channel as it flows. In this way, a stream grows deeper and wider.

The sediment that a stream carries is called its load. Water in the stream lifts and carries some of the lighter sediments. These lighter sediments are called the suspended load, and are shown in the figure below. Larger, heavy particles called the bed load just roll along the bottom, or bed, of the stream.

The sediments scrape against the bottom and sides of the stream like sandpaper. Slowly these sediments wear away the rock in the stream. This process is called abrasion.

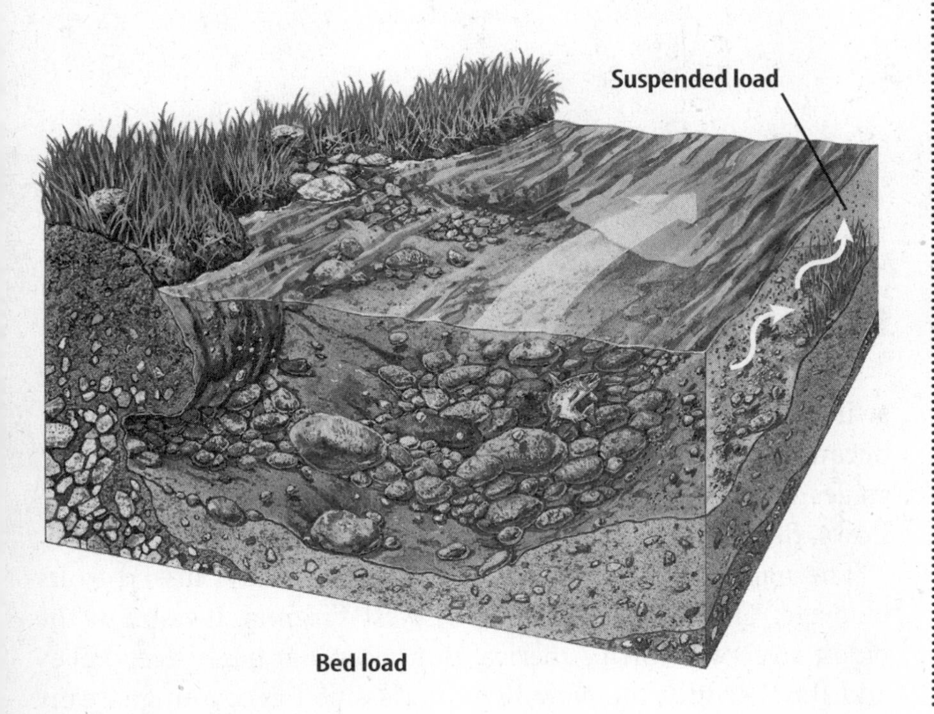


**Picture This**

**4. Identify** Draw a line from the label to the bed load in the stream.

## River System Development

Each day, millions of liters of water flow through rivers. Where does all the water come from? Where is it flowing to?

### What are the parts of a river system?

A tree is a system with twigs, branches, and a trunk. Rivers also are systems with many parts. Water in a river first comes from runoff that enters small streams. Then small streams join together to form larger streams, and these larger streams join to form a river. The river grows larger and carries more water as more streams join. ☑

**Reading Check**

**5. Describe** What makes up a river system?

______________________

______________________

______________________

______________________

### What are drainage basins?

A **drainage basin** is all the land that gathers runoff for a stream or river. Compare a drainage basin to a bathtub. The water in a bathtub flows to one spot—the drain. Likewise, the water in a river system flows to one spot—the main river, or trunk. The largest drainage basin in the United States is the Mississippi River Drainage Basin. The map below shows the area covered by this huge drainage basin.

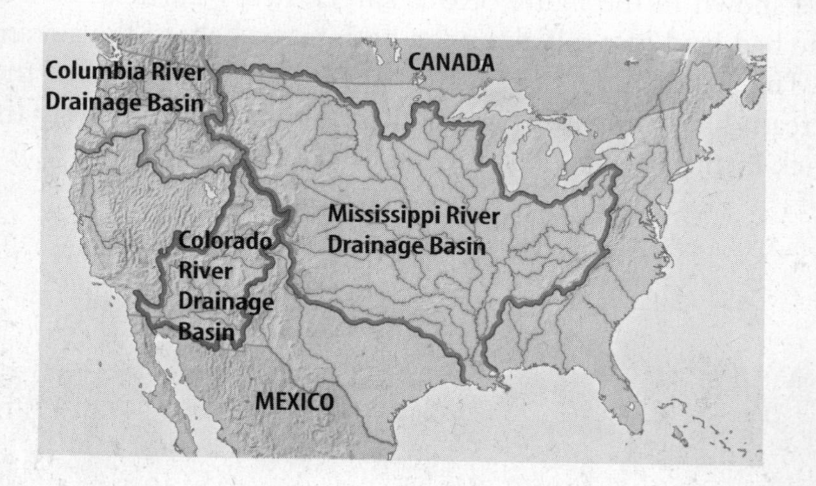


**Picture This**

**6. Determine** What land feature separates the Mississippi River Drainage Basin from the Columbia and Colorado River Drainage Basins?

______________________

## Stages of Stream Development

Streams come in many forms. Some are narrow and move swiftly while others are wide and move slowly. Streams differ because they are in different stages of development. The stages depend on the slope of the ground where the stream flows. Streams are classified as young, mature, or old.

The name of the stage of a stream isn't always linked to its true age. The New River flows in West Virginia. It's one of the oldest rivers in North America. But because it has a steep valley and flows swiftly, the New River is classified as a young stream.

## What are young streams?

A stream that flows swiftly through a steep valley is a young stream. A young stream may have rapids and waterfalls. Water flowing in a steep channel with a rough bottom has a lot of energy. It erodes the stream bottom faster than the sides.

## What are mature streams?

The next stage in the development of a stream is the mature stage. A mature stream flows more smoothly in its valley than a young stream. In time, most of the rocks in the stream that cause waterfalls and rapids are eroded. The rocks are worn down by the running water and the sediments in it.

Erosion doesn't occur just on the bottom of a mature stream. Erosion begins to occur along the sides of the mature stream. In time, as the sides erode, some curves form. The curves form because the speed of the water changes across the width of the stream.

**Meander** In a shallow part of a stream, water moves slower because it drags along the bottom. In the deeper part of the stream, the water flows faster. If the deep part of a stream is at one side of the channel, the water will erode that side making a slight curve. Over time, the curve grows and becomes a broad bend called a **meander** (mee AN dur).

**Floodplain** The broad, flat valley made by a meandering stream is called a floodplain. As you might guess, floods often occur on a floodplain. When a stream floods, it often covers all or part of the floodplain.

## What are old streams?

The last stage in the development of a stream is the old stage. An old stream flows smoothly in a broad, flat floodplain. South of St. Louis, Missouri, the lower Mississippi River is in the old stage.

River systems have young, mature, and old streams. In the upper part of a system, young streams move swiftly down mountains and hills. At the foot of mountains and hills, mature streams start to meander. These streams meet at the trunk of the drainage basin. There, they form a major river.

**FOLDABLES™**

**B Identify Main Ideas** Make a three-tab Foldable to learn the main ideas about stream development.

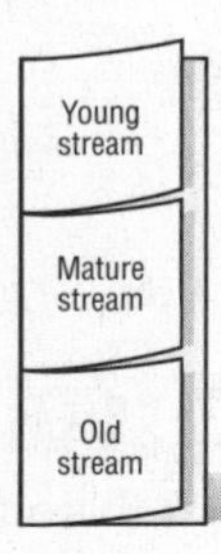


**Think it Over**

7. **Contrast** Name one way in which mature streams are different from old streams.

________________

________________

## Too Much Water

Sometimes, heavy rains or melting snow pour large amounts of water into a river. What happens when a river has more water than it can hold? The water has to go somewhere. It flows over the river banks causing a flood. Floods can bring disaster by destroying homes, bridges, or crops.

### What are dams and levees?

Dams and levees are built to prevent rivers from overflowing their banks. Dams across rivers are used to control the flow of water downstream. They may be built of soil, sand, or steel and concrete. Levees are mounds of earth built along the sides of rivers to help keep water inside the banks. ☑

Dams and levees are built to stop floods. But, they do not stop the worst floods. Dams and levees were not able to stop the floods in 1993. That year heavy rains caused the Mississippi River to overflow its banks. In nine Midwestern states, the flooding resulted in billions of dollars of damage.

**Reading Check**

8. **Explain** the difference between the places where dams and levees are built.

______________________

______________________

______________________

### What are catastrophic floods?

At certain times in Earth's past, many huge floods have changed large areas of Earth's surface. Such floods are called catastrophic floods.

One catastrophic flood formed the Channeled Scablands in eastern Washington State. This large area is shown in the map below. Now the soil is gone and the land is just deeply worn rock.

**Picture This**

9. **Identify** Circle the names of the four states shown on the map.

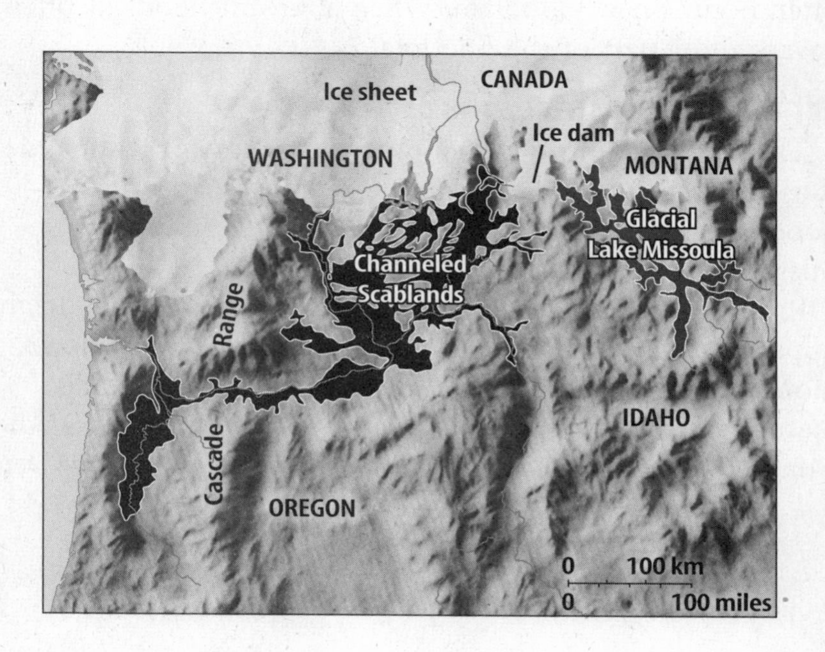

### How did the Channeled Scablands form?

The figure on the previous page shows a time when huge Lake Missoula covered much of western Montana. A natural dam of ice formed this lake. When the dam melted or eroded, huge amounts of water quickly flowed out. Floodwaters poured through what is now the state of Idaho into Washington, removing topsoil and cutting deep channels in the rock.

## Deposition by Surface Water

You know how hard it is to carry a heavy object for a long time. After awhile, you need to set it down. As water moves in a river system, it loses some of its energy. The water can't carry all of its sediment. As a result, the sediment drops, or is deposited, to the bottom of the stream. ☑

Streams carry some sediment just a short way. In fact, they often drop heavy pieces of soil and rock in the stream channel. Streams carry lighter pieces of soil and rock greater distances. The sediments picked up when rills and gullies erode are examples of this.

Water has a lot of energy when it runs down a steep slope. It can carry large amounts of sediment as long as it is moving fast. When water flows onto level land, it slows. As it slows, it drops its load of soil and rock. Water also slows and deposits its load of sediment when it empties into an ocean or lake.

**Reading Check**

10. **Explain** When river water begins to lose its energy, what happens to the sediment it carries?

______________________

______________________

### What are deltas and fans?

When water flows into an ocean or lake, it slows and deposits its sediment. This deposit, shaped like a triangle or fan, is called a delta. When a river flows from a mountain valley onto an open plain, the deposit is called an alluvial (uh LEW vee ul) fan. ☑

**Reading Check**

11. **Describe** How is a delta shaped?

______________________

______________________

### How is the Mississippi River a model river?

The Mississippi River is a model of what you have learned in this section. Runoff causes rills and gullies. Sediment is picked up and carried into bigger streams that eventually flow into the Mississippi River.

As the river flows, it cuts into its banks, picking up more sediment. Where the land is flat, the river deposits some of the sediment in its own channel. At last, the Mississippi River enters the Gulf of Mexico where it slows, drops much of its sediment, and forms the Mississippi River delta.

# After You Read

## Mini Glossary

**channel:** groove created by water moving down the same path

**drainage basin:** land area that gathers runoff for a stream or river

**meander (mee AN dur):** a broad arc or curve in a stream or river formed by erosion of its outer bank

**runoff:** any rainwater that does not soak into the ground or evaporate, but flows over Earth's surface

**sheet erosion:** erosion caused by runoff that occurs when water flowing as sheets picks up sediments and carries them away

1. Review the terms and their definitions in the Mini Glossary. Write one or two sentences telling how runoff can form channels.

_______________________________________________

_______________________________________________

_______________________________________________

2. Fill in the spaces below to compare different types of erosion by surface water.

| Type of Erosion | Location | Cause | Result |
|---|---|---|---|
| Rill | On a slope | | |
| Gully | | Heavy rainfall | |
| Sheet | | | Land is covered by a thin layer of water for a short time; mild loss of soil |
| Stream | | Normal water flow in the channel | |

Visit earth.msscience.com to access your textbook, interactive games, and projects to help you learn more about surface water.

# Water Erosion and Deposition

## section 2 Groundwater

### Before You Read

Where do you think the water you drink every day comes from? Write your thoughts on the lines below.

______________________________________________

______________________________________________

______________________________________________

**What You'll Learn**

- why groundwater is important
- how groundwater erodes and deposits rock

### Read to Learn

**Mark the Text**

**Highlight** As you read this section, highlight each vocabulary term and its definition.

#### Groundwater Systems

What would happen if you spilled milk on a table and it ran onto a carpeted floor? The milk might soak into the carpet. Water that falls on Earth can soak into the ground just like milk soaks into a carpet.

#### What is groundwater?

Water that soaks into the ground becomes part of a system. Soil is made up of many small fragments of rocks and minerals. Even though the fragments are all touching one another, there is some empty space between them. Holes, cracks, and gaps are found in the rock under the soil. Water that doesn't run off, soaks into the ground. Water that soaks into the ground and collects in pores and empty spaces is called **groundwater**. Groundwater is an important source of drinking water. ☑

How much of Earth's water do you think is held in spaces in the ground? One estimate is that 14 percent of all freshwater on Earth is in the ground. That is almost 30 times more water in the ground than in all of Earth's lakes and rivers.

**Reading Check**

1. **Determine** When water soaks into the ground, where does it collect?

______________________

______________________

## How does water pass through the ground?

A groundwater system is similar to a river system. A river has channels that connect the parts of the drainage basin. A groundwater system has connecting pores. If the soil and rock are **permeable** (PUR mee uh bul), they have connecting pores through which water can flow. Sandstone is an example of permeable rock. ☑

Soil or rock that has many big, connecting pores is permeable. Water can pass through it easily. However, if some rock or sediment has few pore spaces or they are not well connected, the flow of water is blocked. **Impermeable** materials do not let water pass through them. Granite has few or no pore spaces at all. Clay has many small pore spaces, but they are not well connected.

**Reading Check**

**2. Define** What term describes soil and rock that have connecting pores that water can pass through?

_______________

## Where does groundwater move?

The figure below shows water moving in the ground through permeable soil and rock. An **aquifer** (AK wuh fur) is a layer of permeable rock that allows water to move freely.

Groundwater keeps going down until it reaches a layer of impermeable rock. At that point, the water stops moving and starts filling up the pores in the rocks above. When all of the pores in the rock are filled with water, the area is called the zone of saturation. The **water table** is the upper surface of the zone of saturation. A stream's surface level is the water table.

**Picture This**

**3. Locate and Label** Write *Impermeable layer* and *Permeable layer* on the lines to show where these are located in the figure.

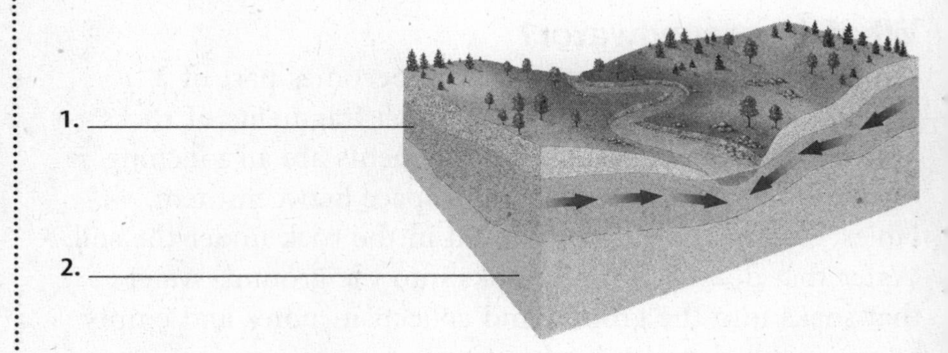


## Water Table

Why is the water table so important? Each person in the United States uses about 626 L of water per day. That could fill nearly two thousand soft drink cans. Many people get their water from wells drilled deep into the ground. There is only so much groundwater. During a drought, the water table drops. That is why you should conserve water.

## How do wells provide water?

A good well goes deep into the ground, past the top of the water table. Water flows into the well and a pump brings it to the surface. In very dry seasons, the water table may drop. Even a good well can go dry. It takes time for the water table to rise again. More water may come from rain or from groundwater flowing from other areas in the aquifer.

In some places, groundwater is the main source of drinking water. It's important to control the number of wells and how much water is pumped out of the ground. There is only so much water in the ground. Wells can go dry if water is pumped out faster than it is replaced.

If too much water has been pumped out of an area, the land level can sink. The weight of the soil above the empty pore spaces causes this.

## Do all wells need pumps?

Most wells need a pump to bring water to the surface. But in an artesian well, the water rises without a pump. Artesian wells are not as common as other types of wells because they require special conditions.

The aquifer for an artesian well has to be between two layers of impermeable soil or rock. The layers must slope. This is shown in the figure of the artesian well below.

Water enters the upper end of the aquifer. The water in the upper end puts pressure on the water below. What happens if a well is drilled? Water flows out quickly. If there's much pressure, the water shoots up into the air like a fountain.

### Think it Over

**4. Infer** Why is it important for a well to be deep?

______

______

______

### Picture This

**5. Explain** Why doesn't an artesian well need a pump?

______

______

______

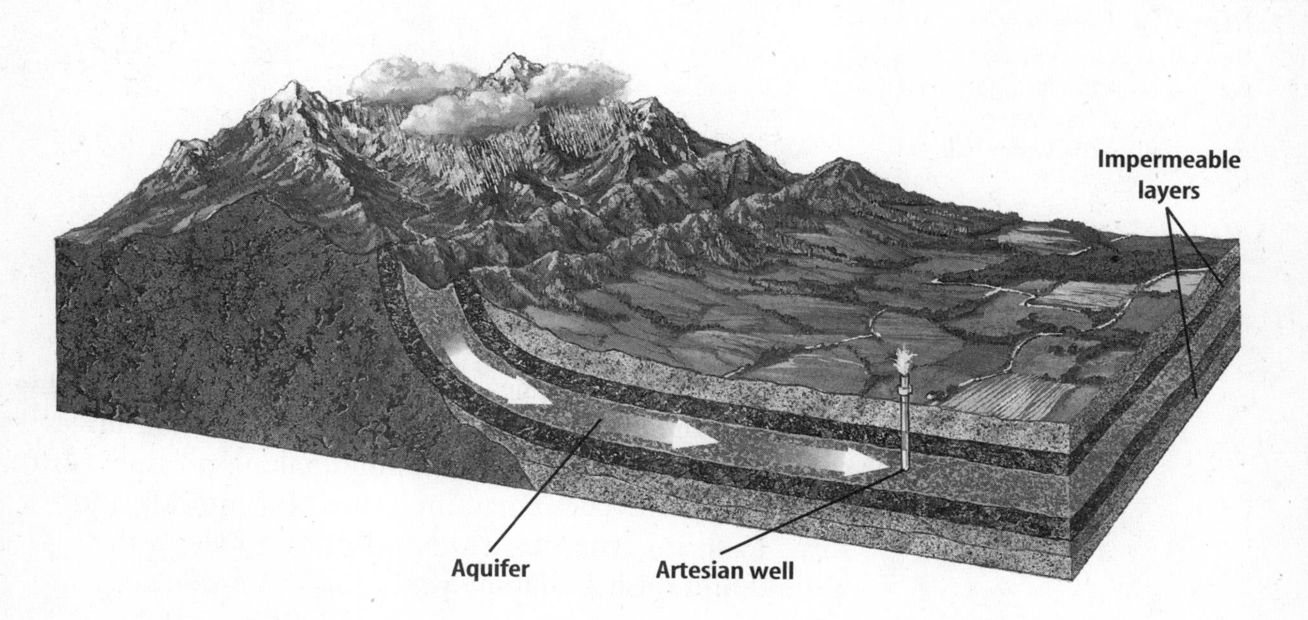

## What are springs?

In some places, the water table is so near Earth's surface that water flows out of the ground. A **spring** forms when the water table meets Earth's surface. Often springs are found on hillsides or other places where the water table meets a sloping surface. Springs often are used as a source of freshwater. ☑

Water from most springs is cool. Soil and rock protect the groundwater from changing temperatures. However, in some places, melted rock rises near Earth's surface and heats the rock around it. Water that touches the hot rock can come to the surface as a hot spring.

**Reading Check**

6. **Identify** What forms when the water table meets Earth's surface?

______________________

## What are geysers?

When water is put into a teakettle to boil, it heats slowly at first. Then steam starts to come out of the cap on the spout. Suddenly the water starts boiling. The teakettle starts to whistle as steam is forced past the cap.

A similar process can occur with water in the ground. It can get so hot, it turns to steam. Yellowstone National Park in Wyoming is a place where ground water is heated. The park has both hot springs and geysers. A **geyser** is a hot spring that erupts periodically and shoots water and steam into the air. The figure below shows a cross section of a geyser.

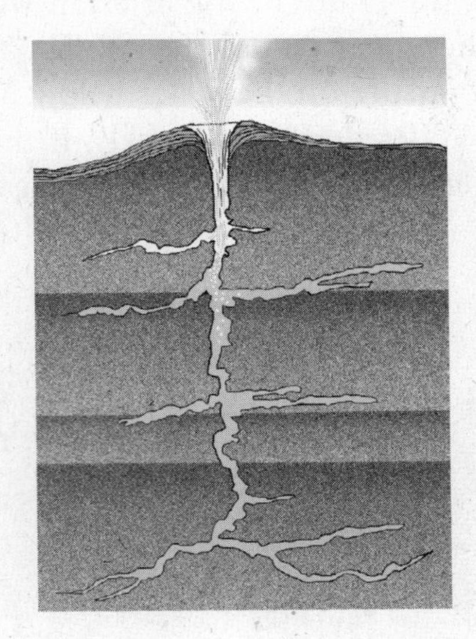

**Picture This**

7. **Identify** Trace the path that hot water and steam will follow in the figure of the geyser. What has to occur before the water will turn to steam?

______________________

______________________

In a geyser, water is heated to high temperatures, making it expand, or spread, underground. This expansion forces some of the water out of the ground, taking pressure off the water that's left. The remaining water boils quickly and turns to steam. The steam shoots out of a hole in the ground and pushes out any water that's still there.

## The Work of Groundwater

Although water is the most powerful cause of erosion on land, it also can have a great effect underground. Water mixes with carbon dioxide gas. This solution of water and carbon dioxide forms a weak acid, called carbonic acid, that can dissolve some rocks. ☑

Carbonic acid easily dissolves limestone. In the soil, acidic water flows through cracks and pores in rock dissolving the limestone. Slowly the cracks in the limestone get larger and larger until a cave is formed. A **cave** is an underground opening that forms when acidic groundwater dissolves limestone.

### How are deposits formed in caves?

You've probably seen a picture of the inside of a cave. Perhaps you've even visited a cave. Groundwater not only dissolves limestone to form caves, it also makes deposits on the insides of the caves.

Water often drips from cracks in the walls and ceilings of caves. The water carries calcium from the limestone. If this water evaporates while hanging from the ceiling of a cave, it leaves behind a deposit of calcium carbonate. Over time, stalactites form from the buildup of calcium. Stalactites hang from the cave's ceiling and grow larger as the calcium continues to build up. If the drops of water fall to the floor of the cave, a stalagmite forms. ☑

### How does a sinkhole form?

A sinkhole can form when underground rock near the surface dissolves. A sinkhole is a low spot on the surface of the ground that forms when a cave collapses. A sinkhole also can form when material near the surface dissolves. Sinkholes are common in states like Florida and Kentucky that have lots of limestone and enough rain to keep the ground full of water. Sinkholes can cause damage if they form where people live.

### What happens to groundwater?

When rain soaks into the ground it becomes part of groundwater systems. It might dissolve limestone and form a cave. It might erupt from a geyser or it might be pumped from a well at your home.

**Reading Check**

**8. Recall** What gas mixes with water to form carbonic acid?

______________________

**Reading Check**

**9. Identify** What two cave formations are caused by water evaporating and leaving behind calcium deposits?

______________________

______________________

**FOLDABLES™**

**C Compare and Contrast** Make a half-book Foldable to record information about groundwater erosion.

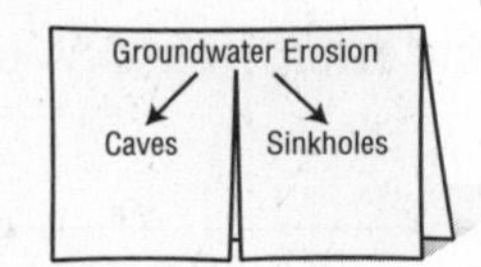

# After You Read

## Mini Glossary

**aquifer:** layer of permeable rock that allows water to flow through it

**cave:** underground opening that forms when acidic groundwater dissolves limestone

**geyser:** hot spring that erupts periodically and shoots water and steam into the air

**groundwater:** water that soaks into the ground and collects in pores and empty spaces

**impermeable:** material that water can't pass through

**permeable:** material that has connecting pores through which water can flow

**spring:** forms where the water table meets Earth's surface

**water table:** top of the zone of saturation where the pores in the soil and rock are filled with water

1. Review the terms and their definitions in the Mini Glossary. Write one or two sentences that tell how groundwater and aquifers are related.

_______________

_______________

_______________

2. Fill in the concept map about cave formation with these terms: *stalagmites, water, carbon dioxide, stalactites.*

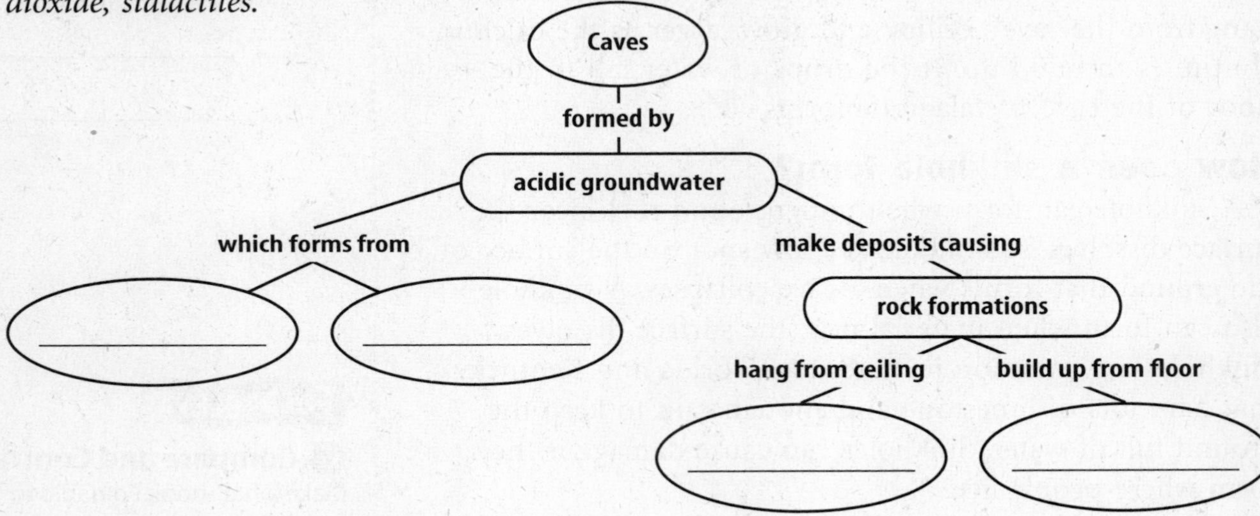


3. In this section you highlighted words and their definitions. Did this strategy help you learn their meanings? What other strategy would help you learn the meanings of new words?

_______________

Science Online Visit **earth.msscience.com** to access your textbook, interactive games, and projects to help you learn more about groundwater.

# Water Erosion and Deposition

## section 3 Ocean Shoreline

## Before You Read

Have you ever visited an ocean beach or seen a picture of one? On the lines below, describe what a beach is like.

___

___

___

### What You'll Learn

- about different types of shoreline
- what causes shoreline erosion
- how sand is eroded and deposited

## Read to Learn

### The Shore

Imagine sitting on a beautiful, sandy beach. Nearby, palm trees sway in the breeze. The waves are quiet and children play at the water's edge. It's hard to imagine a more peaceful place. Now imagine sitting on a high cliff by a different shore where waves are crashing into the rocks below. Both places are shorelines. An ocean shoreline is where land meets the ocean.

Both shorelines have waves, tides, and currents. Yet, the two shores are not the same. The action of water changes shorelines all the time. You may see changes from hour to hour. Why are shorelines so different? They are different because of the forces that shape them.

### What forces shape shorelines?

Waves constantly pound the beach, breaking rocks into smaller and smaller pieces. Currents move tons of sediment along the shoreline. The tiny grains of sediment grind against each other like sandpaper. Sometimes the tide carries sediment to deeper water. Then the tide returns, bringing new sediment with it. These three forces—waves, currents, and tides—are at work slowly changing the shape of the shoreline.

Study Coach

**Identify Main Ideas** As you read this section, write the main idea of each topic on a sheet of paper. Use your notes to review at the end of the section.

FOLDABLES™

**D Classify** Make a three-tab Foldable to learn about the main forces that shape shorelines.

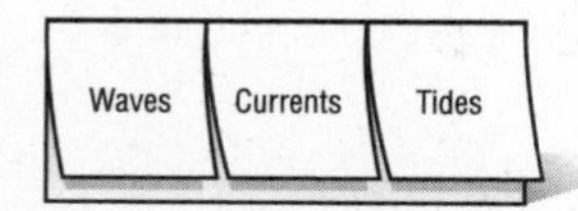

## Picture This

1. **Locate and Label** In the figure, write *Tidal current* by the arrows showing the movement of incoming and outgoing tides. Write *Longshore current* by the arrows showing this type of current. Now read about these currents below.

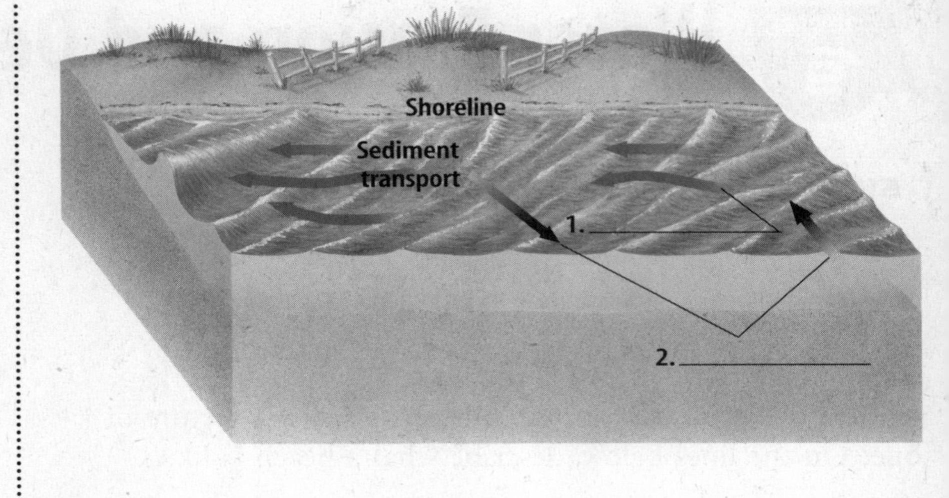


## How do waves and tides affect shorelines?

Waves are a powerful force as they crash against the shoreline. Often waves hit the shore at slight angles creating a longshore current. **Longshore currents** run parallel to the shoreline and move tons of loose sediment.

Tides create currents that move directly to or away from the shore at right angles. These are tidal currents. Outgoing tides carry sediment away from the shores while incoming tides bring new sediment toward the shore. Tides work with waves to shape the shoreline. Longshore currents and tidal currents are shown in the figure above.

These forces affect all shorelines. There are other forces that make one shore a flat, sandy beach and another shore a steep, rocky cliff.

**Applying Math**

2. **Calculate** About 14,000 waves hit the shore each day. How many waves hit the shore each hour? Show your work.

## Rocky Shorelines

Rocks and cliffs are the most common features along rocky shores. Waves crash against the rocks and cliffs. Sediments in the water grind against the cliffs. Slowly, over time, they wear away the rock. Waves grind up pieces of rocks that break from the cliffs. The endless motion of the waves grinds rocks into sediment and longshore currents transport it.

Softer rocks erode before harder rocks do. At first, erosion leaves islands of harder rocks by the shore. Over time, the harder rocks wear away too. This process takes thousands of years. Remember, the ocean never stops. In one day, about 14 thousand waves crash into shore.

## Sandy Beaches

Smooth, gently sloping shorelines are different from steep, rocky ones. Beaches are found here. A **beach** is a deposit of sediments that is parallel to the shore. Beaches are made up of different materials. Some beaches are made of rock fragments from the shore. Many beaches are made of grains of sand and pieces of seashells. ☑

Rock fragments may be larger than your hand. Sand grains range from 0.06 to 2 mm wide. Why do many beaches have particles this small? Waves grind rocks and seashells into tiny pieces. The back and forth motion of waves bumps the pieces of rocks, shells, and sand together. That breaks the pieces into smaller pieces. The bumping motion also polishes their jagged corners and makes them smoother.

Some beaches have sand that is made of other things. For example, Hawaii's black sands are made of basalt, and its green sands are made of the mineral olivine. Jamaica's white sands are made of bits of coral and shells.

## Sand Erosion and Deposition

Beaches are fragile, short-term land features because they are constantly changing. Tides and currents carry the sand. Storms and waves damage beaches. Sometimes, human activity such as construction changes beaches. Beach erosion is a problem that affects many coastal areas. Places as far apart as Long Island, New York; Malibu, California; and Padre Island, Texas, have problems because of beach erosion. ☑

### What are barrier islands?

Barrier islands are deposits of sand that run parallel to the shore but are separate from the mainland. These islands start as underwater sand ridges formed by breaking waves. Hurricanes and storms add sediment to them. This raises some of these islands to sea level. When a barrier island becomes large enough, the wind blows the loose sand into dunes. The dunes keep the new island above sea level. Yet, these islands don't last long.

The forces that build barrier islands also can erode them. Storms and waves carry sand away. Barrier islands are short-term places that can last from a few years to a few centuries.

✔ Reading Check

3. **Identify** Where are beaches usually found?

_______________

_______________

✔ Reading Check

4. **Explain** Name one reason beaches are always changing.

_______________

_______________

# After You Read

## Mini Glossary

**beach:** deposit of sediments whose materials vary in size, color, and composition and is usually found on a smooth, gently sloped shoreline

**longshore current:** current that runs parallel to the shoreline and moves tons of loose sediment; caused by waves hitting the shoreline at an angle

1. Review the terms and their definitions in the Mini Glossary. Write one or two sentences that tell how longshore currents and tides are alike and different.

___

___

___

2. Complete the concept map below.

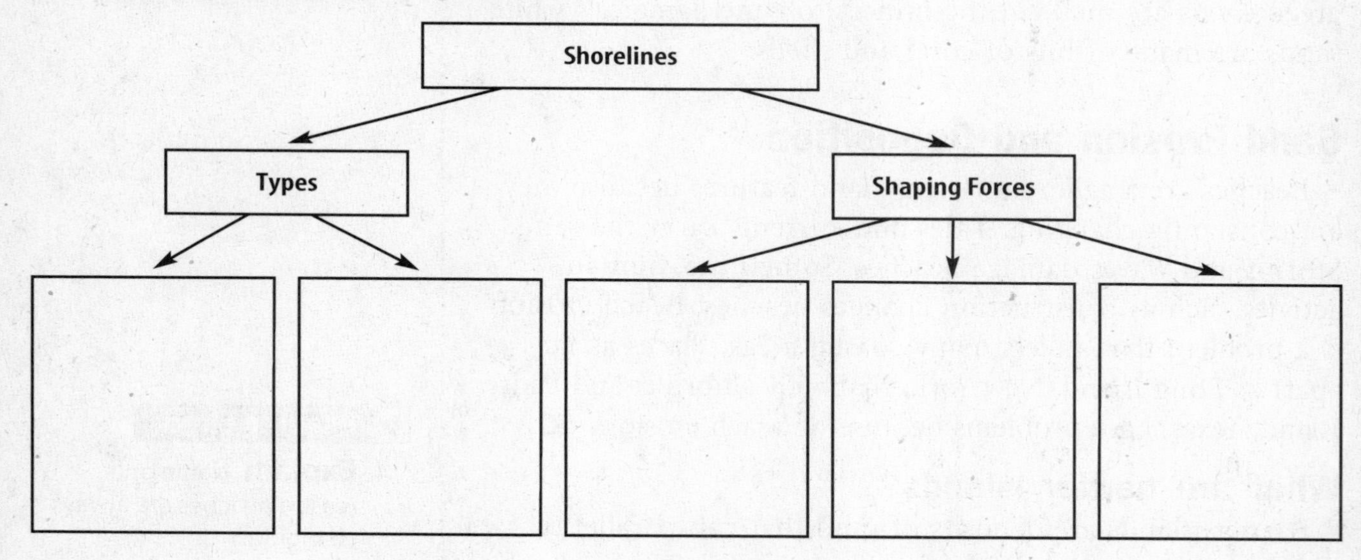


3. In this section you wrote the main idea of each topic and then reviewed. Would you use this method again to learn new information? Explain why or why not.

___

___

Science Online Visit **earth.msscience.com** to access your textbook, interactive games, and projects to help you learn more about the ocean shoreline.

# Plate Tectonics

## section 1 Continental Drift

### Before You Read

After you peel an orange, can you fit the pieces of peel back together again like a puzzle? Why or why not?

_______________________________________________

_______________________________________________

**What You'll Learn**
- what continental drift is
- what evidence supports continental drift

### Read to Learn

## Evidence for Continental Drift

Take a look at a map of Earth's surface. Look carefully at the shape of each continent. The continents look like they might fit together like pieces of a puzzle, don't they? People throughout history have noticed this and wondered what it meant. For example, over 400 years ago Abraham Ortelius, a Dutch mapmaker, noticed that the coastlines of South America and Africa fit together.

**Study Coach**

**Two-column Notes** As you read, organize your notes in two columns. In the left column, write the main idea of each paragraph. In the right column, write details about the main idea.

### Have you ever been to Pangaea?

In 1912, a German scientist named Alfred Wegener (VEG nur) proposed that the continents did at one time fit together. He suggested that all the continents were once connected. Wegener said they were once one big landmass. He called this landmass **Pangaea** (pan JEE uh), which means "all land."

According to Wegener, about 200 million years ago, Pangaea broke into pieces. The pieces drifted away from each other. We call these pieces continents. Wegner's hypothesis is called continental (kahn tuh NEN tul) drift. **Continental drift** is the hypothesis that Pangaea broke apart into continents that moved slowly to where they are today. So even though you haven't visited Pangaea, the area where you live probably was part of Pangaea millions of years ago.

**FOLDABLES™**

**A Organize** Make a four-door Foldable as shown below to help organize facts you learn about Alfred Wegener and continental drift.

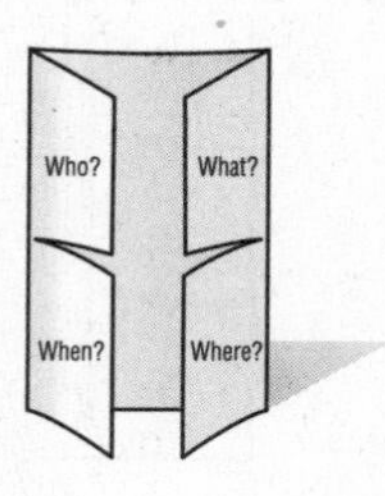

## Picture This

1. **Identify** Trace over the boundaries separating the continents on the map.

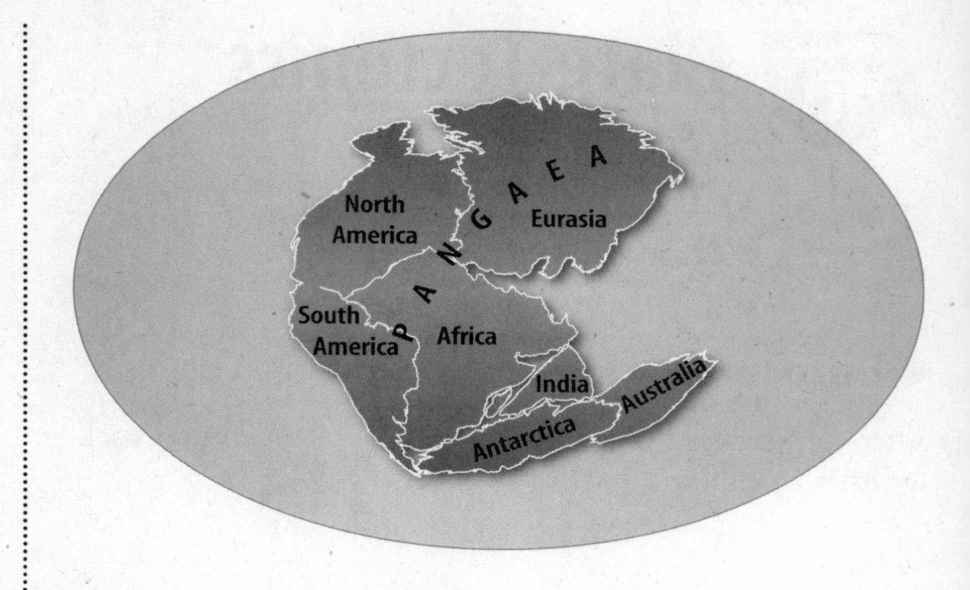


**Reading Check**

2. **Explain** Why didn't Wegener convince people his ideas were right?

______________________

______________________

______________________

## Did everyone accept Wegner's idea?

Wegener's hypothesis on continental drift was a controversial idea. It caused a lot of argument. Wegener didn't convince people that his ideas were right during his lifetime. He didn't have enough evidence. For example, he couldn't explain what made the continents drift. He thought Earth spinning on its axis might cause the continents to plow through the ocean floor. But geologists and physicists of that time rejected this explanation. Wegener's basic idea wasn't accepted until long after his death in 1930. More evidence came later to support his ideas. ☑

## How do animal fossils support continental drift?

Animal fossils offer one clue that the continents might have been joined together millions of years ago. Fossils are the remains, imprints, or traces of prehistoric organisms. Fossils can tell when and where organisms once lived and how they lived. For example, fossils of *Mesosaurus* have been found in South America and in Africa. *Mesosaurus* is a reptile that lived on land and in fresh water. How could this reptile move between two continents separated by a salty ocean? It is not likely that it swam across the Atlantic Ocean. Wegener's hypothesis of continental drift proposes that *Mesosaurus* lived on both continents when they were joined together. The map on the next page shows where fossil remains of different animals have been found on different continents.

**FOLDABLES™**

**B Record Information** Make a three-tab Foldable to record facts about continental drift. You will complete your Foldable in the next two sections.

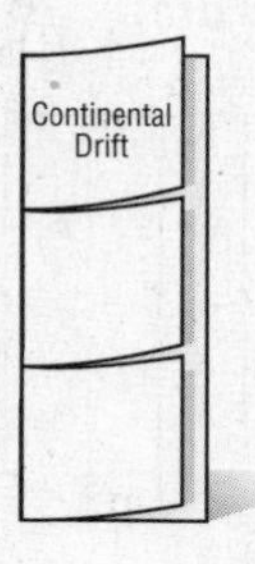

## Fossils Support Continental Drift

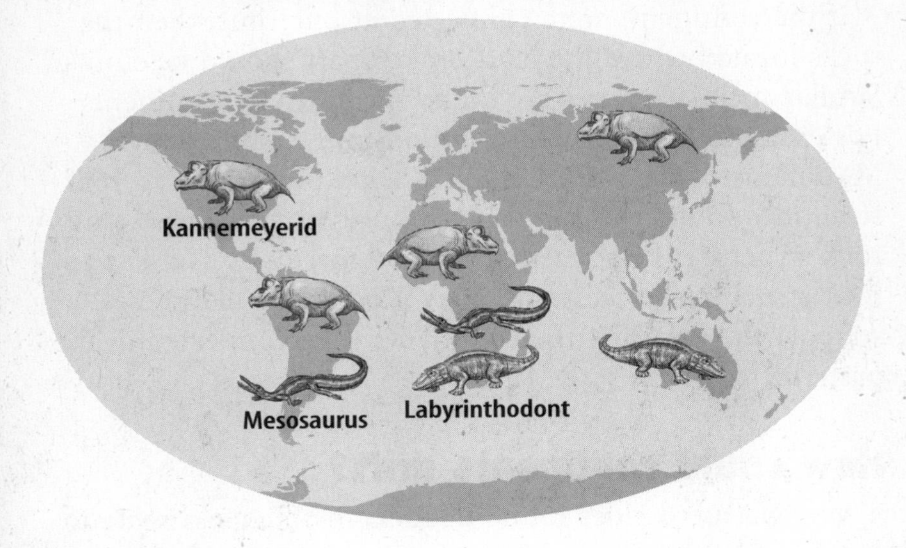


## How do plant fossils support continental drift?

Another fossil that supports continental drift is *Glossopteris* (glahs AHP tur us). Fossils of this plant are found on five continents—Africa, Australia, Asia, South America, and Antarctica. Finding *Glossopteris* in so many areas supported the idea that all of these regions once were connected and had similar climates.

## What do climates tell us about continental drift?

Scientists have found fossils of warm-weather plants on the island of Spitsbergen in the Arctic Ocean. This is one of the coldest places on Earth. How did this happen? Wegener's hypothesis of continental drift proposes that Spitsbergen Island drifted to the Arctic from a tropical region of Earth.

## How do rocks support continental drift?

Glaciers are large, slow moving bodies of ice on land. Glaciers leave tracks as they move, scouring and polishing rock surfaces under them. They also leave deposits of glacial rock and sediment. Glaciers are found in cold areas. Yet there are traces of glaciers and deposits in many tropical and temperate regions of Earth. How is this possible? ☑

Wegener's hypothesis of continental drift says that millions of years ago, some tropical and temperate lands had cold climates and were located near Earth's south pole. These continents were joined together and partly covered with ice. When the continents drifted apart into warmer climates, the glaciers melted away from these areas. The glacial deposits left behind add more evidence to support continental drift.

### Picture This

3. **Apply** Fossil remains of the extinct plant *Glossopteris* have been found in Asia, Australia, Antarctica, Africa, and South America. Add a leaf symbol to these areas on the map to represent *Glossopteris.*

### Reading Check

4. **Identify** What do glaciers leave behind when they melt?

______________________

______________________

## What do rocks tell us about continental drift?

If the continents were connected at one time, then the rocks located where the land broke apart should be similar. Similar rock structures are found on different continents. For example, rock structures found in the Appalachian Mountains of the eastern United States are similar to rock structures found in Greenland and western Europe. Some rock structures from eastern South America are similar to rock structures in western Africa. Rock clues like these support the idea that the continents were connected in the past. ☑

**Reading Check**

**5. Explain** How do rocks provide evidence for continental drift?

______________________

______________________

______________________

## How could continents drift?

Wegener used clues found in rock, fossil, and climate to support his hypothesis of continental drift. The computer model below shows how the continents might have drifted over millions of years. But Wegener was not able to explain why the continents broke apart. Most importantly, he could not explain what caused the continents to drift. Wegener hypothesized that the continents plowed through the ocean floor. He thought that the spinning of Earth on its axis might have been the cause. He couldn't prove his ideas with facts. Physicists of his day thought he was wrong. The idea was rejected because it was too unusual.

Today, there is more evidence for continental drift. After Wegener's death, new technology provided new evidence. New ideas about continental drift were developed. One of these ideas is called seafloor spreading. You'll read about this in the next section.

**Picture This**

**6. Label** Find North America on each of the three drawings. Label it.

**How Continents May Have Drifted**

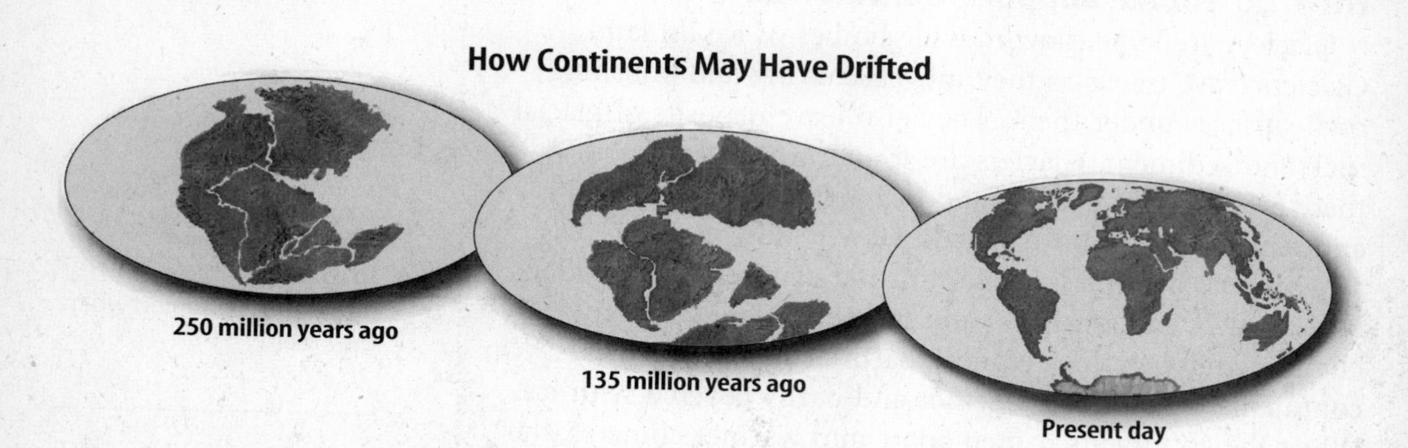

# After You Read

## Mini Glossary

**continental (kahn tuh NEN tul) drift:** Wegener's hypothesis that all continents were once connected in a single large landmass that broke apart and drifted slowly to their current positions

**Pangaea:** a single landmass composed of all the continents joined together

1. Review the terms and their definitions in the Mini Glossary. Write a sentence that explains how Pangaea relates to the hypothesis of continental drift.

2. According to the continental drift hypothesis, in what order did these three events take place? Write the events in the correct order.

**Climates on continents changed.**
**Pangaea broke apart.**
**Continents began to drift.**

**First**

**Second**

**Third**

3. In this section you read about later evidence that supports continental drift. Choose one piece of evidence. Write one or two sentences explaining how this evidence supports Wegener's hypothesis.

**Science Online** Visit **earth.msscience.com** to access your textbook, interactive games, and projects to help you learn more about continental drift.

**End of Section**

# Plate Tectonics

## section 2 Seafloor Spreading

**What You'll Learn**
- seafloor spreading
- what evidence supports seafloor spreading

## Before You Read

Have you ever seen the rings inside a tree that has been cut down? Do you think the rings closest to the center are the youngest or the oldest? Why?

_______________________________________________

_______________________________________________

Study Coach

**Sticky-note** As you read, use sticky-notes to mark pages you find interesting or you have a question about. Share the interesting fact with another student in your class. Ask your teacher to answer your questions.

## Read to Learn

### Mapping the Ocean Floor

If you were to lower a rope from a boat until it reached the seafloor, you would know how deep the ocean was in that spot. This is the first way people mapped the ocean floor.

#### How did technology improve seafloor mapping?

The rope method was used until German scientists discovered how to use sound waves to locate submarines. Later, sound waves on ships were used to map the seafloor. Sound waves travel through the water. When they hit the seafloor, they bounce back. The longer it takes for the sound waves to bounce back to the ship, the deeper the water is. The new sound wave technology made it easier to make better maps of the seafloor.

FOLDABLES™

**B Record Information** Use your Foldable to record facts about seafloor spreading.

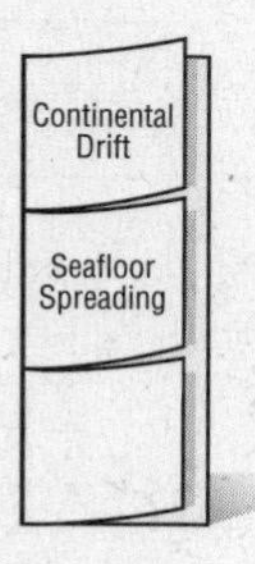


#### What did sound waves help discover?

By using sound waves, scientists found that the seafloor had an underwater system of ridges. These ridges are like mountain ranges and valleys on land. In some of these ridges are long rift valleys. These rift valleys are like rips in the ocean floor. Volcanic eruptions and earthquakes occur in the rift valleys from time to time. Underwater volcanic eruptions create underwater mountains. When these mountains push out of the water, they create islands.

## What are mid-ocean ridges?

The mid-ocean ridges are a chain of ridges and valleys stretching along Earth's ocean floor. Many of these ridges are connected. They circle Earth much like the stitching on a baseball.

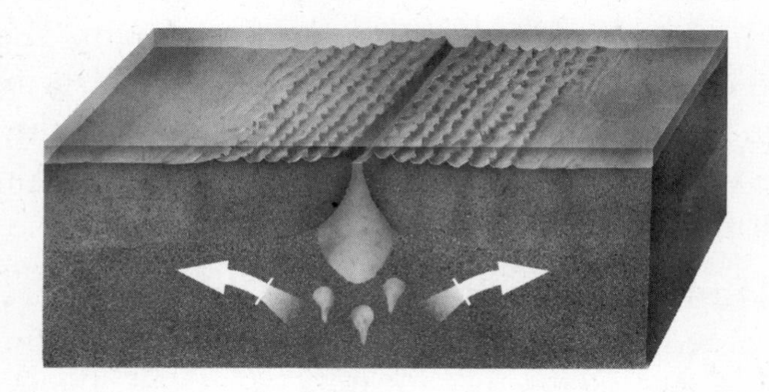

### Picture This

1. **Interpret** Circle the mid-ocean ridge in the figure. Trace over the arrows that show the direction in which the old seafloor is moving.

## Is the seafloor spreading?

In the 1960s, Harry Hess, an American scientist, proposed that the ocean floor moves. He called his theory seafloor spreading. The theory of **seafloor spreading** proposes that magma, or melted rock, under Earth's crust is forced up toward the surface at the mid-ocean ridges, forming new seafloor. When the less dense magma hits Earth's crust, it flows sideways. The magma carries the seafloor away from the central ridge in both directions. New seafloor is continuously being created. Older sea floor is pushed away from the central ridge as shown in the figure above.

# Evidence for Spreading

In 1968, scientists began studying rocks on the seafloor. They took rock samples from the mid-ocean ridges. They also took rock samples farther away from the ridge. They found that rocks near the mid-ocean ridge were the youngest rocks. Rocks farther away from the ridge were older.

According to Hess's theory of seafloor spreading, the seafloor near the ridge has formed more recently from magma. The older seafloor is pushed away from the ridge. Like tree rings, the further away the rocks, the older they are. The age of the rocks and their distance from the mid-ocean ridge supports the theory of seafloor spreading.

New life-forms have been discovered near the mid-ocean ridges. These giant clams, mussels, and tube worms, get heat and chemicals from magma pouring out of rifts in mid-ocean ridges.

2. **Distinguish** Are the oldest seafloor rocks located close to or far from the mid-ocean ridges?

_______________

## Does Earth's magnetic field change?

Earth's magnetic field has a north pole and a south pole. Invisible lines of magnetic force leave Earth near the south pole and enter Earth near the north pole. At this time, Earth's magnetic field travels from south to north. This is not always true. At times, the lines of magnetic force have traveled in the opposite direction, north to south. These direction changes are called magnetic reversals. During a magnetic reversal, the lines of magnetic force run the opposite way. All of these magnetic reversals are recorded in rocks forming along mid-ocean ridges.

**Think it Over**

3. **Apply** If Earth had a magnetic reversal, which way would the invisible lines of magnetic force run?

______________________

______________________

## How does the seafloor record history?

Minerals containing iron, such as magnetite, are found in rocks on the seafloor. Iron in the rock records the magnetic reversal. A device called a magnetometer (mag nuh TAH muh tur) tells scientists what direction a magnetic field has.

How do scientists know when a magnetic reversal happened and when it changed back? A strong magnetic reading is recorded when the polarity of a rock is the same as the polarity of Earth's magnetic field. Look at the figure below. Normal polarities in rocks show up as large peaks. After the magnetic reversal, the magnetometer records a weak reading. Over time, the reversals are shown in strips parallel to mid-ocean ridges.

Changes in Earth's magnetic field can be seen on both sides of mid-ocean ridges. This discovery adds to the evidence that the seafloor is spreading. The magnetic reversals showed that new rock was being formed at mid-ocean ridges. This helped explain how Earth's crust could move. It gives evidence that the continental drift hypothesis did not provide.

**Think it Over**

4. **Explain** How do rocks that form at mid-ocean ridges show that magnetic reversals have occurred?

______________________

______________________

______________________

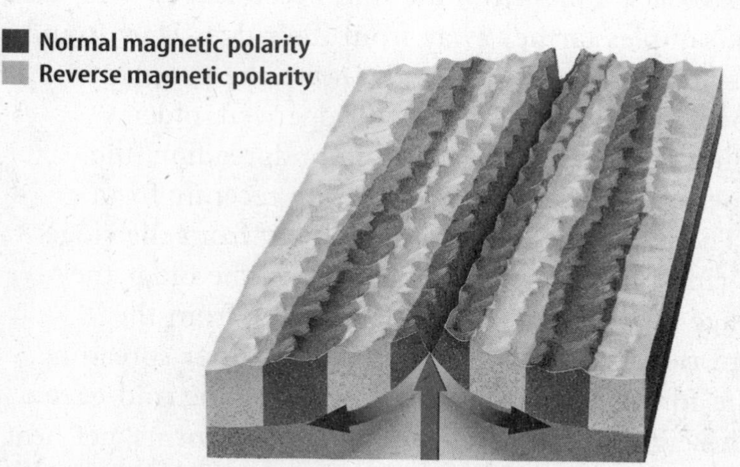


Changes in Earth's magnetic field are recorded in rock.

# After You Read

## Mini Glossary

**seafloor spreading:** Hess's theory that magma under Earth's crust is forced up toward the surface at the mid-ocean ridges, forming new seafloor

1. Review the term and its definition in the Mini Glossary. Write a sentence explaining where the magma at the mid-ocean ridge comes from.

_______________________________________________

_______________________________________________

2. Find the effect of this cause. Write a sentence describing the effect in the box.

**CAUSE:**

| **According to the theory of seafloor spreading, magma pushes up through cracks in the seafloor.** |
|---|

**EFFECT:**

| |
|---|

3. You put sticky notes on pages that you found interesting or that you had a question about. Did you discuss the facts or questions? How did this strategy help you understand more about seafloor spreading?

_______________________________________________

_______________________________________________

_______________________________________________

Visit **earth.msscience.com** to access your textbook, interactive games, and projects to help you learn more about seafloor spreading.

# Plate Tectonics

## section 3 Theory of Plate Tectonics

**What You'll Learn**
- about plate boundaries
- about features caused by plate tectonics
- how heat inside Earth causes plate tectonics

## Before You Read

Have you ever been swimming and noticed some areas in the water are colder and other areas are warmer? Why do you think this happens?

______________________________

______________________________

______________________________

Mark the Text

**Understand Key Terms** Underline meanings of the key terms as you read this section.

## Read to Learn

### Plate Tectonics

In the 1960s, scientists developed a new theory that combined continental drift and seafloor spreading. According to the theory of **plate tectonics** (tek TAH nihks), Earth's crust and part of the upper mantle are broken into plates, or sections, that move around on a plasticlike layer of the mantle.

### What are plates?

**Plates** are large sections of Earth's crust and upper mantle. These plates float and move around on a plasticlike layer of the upper mantle. You can think of the plates as rafts that float on this layer.

FOLDABLES™

**C Cause and Effect** Use three half-sheets of paper to record what happens at plate boundaries.

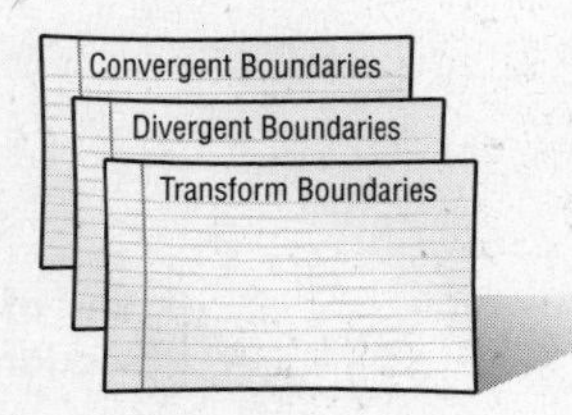


### What are the lithosphere and the asthenosphere?

Together, the crust and the rigid upper mantle form the **lithosphere** (LIH thuh sfihr). The lithosphere is about 100 km thick. The layer below is called the **asthenosphere** (as THE nuh sfihr). The asthenosphere is plasticlike. The rigid plates of the lithosphere float and move around on the plasticlike asthenosphere.

## Major Plates of the Lithosphere

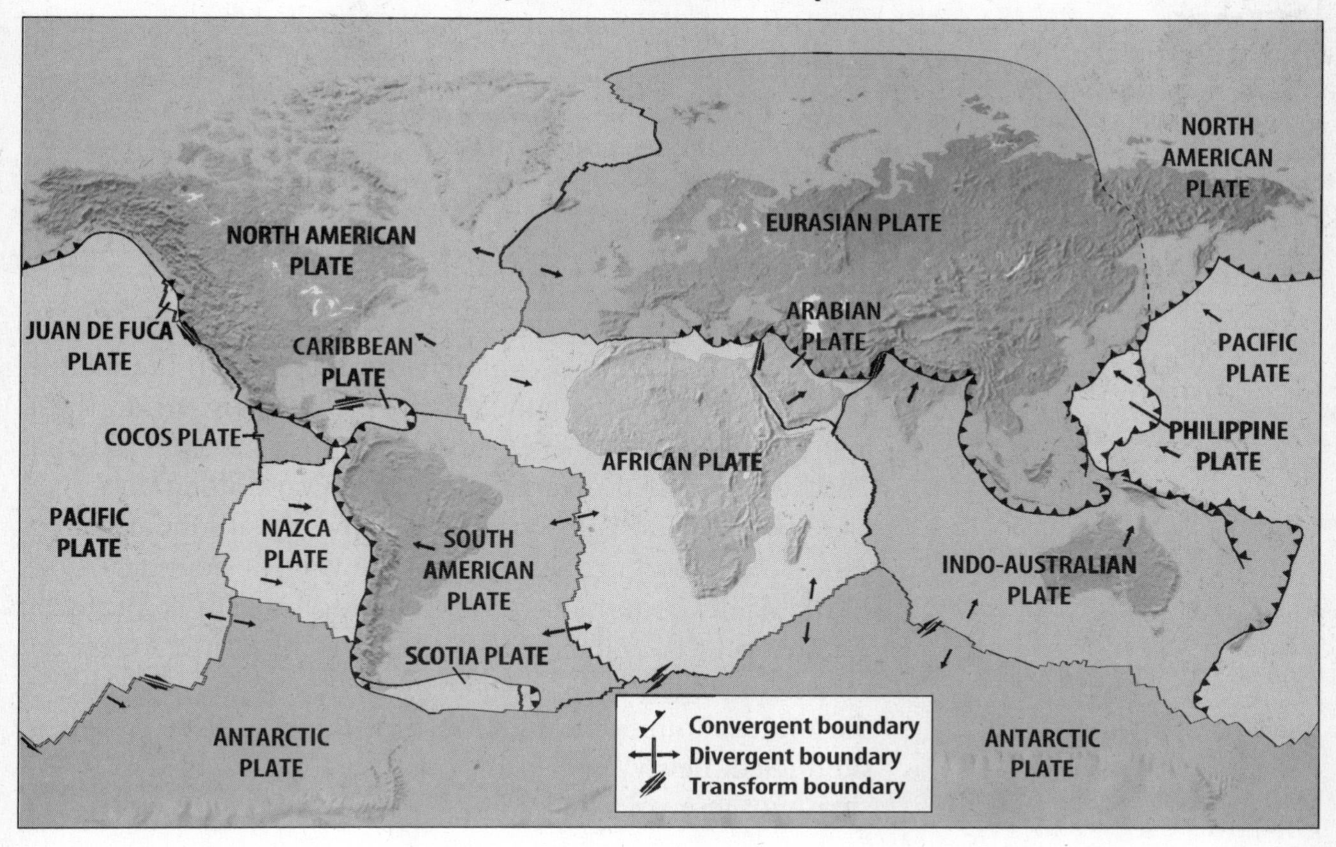


## Plate Boundaries

When plates move, several things can happen. They can move closer together and converge, or collide. They also can pull apart or slide by one another. When plates move, the result of these movements shows up at plate boundaries. Movements at one boundary means that changes must happen at other boundaries. The figure above shows the major plates, the way these plates are moving, and plate boundaries.

### What is a divergent boundary?

The area where two plates meet is called a boundary. When two plates are moving apart, the area between them is called a divergent boundary. In the figure above, find the North American Plate. Now find the African and Eurasian plates. The arrows show the North American Plate moving away from the Eurasian and African Plates. The divergent boundary between these plates is called the Mid-Atlantic Ridge. It is a mid-ocean ridge. As the plates pull apart, magma pushes up and becomes new seafloor. At some divergent plate boundaries, rift valleys form as the plates pull apart and crust sinks. ☑

**Picture This**

1. **Locate** Notice the small arrows on the figure above. Find and circle two arrows pointing in opposite directions. These arrows show two plates moving away from each other.

**Reading Check**

2. **Define** What type of boundary occurs when two plates move away from each other?

_______________

FOLDABLES™

**B Record Information** Complete your Foldable by adding facts about plate tectonics.

## What is a convergent boundary?

When two plates converge, or come together, they form a convergent boundary. What happens to the plates when they come together? One plate can sink and disappear under the other plate. For example, oceanic plates are denser than continental plates. When an oceanic plate converges with a less dense continental plate, the denser oceanic plate sinks under the continental plate.

## What is subduction?

The area where an oceanic plate subducts into the mantle is called a subduction zone. As the plate subducts into the mantle, it begins to melt. The melting rock becomes magma. The newly formed magma is forced upward along these plate boundaries. Volcanoes form above these subduction zones. Subduction zones occur at convergent boundaries.

The Andes mountain range in South America is located at a convergent boundary. The Nazca and the South American Plates converged to form them. There are many volcanoes in the Andes mountain range. When the Nazca Plate subducted, newly formed magma was forced upward, creating these volcanoes.

**Picture This**

3. **Explain** As the oceanic plate sinks into the mantle, it gets smaller. What is happening to the continental plate that balances this action?

______________________

______________________

______________________

**Subduction Zone**

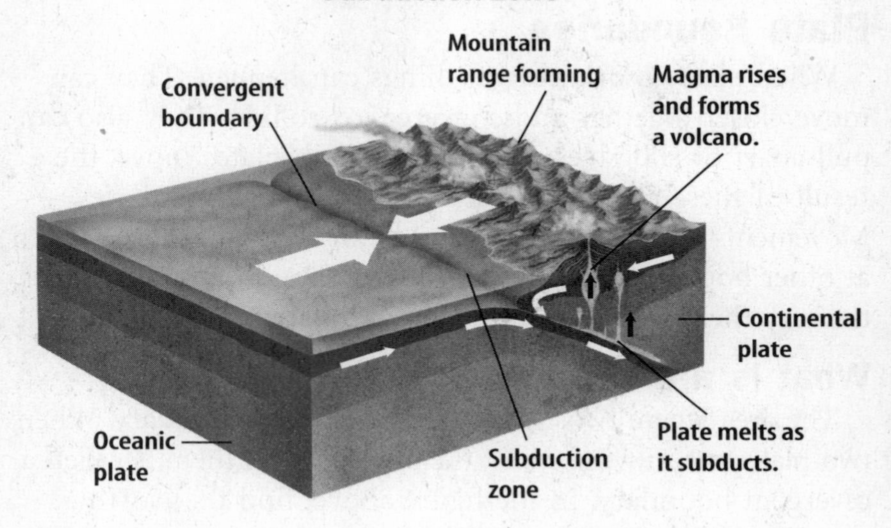


The figure above shows a subduction zone and what can occur at a convergent boundary between oceanic and continental plates. The denser oceanic plate is sinking under the less dense continental plate. High temperatures cause the rock to melt around the subducting slab as it moves under the other plate.

## What happens when oceanic plates converge?

When two oceanic plates converge, the colder, older, denser plate bends and sinks down into the mantle. A subduction zone forms where these plates collide. Volcanoes can form and, over time, some volcanoes form islands. The Mariana Islands in the western Pacific Ocean are a chain of volcanic islands that formed where two oceanic plates collided.

## What occurs if continental plates collide?

When two continental plates collide or converge, neither of the plates sinks under the other. Subduction usually doesn't occur. The continental plates are less dense than the asthenosphere below them. As a result, when these two plates collide, they fold and crumple to form mountain ranges. Earthquakes are common at these convergent boundaries. Volcanoes do not form because there is no, or little, subduction. ☑

The Indo-Australian Plate is colliding with the Eurasian Plate. These converging plates are forming the Himalaya in Asia.

**Reading Check**

**4. Describe** What usually happens when two continental plates converge?

______________________

______________________

## What is a transform boundary?

The third type of plate boundary is called a transform boundary. Transform boundaries occur when two plates slide past one another. In one type of transform boundary, two plates slide past each other in opposite directions. In another type, two plates are moving in the same direction, but at different rates. When one plate moves past another, earthquakes occur.

As shown below, the San Andreas (an DRAY us) Fault in California is part of a transform plate boundary. The Pacific Plate is sliding past the North American Plate. Both plates are moving in the same direction, but at different rates. As a result, this area has many earthquakes.

**Transform Boundary**

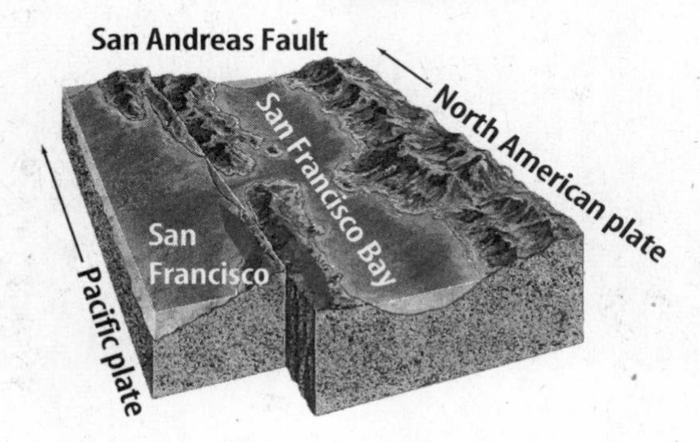


**Picture This**

**5. Identify** Highlight the arrows showing that each plate is moving in the same direction.

## Causes of Plate Tectonics

Scientists don't know exactly why Earth's plates move. They hypothesize that plates move by the same basic process that occurs when soup is heated in a pan.

### What is convection current?

Convection (kun VEK shun) currents can be found in a pan of soup that is cooking. As it heats, some of the soup becomes hotter and less dense. Some of the soup is cooler and more dense. This difference in temperature causes movement in the soup.

The cooler soup sinks and forces the hotter soup to rise to the top of the pot. As the hot soup reaches the surface, it cools and sinks back down into the pan. This happens in a cycle, over and over. This cycle of heating, rising, cooling and sinking is called a **convection current**. ☑

**Reading Check**

6. **Define** What is the cycle of heating, rising, cooling, and sinking called?

_______________

### Are there convection currents inside Earth?

What causes Earth's plates to move? A type of convection current is occuring inside Earth. Materials deep inside Earth have different amounts of heat and density. The colder, denser materials force the hotter, less dense materials towards Earth's surface. The arrows in the figure below show the rise and fall of materials in Earth's mantle.

One hypothesis suggests the transfer of heat inside Earth provides the energy to move plates and causes many of Earth's surface features. All of the hypotheses use convection currents in some way to explain the movements of plates.

**Picture This**

7. **Summarize** Choose one convection current in the figure and label the arrows *heating, rising, cooling,* and *sinking.*

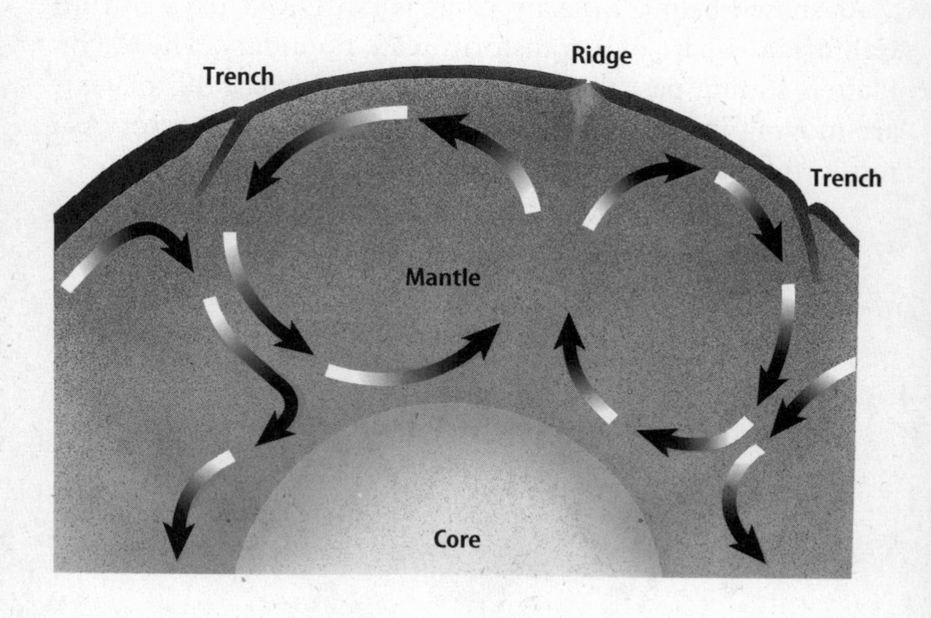

## Features Caused by Plate Tectonics

Earth is an active planet with a hot interior. The heat inside Earth causes convection that powers the movement of Earth's plates. When the plates move and interact, they produce forces that cause Earth's surface to change. These changes may happen over millions of years.

### How do normal faults and rift valleys form?

If forces are pulling Earth's crust in opposite directions, the crust will stretch. These pull-apart forces are called tension forces. As the crust stretches, large blocks of crust will break and slip down the broken surface of the crust.

When rocks break and move along surfaces, a fault forms. Faults move rock layers out of place. In the process, mountains can form. Usually faults that form this way are called normal faults. In normal faults, the rock layers above the fault move down when compared with the rock layers below the fault. Look at the figure below. The arrows show how tension forces stretch Earth's crust causing the movement of rock along normal faults. A range of mountains, called fault-block mountains, can form in the process.

**Tension Forces Cause Normal Faults**

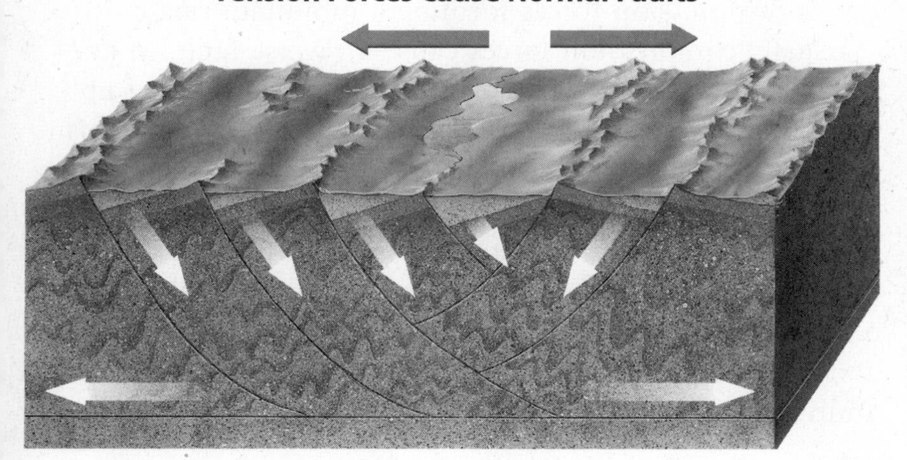

Tension forces also cause rift valleys and mid-ocean ridges. Rift valleys and mid-ocean ridges are large cracks that form where Earth's crust separates. One example of a rift valley is the Great Rift Valley in Africa. Valleys also occur in the middle of mid-ocean ridges. The Mid-Atlantic Ridge and the East Pacific Rise are two examples of mid-ocean ridges.

**FOLDABLES™**

**D Record Information** Make the Foldable shown below to record information about the features caused by plate tectonics.

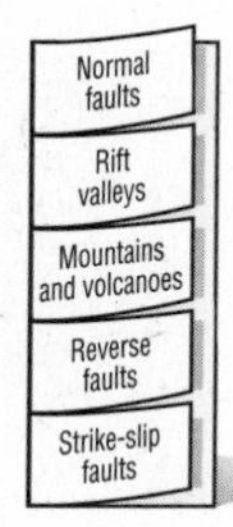


**Picture This**

8. **Analyze** Circle arrows that show the direction of the tension forces. Color the arrows blue that show the direction of fault movement.

## Picture This

**9. Locate** Circle the arrows that show the direction of the compression forces.

### Compression Forces Cause Reverse Faults

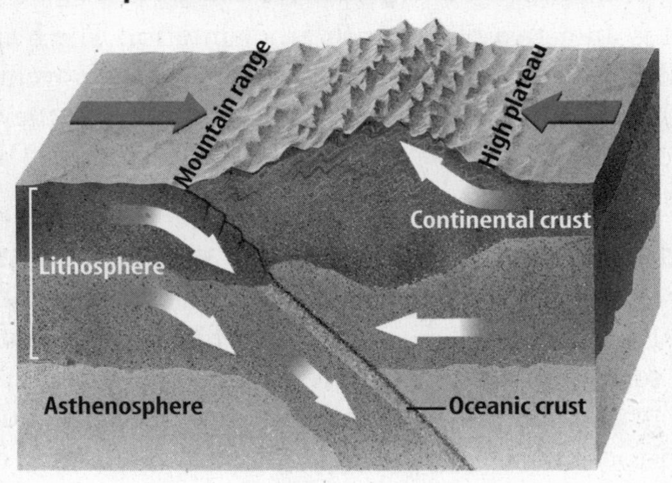


## How do mountains, reverse faults, and volcanoes form?

Compression forces squeeze objects together. Where Earth's plates come together, compression forces produce several effects. As continental plates collide, compression forces cause rock layers to fold and fault. Mountains can form. The Himalaya (hih muh LAY uh) are mountains being formed where two plates are colliding and forcing huge sections of rock to fold and break. The figure above shows compression forces forming a mountain range.

Usually compression forces cause a reverse fault. In a reverse fault, rock layers above the fault surface move up when compared with the rock layers below the fault. This is the opposite of a normal fault.

As you read earlier, when two oceanic plates converge, the denser plate will sink under the less dense plate. If an oceanic plate converges with a continental plate, the denser oceanic plate slides under the continental plate. Mountains and volcanoes can form as a result of the folding and faulting that occurs at the plate boundaries.

## Think it Over

**10. Identify** Name two land features caused by compression forces.

______________________

______________________

## What are strike-slip faults?

Strike-slip faults occur where two plates stick, or strike, and then slip by one another. Strike-slip faults occur at transform boundaries. A transform boundary is where two plates slide past one another. The plates can slide by in opposite directions or they may slide by in the same direction, but at different rates. When the plates move suddenly, they cause vibrations inside Earth that we feel as earthquakes. The San Andreas Fault is a strike-slip fault.

## Testing for Plate Tectonics

Only recently have scientists been able to measure exact movements of Earth's crust. All the early methods to check for plate movements were indirect.

### What are indirect methods of testing?

You have been reading about indirect methods of testing plate movements in this chapter. One method is studying the magnetic properties of rocks on the ocean floor. Scientists also could study volcanoes and earthquakes. These methods supported the theory that plates have moved and are still moving. However, these methods did not prove that plates are moving. ☑

**Reading Check**

11. **Infer** Is an earthquake an example of direct or indirect evidence of plate movement?

______________________

### How is plate movement measured?

There is a new method of measuring small amounts of plate movement that uses lasers and satellites. The figure below shows the Satellite Laser Ranging System. From the ground, scientists aim laser pulses at a satellite in orbit. The pulse reflects off of the satellite and returns to Earth. With this new technology, scientists can measure exact amounts of movement of Earth's plates. This new method shows that the plates move at rates between 1 cm to 12 cm per year. For example, Hawaii is moving toward Japan at a rate of 8.3 cm per year.

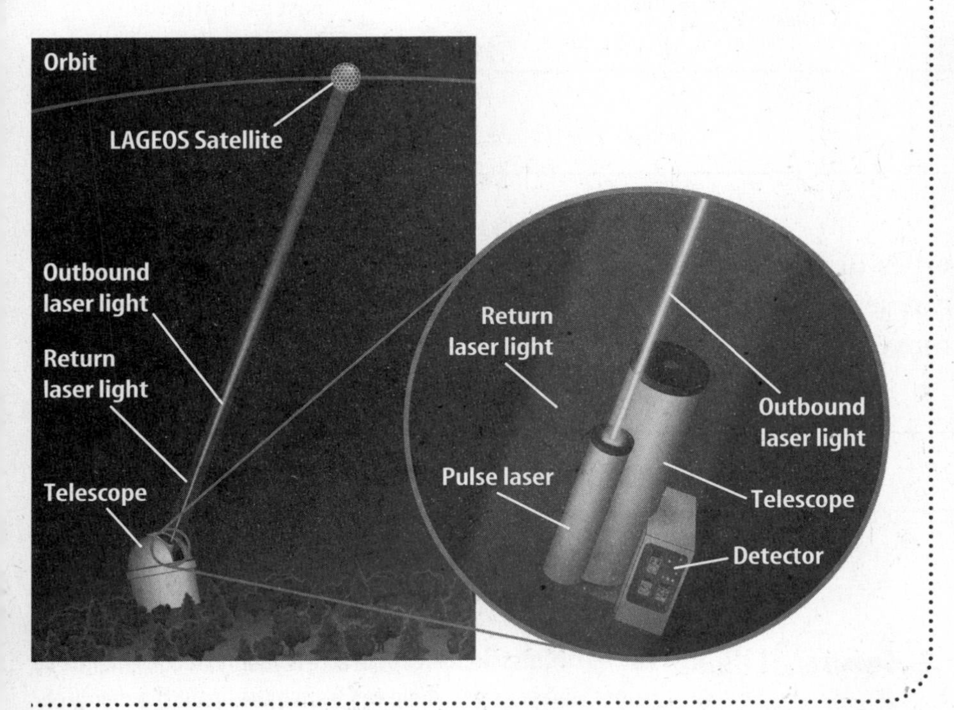


**Picture This**

12. **Interpret** What happens to the laser light when it reaches the satellite in the ranging system?

______________________

______________________

______________________

# After You Read

## Mini Glossary

**asthenosphere (as THE nuh sfihr):** plasticlike layer of Earth on which the lithosphere plates move

**convection current:** cycle of heating, rising, cooling, and sinking

**lithosphere (LIH thuh sfihr):** rigid layer of Earth made of the crust and upper mantle that move on the asthenosphere

**plate:** large section of Earth's crust layer and upper mantle layer that moves around on the asthenosphere

**plate tectonics (tek TAH nihks):** theory that Earth's crust and upper mantle are broken into plates that move around on a plasticlike layer of mantle.

1. Review the terms and their definitions in the Mini Glossary. Write a sentence explaining how the asthenosphere and the lithosphere interact according to the theory of plate tectonics.

______________________________________________________________

______________________________________________________________

2. Complete the concept map with the name of the plate boundary.

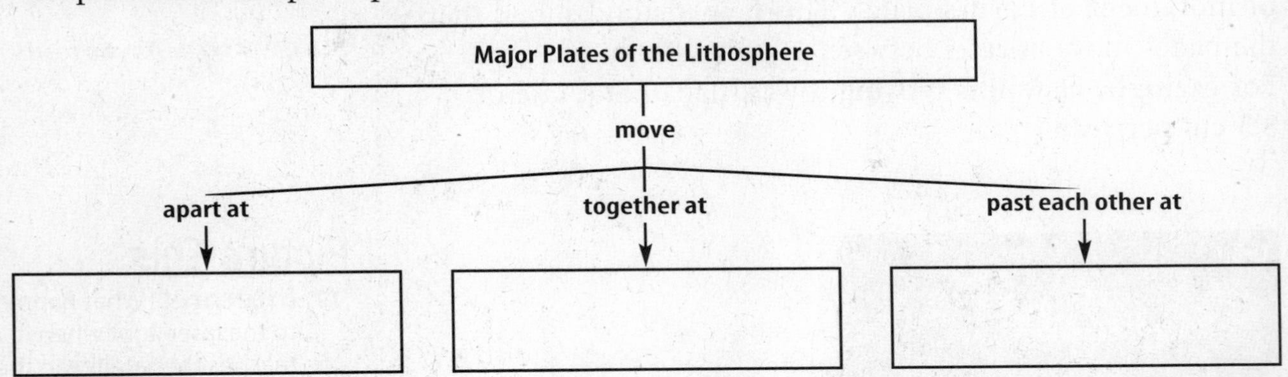


3. You learned about three types of faults in this section—normal, reverse, and strike-slip. Choose one fault. Write two sentences describing the force that causes the fault and what occurs as a result of the movement.

______________________________________________________________

______________________________________________________________

______________________________________________________________

Science Online Visit **earth.msscience.com** to access your textbook, interactive games, and projects to help you learn more about the theory of plate tectonics.

# Earthquakes

## section 1 Forces Inside Earth

### Before You Read

Have you ever stretched a rubber band so far that it broke? On the lines below, tell why you think a rubber band breaks when it is stretched too much.

______________________________________________

______________________________________________

______________________________________________

**What You'll Learn**
- how earthquakes result from buildup of energy
- how different types of faults form

### Read to Learn

#### Earthquake Causes

Rubber bands are elastic so they can stretch and then return to their original shape. But if the rubber band is stretched too far, it will break. It has reached its elastic limit. Rocks also have an elastic limit. Forces in Earth bend or stretch rocks. Rocks can bend and stretch up to a point. But once a rock's elastic limit is passed, the rock breaks.

When rocks break in this way, they move along surfaces, or faults. A **fault** is the surface along which rocks move when they pass their elastic limit and break. A great deal of force is needed to move rocks along a fault. Rock on one side of a fault may move up, down, or sideways in relation to rock on the other side of the fault.

Study Coach

**Summarize** As you read each paragraph, write a one or two sentence summary of its main idea.

FOLDABLES™

**A Define** Make a seven-tab Foldable to help you remember the terms in this chapter.

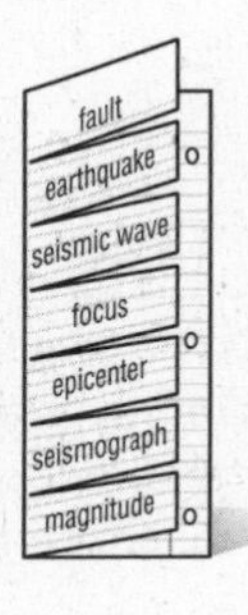


#### What causes faults?

The surface of Earth is in constant motion because of forces deep inside the planet. These forces cause sections of Earth's surface, called plates, to move. The movement puts stress on rocks near the edges of the plates. To relieve this stress, rocks bend, stretch, or compress. If the force is great enough, the rocks will break. An **earthquake** is the vibrations, or shaking, produced when rocks break along a fault.

## What causes earthquakes to occur?

As rocks along a fault move past each other, the rough surfaces catch, or grind, to a temporary halt. However, the forces inside Earth are so strong, they do not let the rocks stop moving. The forces keep driving the rocks, and stress builds up along the fault. When the rocks are stressed beyond their elastic limit, they break, move along the fault, and then return to their original shapes. An earthquake results. Earthquakes may be small vibrations that no one notices. Or they may be enormous vibrations that cause a great deal of damage. No matter what their strength, most earthquakes result from rocks moving over, under, or past each other along fault surfaces.

# Types of Faults

Three types of forces act on rocks—tension, compression, and shear. Tension is the force that pulls rocks apart. Compression is the force that squeezes rocks together. Shear is the force that causes rocks on either side of a fault to slide past each other. ☑

**Reading Check**

1. **Identify** Name three types of forces that act on rocks.

______________________

______________________

______________________

## What are normal faults?

Forces of tension inside Earth cause rocks to be pulled apart. When rocks are stretched by these forces, a normal fault can form. Along a **normal fault**, rocks are pulled apart, and the rock above the fault moves downward in relation to rock below the fault.

The figure below shows how rocks move along a normal fault. Arrows show where tension forces pull the rocks apart. Find the arrows that point out direction of motion. These arrows show how rock above the normal fault surface moves downward in relation to rock below the fault surface.

**Picture This**

2. **Identify** Trace over the arrows in the figure that show the direction that rock moves in a normal fault.

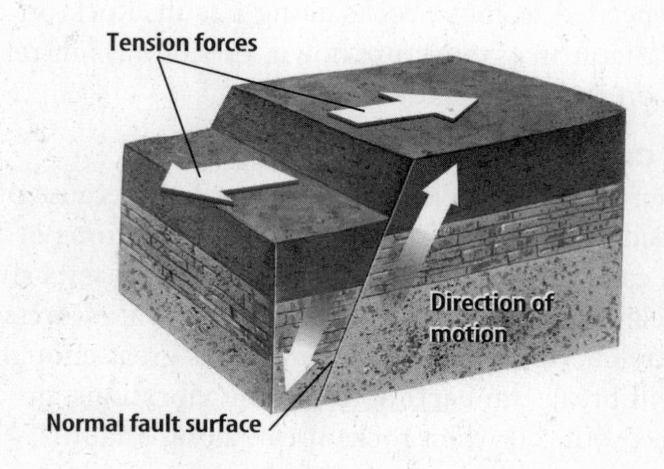

## What are reverse faults?

Some faults occur when forces of compression squeeze rocks together. In a **reverse fault**, rock breaks from forces of compression pushing on it from opposite directions. The force of compression along a reverse fault pushes the rock that is above the fault up and over the rock that is below the fault.

In the figure below, arrows show forces of compression pushing the rock together. The rock above the reverse fault surface moves upward in relation to the rock below the fault surface.

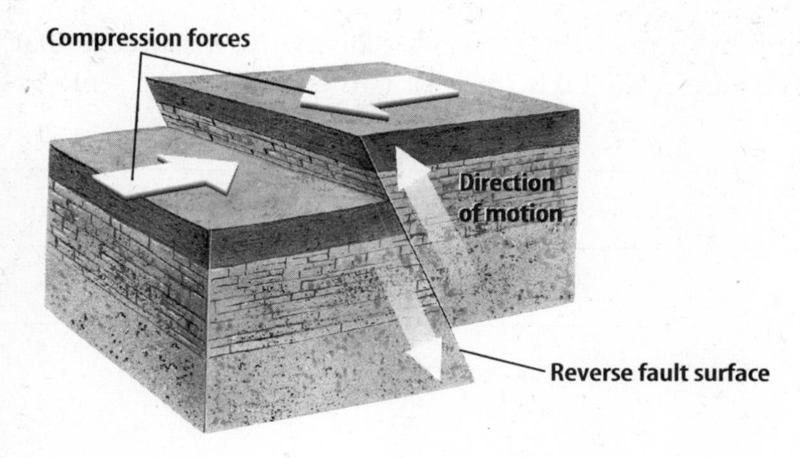


## What are strike-slip faults?

In a **strike-slip fault**, rocks move past each other without much movement upward or downward. The figure below shows that shear forces push rocks past each other on either side of the strike-slip fault surface. The San Andreas Fault in California is a strike-slip fault that extends for more than 1,100 km through the state. The San Andreas Fault is the boundary between two of Earth's plates that are moving sideways past each other.

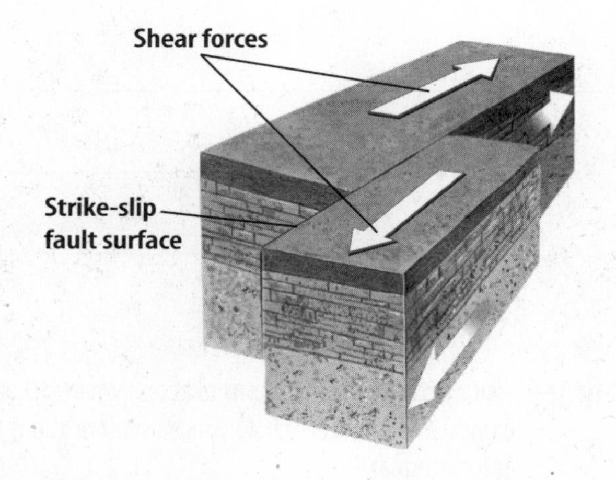


FOLDABLES™

**B Classify** Make a three-tab Foldable to record information about normal, reverse, and strike-slip faults.

**Picture This**

**3. Compare** Look at the diagrams of a reverse fault and a strike-slip fault. How does the direction of force differ in these two faults?

______________________

______________________

______________________

## After You Read

### Mini Glossary

**earthquake:** vibrations, or shaking, produced when rocks break along a fault

**fault:** surface along which rocks move when they reach their elastic limit and break

**normal fault:** break in rock caused by tension forces where rock above the fault surface moves down in relation to rock below the fault surface

**reverse fault:** break in rock caused by compression forces where rock above the fault surface moves upward in relation to rock below the fault surface

**strike-slip fault:** break in rock caused by shear forces where rocks move past each other without much movement upward or downward

**1.** Review the terms and their definitions in the Mini Glossary. Then write a sentence that explains how earthquakes and faults are related. Use at least two terms in your sentence.

______________________________________________

______________________________________________

______________________________________________

**2.** Write the correct term in each box.

**Types of Faults**

| ________ | ________ | ________ |
|---|---|---|
| **caused by shear forces** | **caused by tension forces** | **caused by forces of compression** |

**3.** How did summarizing the main idea in each paragraph help you understand the information in this section?

______________________________________________

______________________________________________

______________________________________________

**End of Section**

# Earthquakes

## section 2 Features of Earthquakes

## Before You Read

Imagine you and a friend are holding opposite ends of a long rope. You shake one end. On the lines below, describe how you think the rope would move.

______________________________________________

______________________________________________

______________________________________________

### What You'll Learn

- what seismic waves are
- how seismic waves are measured and studied
- the structure of Earth's interior

## Read to Learn

### Seismic Waves

When one end of a rope is shaken, the rope moves up and down, or side to side. Energy travels through the rope in the form of waves. **Seismic** (SIZE mihk) **waves** are generated by earthquakes. Seismic waves travel through Earth just as waves travel through rope. When an earthquake occurs, the ground moves forward and backward, up and down, or side to side. Sometimes, earthquakes cause the surface of Earth to ripple like the waves on the ocean.

Mark the Text

**Main Ideas** Highlight the main idea in each paragraph as you read. Use a different color to highlight the vocabulary words and their definitions.

### What is the origin of seismic waves?

Rocks move past each other along faults. Stress builds up along rock surfaces that catch on each other. The stress continues to build up until the rocks' elastic limit is passed. Then the built up energy is released as seismic waves. The **focus** (plural, *foci*) of an earthquake is the point where this energy is first released.

The foci of most earthquakes are within 65 km of Earth's surface. A few earthquakes have occurred as much as 700 km beneath the surface. Seismic waves are produced and travel outward from the earthquake focus.

FOLDABLES™

**C Classify** Make a three-tab Foldable to organize information about primary, secondary, and surface waves.

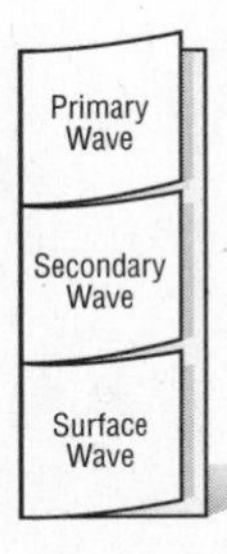

## What are primary waves?

Earthquakes produce three different types of seismic waves—primary, secondary, and surface waves. All three types of seismic waves are produced at the same time. But each type of wave behaves in a different way within Earth. ☑

**Reading Check**

1. **Classify** What are the three types of seismic waves?

______________________

______________________

______________________

**Primary waves** (P-waves) cause particles in rock to move back and forth in the same direction that the wave is traveling. Primary waves occur when particles in rocks are compressed and then stretch apart. This motion sends primary waves traveling through the rock.

## What are secondary and surface waves?

**Secondary waves** (S-waves) move through Earth by causing particles in rock to move at right angles to the direction in which the wave is traveling.

**Surface waves** move particles in rock in a backward, rolling motion and also in a side-to-side, swaying motion. The arrows in the figure below show the different directions of movement made by surface waves. Surface waves cause most of the damage from earthquakes. Many buildings are made with stiff materials that crack when surface waves shake the ground. The buildings fall apart or collapse when surface waves cause different parts of the building to move in different directions.

**Picture This**

2. **Explain** Why do surface waves cause much more damage than other waves?

______________________

______________________

______________________

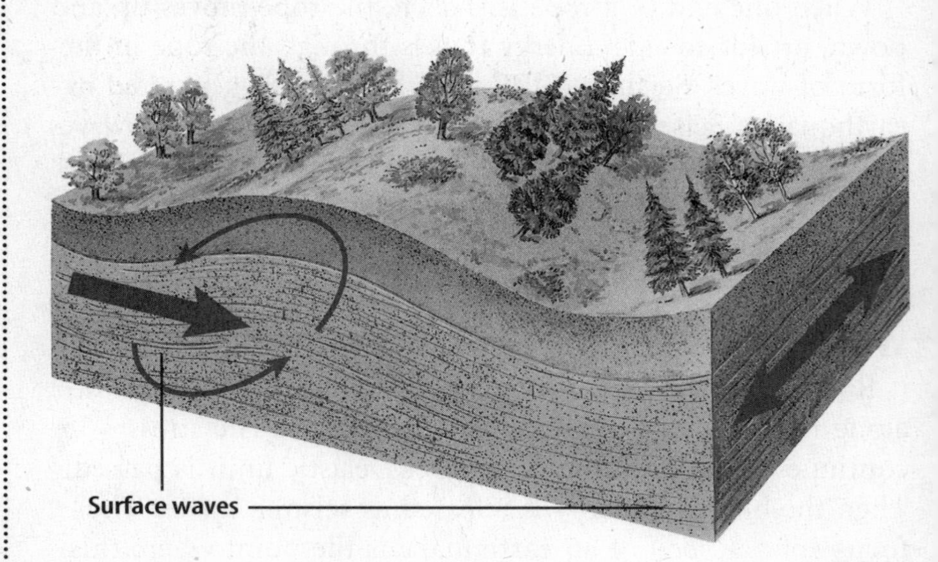


Surface waves are produced when earthquake energy reaches Earth's surface. Surface waves travel outward from the epicenter of an earthquake. An earthquake's **epicenter** (EH pih sen tur) is the point on Earth's surface directly above the earthquake focus.

## Locating an Epicenter

Different seismic waves travel through Earth at different speeds. Primary waves are the fastest; secondary waves are slower. They travel about half the speed of primary waves. Surface waves are the slowest seismic waves. ☑

Scientists have learned to use the different speeds of seismic waves to figure out the distance to an earthquake's epicenter. When an earthquake's epicenter is far from a location, the primary wave has more time to get farther ahead of the secondary and surface waves. It reaches the monitoring center first, ahead of the other seismic waves.

### How are seismic waves measured?

A **seismograph** is the instrument scientists use to measure seismic waves. A seismograph registers the waves and records the time that each wave arrived.

A seismograph is made up of a rotating drum of paper and a hanging weight, or pendulum, with a pen attached to it. When seismic waves reach the seismograph, the drum vibrates, but the pendulum remains at rest. The unmoving pen traces a record of the vibrations on the moving drum of paper. The paper record of a seismic event is called a seismogram. The figure below shows two different seismographs recording two different ground movements—horizontal and vertical.

**Reading Check**

3. **Interpret** What are the fastest seismic waves?

______________________

**Think it Over**

4. **Explain** Why are two different seismographs used to record Earth's movements?

______________________

______________________

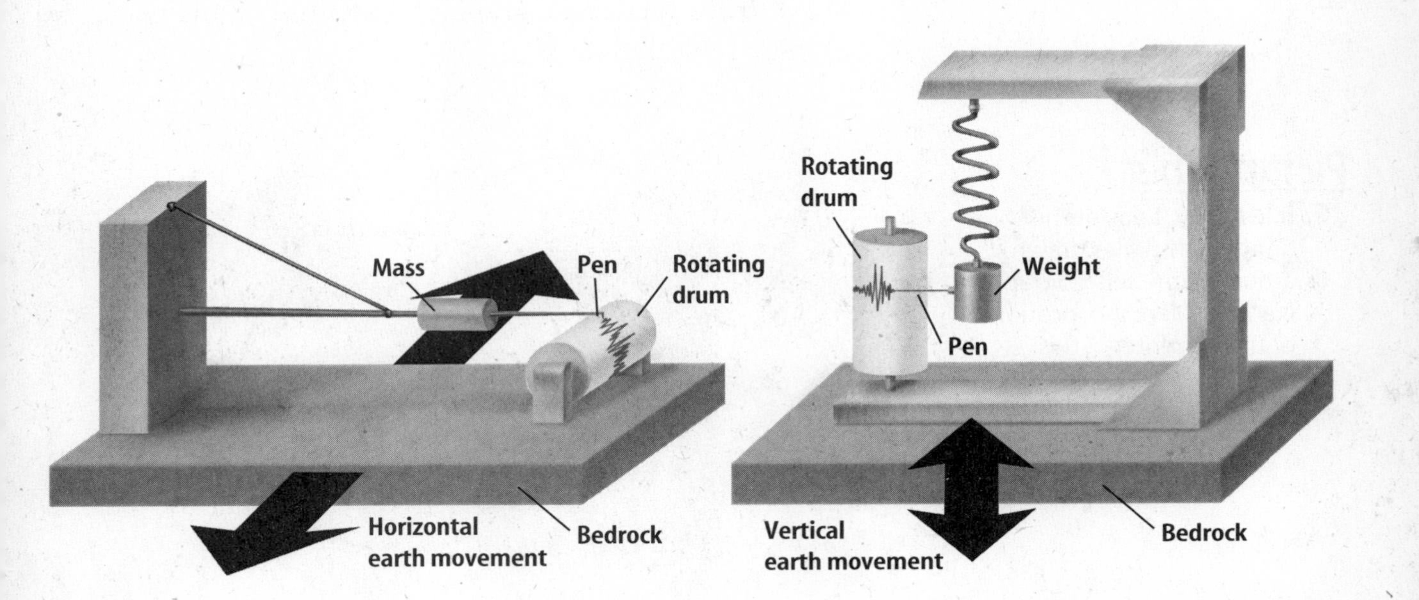

## What are seismograph stations?

Seismographs are located at seismograph stations, where scientists record and monitor earthquake activity. Each type of seismic wave reaches the seismograph station at a different time, depending on its speed. Primary waves reach the seismograph station first. Secondary waves reach the station second, and the slow surface waves reach the seismograph station last. Scientists use the time difference to figure out the distance between the seismograph station and the earthquake epicenter. For example, if a seismograph station is located 4,000 km from an earthquake epicenter, primary waves will reach the station about 6 minutes before secondary waves. ☑

**Reading Check**

**5. Analyze** How is the speed of different seismic waves used to locate an earthquake epicenter?

_______________

_______________

_______________

_______________

## How is an epicenter located?

Seismic waves must be recorded at three or more stations to determine an earthquake's epicenter. To locate an epicenter, scientists draw a circle around each station's position on a map, as shown in the figure below. The size of each circle is based on how far each seismograph station is from the epicenter of the earthquake. The point at which the three circles meet is the location of the earthquake epicenter.

Scientists usually describe earthquakes based on their distance from the seismograph. Local earthquakes occur less than 100 km away. Regional events occur from 100 km to 1,400 km from the seismograph. Teleseismic events occur more than 1,400 km away.

**Picture This**

**6. Identify** Look at the figure with circles drawn around three seismograph stations. Mark the location of the epicenter of this earthquake.

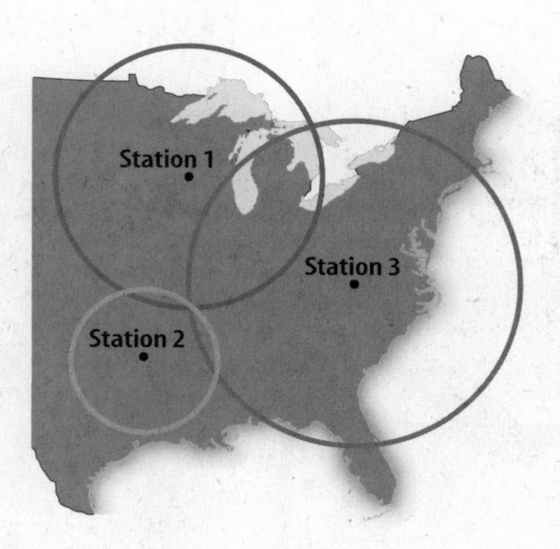

## Basic Structure of Earth

Earth's different layers are shown in the figure below. The center of Earth is a very dense, solid core made mostly of iron, with smaller amounts of nickel, oxygen, silicon, and sulfur. The core is solid and dense because of the intense pressure of all the layers above it. Above the inner core lies the liquid outer core, which also is made mostly of iron.

Earth's mantle is the largest layer. The mantle lies above the liquid outer core. The mantle is made up mainly of silicon, oxygen, magnesium, and iron. The mantle is divided into an upper mantle and a lower mantle, based on the different speeds at which seismic waves move through them. One part of the upper mantle, called the asthenosphere (as THE nuh sfihr), contains weak rock that can flow slowly, like a thick liquid.

FOLDABLES™

**D Describe** Make a half-book Foldable showing a cross section of Earth. Label Earth's layers. Describe each layer inside the book.

### What is Earth's crust?

The outermost layer of Earth is called the crust. The crust contains more silicon and aluminum, but less magnesium and iron, than the mantle. The crust and the part of the mantle just beneath it make up the lithosphere (LIH thuh sfihr). The lithosphere is broken up into a number of plates. The plates move over the asthenosphere that lies just below them.

The thickness of Earth's crust varies. In some mountain regions, it is more than 60 km thick. Under some parts of the ocean, it is only 5 km thick. Earth's crust is generally less dense than the mantle beneath it.

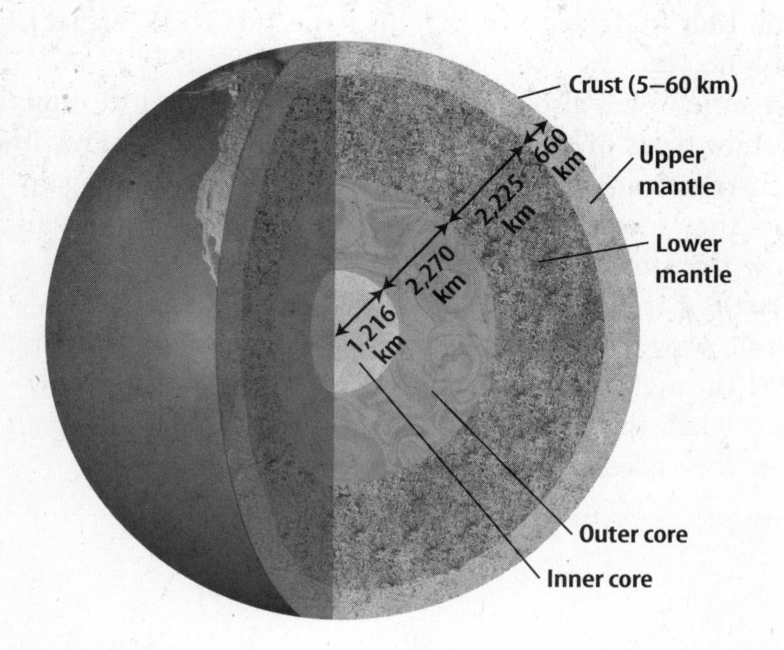


### Picture This

**7. Locate** Use a colored pencil or marker to highlight the lithosphere in the figure.

## How can scientists know about Earth's internal structure?

The speeds and paths of seismic waves change as they travel through materials with different densities. By studying seismic waves, scientists have concluded that different layers of Earth are made of different materials with different densities. In general, the deeper inside Earth a layer is, the denser it is. Studying how seismic waves travel through Earth has allowed scientists to map Earth's internal structure without being there.

**Think it Over**

**8. Infer** What part of Earth has the greatest density?

______________________

## Do seismic waves travel through liquid?

Scientists also discovered that seismic waves do not travel through some regions of Earth. Seismic waves are not detected in the region of Earth called the shadow zone. The shadow zone occurs because secondary waves cannot travel through liquid, so they stop moving when they reach the liquid outer core. Primary waves are slowed and bent, but not stopped, by the liquid outer core. From these findings, scientists concluded that the liquid outer core and the mantle are made of different materials with different densities.

## How do layer boundaries affect seismic waves?

Seismic waves change speed when they travel through different layers of Earth. Seismic waves speed up when they pass through the bottom of the crust and enter the upper mantle. Seismic waves change speed at this boundary because the crust and the upper mantle have different densities. This boundary is called the Mohorovicic discontinuity (moh huh ROH vee chihch • dis kahn tuh NEW uh tee), or Moho. ☑

**Reading Check**

**9. Identify** What causes seismic waves to change speeds as they move through Earth's layers?

______________________

______________________

Seismic waves also change speed as they travel through different parts of the mantle. For example, waves slow down when they reach the asthenosphere. They speed up again when they move through the more solid part of the mantle below the atmosphere.

Earth's core is divided into two regions based on how seismic waves travel through it. Secondary waves cannot travel through the liquid outer core. Primary waves slow down when they reach the outer core, but speed up again when they reach the solid inner core.

# ● After You Read

## Mini Glossary

**epicenter:** point on Earth's surface directly above the earthquake focus

**focus:** point below Earth's surface where energy is first released in the form of seismic waves

**primary wave:** seismic wave that causes rock particles to move back and forth in the same direction that the wave is traveling

**secondary wave:** seismic wave that moves rock particles at right angles to the direction of the wave

**seismic wave:** wave generated by an earthquake

**seismograph:** instrument used to register seismic waves and record the time that each arrived

**surface wave:** seismic wave that moves rock particles in a backward, rolling motion and also in a side-to-side, swaying motion

**1.** Review the terms and their definitions in the Mini Glossary. Then write two sentences about primary, secondary, and surface waves. Explain how they are alike and how they are different.

______________________________________________

______________________________________________

______________________________________________

**2.** Fill in the blank boxes below to explain how seismic waves are used.

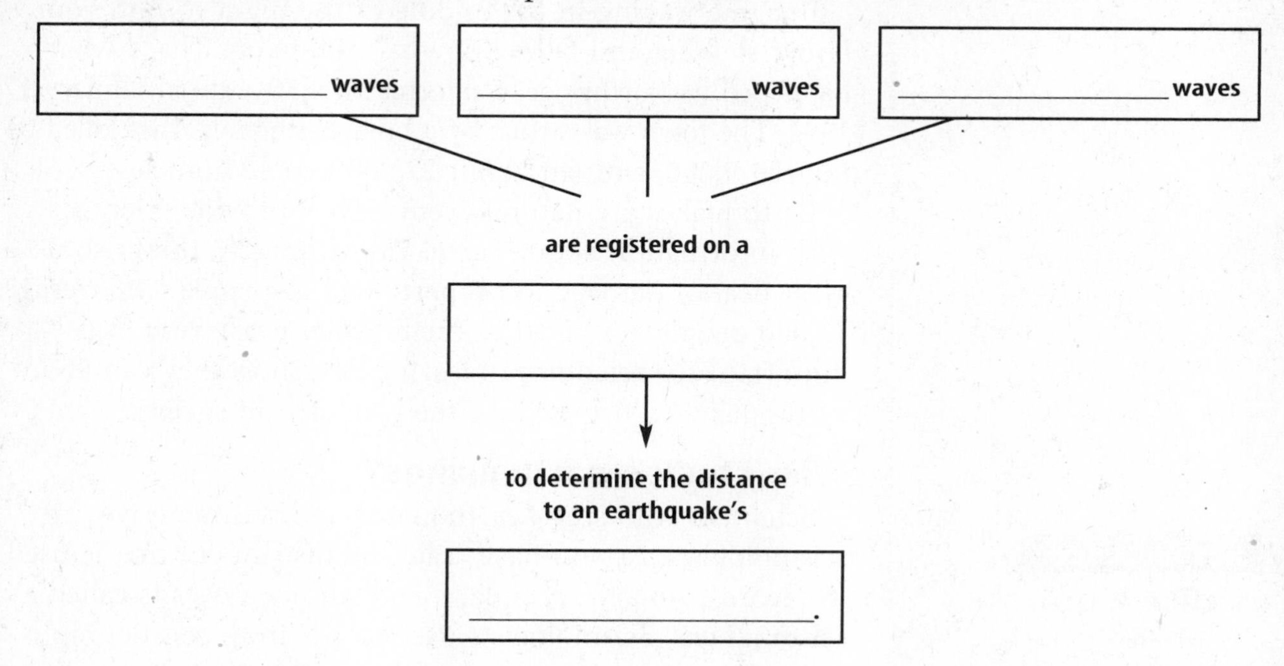


**Science Online** Visit **earth.msscience.com** to access your textbook, interactive games, and projects to help you learn more about earthquakes and the structure of Earth.

End of Section

# Earthquakes

## section 3 People and Earthquakes

### What You'll Learn

- how earthquakes are measured
- how to make your home or school earthquake-safe

### Before You Read

Think about pictures you have seen of a city that has been hit by a powerful earthquake. On the lines below, describe about what the city looked like.

______________________________

______________________________

______________________________

Mark the Text

**Underline Key Ideas** As you read this section, underline key ideas and terms in each paragraph.

### Read to Learn

#### Earthquake Activity

Imagine waking in the middle of the night to find your house shaking and falling down around you. That's what happened in Northridge, California, at 4:30 A.M. on January 17, 1994. The town was struck by a huge earthquake that killed 51, injured 9,000, and left about 22,000 people homeless.

Earthquakes are natural events. They provide scientists with information about Earth. Unfortunately, they also do a great deal of damage to property and to people. On average, 10,000 people are killed in earthquakes every year. It is important for scientists to learn as much as they can about earthquakes to help reduce their impact on society.

#### Who studies earthquakes?

Scientists who study earthquakes and seismic waves are seismologists. As you have read, the instrument that is used to record primary, secondary, and surface waves is called a seismograph. Seismologists use records from seismographs to study earthquakes. There are seismograph stations all over the world that help determine where earthquake epicenters are located. ☑

**1. Define** What are seismologists?

______________________________

______________________________

## What is an earthquake's magnitude?

A seismograph record of an earthquake is called a seismogram. A seismogram is shown in the figure below. The height of the lines on a seismogram is a measure of energy the earthquake released. The **magnitude** of an earthquake is the measure of energy released. The taller the lines on the seismogram, the greater the magnitude of the earthquake.

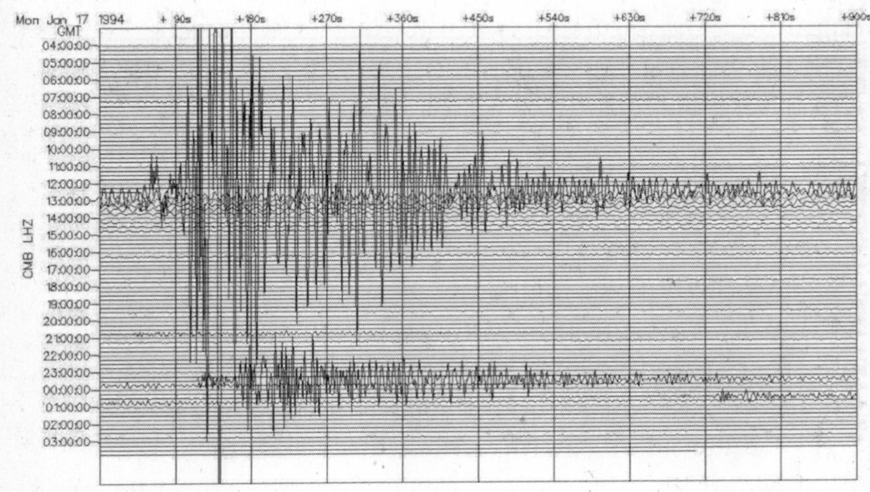

Berkeley Seismological Laboratory

### Picture This

**2. Interpret** Look at the seismogram. Circle the area on the seismogram that recorded the earthquake.

## What is the Richter scale?

The Richter magnitude scale is used to describe the magnitude of earthquakes. The Richter scale is based on the height of the lines on a seismogram. The Richter scale has no upper limit. However scientists think that a Richter scale value of 9.5 is the greatest magnitude an earthquake could register. ☑

### ✓ Reading Check

**3. Identify** What scale is used to describe an earthquake's magnitude?

______________________

For each increase of 1.0 on the Richter scale, the height of the line on a seismogram is ten times taller. For example, the line on a seismogram that shows a magnitude 7.0 earthquake is ten times taller than the line on a seismogram that shows a magnitude 6.0 earthquake.

The difference in the energy released by earthquakes of different magnitudes is even greater. About 32 times as much energy is released for every increase of 1.0 on the Richter scale. An earthquake with a Richter scale magnitude of 8.5 releases about 32 times more energy than an earthquake with a magnitude of 7.5.

## What do we know from past earthquakes?

Thousands of earthquakes occur on Earth every day. Most are so small that no one is aware of them. Earthquakes with a magnitude of 3.0 or below are generally not noticed and cause no damage. Earthquakes with a magnitude between 3.0 and 4.9 can usually be felt by humans. There are about 55,000 earthquakes each year that are felt, but that cause little or no damage. ☑

Most of the earthquakes you hear about are large ones that cause great damage. Destructive earthquakes have a magnitude of 5.0 or more on the Richter scale. The earthquakes that cause the most damage and take the most lives have a magnitude of 6.8 or higher.

**Reading Check**

**4. Determine** At about what magnitude are earthquakes first noticed?

______________________

## How is earthquake intensity described?

Earthquakes also can be described by the amount of damage they cause. The modified Mercalli intensity scale describes earthquake intensity based on the amount of damage an earthquake does to rock formations, buildings, and other structures at a specific location. The amount of damage an earthquake does depends on its strength, the kind of material at Earth's surface, how structures are designed, and how far a location is from the epicenter of the earthquake.

On the Mercalli scale, an intensity-I earthquake would be felt by few people and do no damage. An intensity-IV earthquake would be felt by everyone indoors, and it might do some damage to buildings. Everyone would feel an intensity-IX earthquake, which would cause serious damage to buildings and open cracks in the ground. An intensity-XII earthquake would result in total destruction of structures. The 1994 6.8-magnitude earthquake in Northridge, California, was listed at an intensity of IX because of the damage it caused.

## What is liquefaction?

Have you ever tried to get thick ketchup out of a bottle? Sometimes you must shake the bottle to make the liquid ketchup flow. Something similar can happen to wet soil during earthquakes. The intense shaking from an earthquake can cause wet soil to act like a liquid. **Liquefaction** occurs when wet soil acts more like a liquid during an earthquake. When liquefaction occurs in soil under buildings, the buildings can sink into the soil and collapse.

**FOLDABLES™**

**E Describe** Use two quarter-sheets of notebook paper to write down facts about liquefaction and tsunamis.

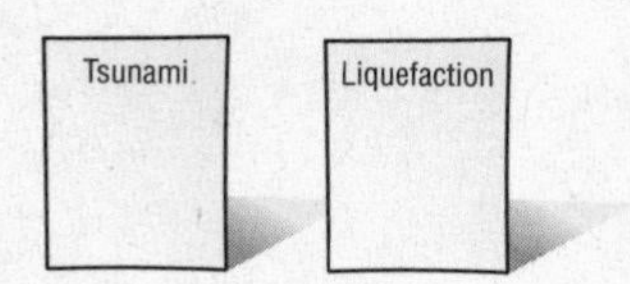

## What are tsunamis?

Some earthquakes occur under the ocean. These earthquakes cause a sudden movement of the ocean floor, which pushes against the water. The powerful wave that results can travel thousands of kilometers across the ocean in all directions, and it may finally crash into a coast. **Tsumamis** (soo NAH meez) are seismic sea waves caused by undersea earthquakes. On the open ocean, tsunamis may not be noticed but when they reach land, tsunamis may form a wall of water up to 30 m high. ☑

**Reading Check**

**5. Identify** What are seismic sea waves caused by underground earthquakes called?

______________________

## What are ways to prepare?

Most tsunamis occur around the Pacific Ocean. The Pacific Tsunami Warning Center in Hawaii monitors undersea earthquakes. People are warned if a tsunami is likely to occur so they have time to leave the danger area.

# Earthquake Safety

Earthquakes may occur anywhere. The map below shows where in the United States they are most likely to occur.

Today buildings can be built to resist earthquake damage. In California, some new buildings are supported by flexible moorings made of rubber and steel. The rubber acts like a cushion to absorb the wave motion of an earthquake. In older buildings, steel rods can be installed to make the walls stronger.

There are some steps you can take before an earthquake to make your home safer. Move heavy objects from high shelves to lower shelves. Make sure water heaters and gas appliances are held securely in place. New sensors can now be installed on gas lines. These sensors automatically shut off the gas when earthquake vibrations are felt.

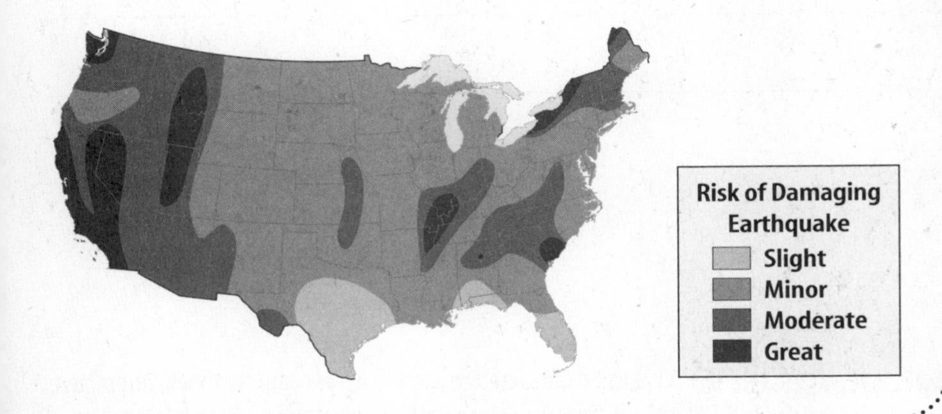


**Think it Over**

**6. Infer** If you live in an earthquake zone, why shouldn't you build a house on loose soil?

______________________

______________________

______________________

# After You Read

## Mini Glossary

**liquefaction:** occurs when wet soil acts more like a liquid during an earthquake

**magnitude:** the measure of energy released during an earthquake

**tsunami (soo NAH mee):** seismic sea waves caused by undersea earthquakes

1. Review the terms and their definitions in the Mini Glossary. Then write one sentence about the effects of a strong undersea earthquake.

_______________

_______________

_______________

2. Fill in the blanks with the correct scale.

**Seismologists use two scales to measure earthquakes.**

| _______________ | _______________ |
|---|---|
| • measures intensity<br>• scale of I to XII | • measures magnitude<br>• scale of 1.0 to about 9.5 |

3. How did underlining main ideas and key terms help you understand the information in this chapter?

_______________

_______________

_______________

# chapter 12 Volcanoes

## section 1 Volcanoes and Earth's Moving Plates

### Before You Read

Think about what happens to a wax candle after you light it. On the lines below, describe what happens to the wax and what it does. Then describe what happens to the wax after the flame is blown out.

______________________________________________

______________________________________________

______________________________________________

**What You'll Learn**

- how volcanoes affect people
- what causes volcanoes
- the relationship between volcanoes and Earth's plates

### Read to Learn

**Mark the Text**

**Highlight** Identify the key terms and their meanings as you read this section.

#### What are volcanoes?

A **volcano** is an opening in Earth's surface that erupts gases, ash, and lava. These materials pile up in layers around the opening, forming volcanic mountains. Today, Earth has more than 600 active volcanoes. An active volcano is one that has erupted within recorded history.

#### Which volcanoes erupt most often?

The most active volcano on Earth is in Hawaii. It is called Kilauea (kee low AY ah). This volcano has been erupting for hundreds of years, but its eruptions are slow, not explosive. Since 1983, Kilauea has had a series of eruptions that continue today. In May of 1990, it destroyed most of the town of Kalapana Gardens. Because its lava flowed slowly, people had time to escape and no one was hurt.

The country of Iceland is also famous for its active volcanoes. This island country is located in an area where Earth's plates move apart. Because of its northern location and active volcanoes, it is known as the land of fire and ice.

**FOLDABLES™**

**A Cause and Effect** Make a two-tab Foldable as shown. As you read, take notes on the causes and effects of volcanoes.

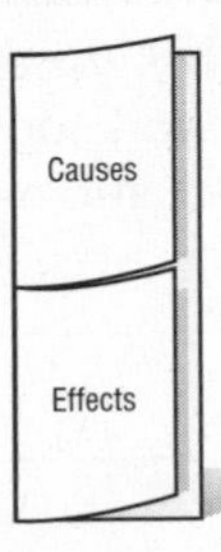

## Effects of Eruptions

Volcanic eruptions have a serious impact on people who live nearby. Their lives may be in danger. Even if people are able to evacuate or escape, their property is often damaged or destroyed. The lava flowing from a volcano destroys everything in its path. Volcanic ash and dust falling from the sky can collapse buildings and block roads. Ash can cause lung diseases in people and animals.

### What is a pyroclastic flow?

Sometimes volcanic ash and other matter rush down the side of a volcano. This is called a pyroclastic (pi roh CLAS tihk) flow. Temperatures inside a pyroclastic flow can be hot enough to catch wood on fire. If the flow is heavy, people in nearby towns are forced to abandon their homes. Buildings, roads and crops may be destroyed by the pyroclastic flow.

**Think it Over**

1. **Explain** two effects of volcanic eruptions.

______________________

______________________

### How do volcanoes affect humans and the environment?

The Soufrière (sew FREE er) Hills volcano on the island of Montserrat erupted in July of 1995. Geologists knew it was about to erupt. They warned people living nearby to evacuate. Two years after the eruption began, large pyroclastic flows swept down the sides of the volcano. Cities and towns were buried. Plant life was destroyed. Twenty people who didn't evacuate were killed. This eruption was one of the largest recent volcanic eruptions near North America.

Sulfurous gas is released during volcanic eruptions. When these gases mix with water vapor in the atmosphere, acid rain forms. On the island of Montserrat, the acid rain destroyed the vegetation. Acid rain fell into the lakes and streams and killed fish. As the vegetation died, the animals living in the forests left or died. When a volcano erupts, it is a danger to all living organisms and to the environment.

## How do volcanoes form?

What happens inside Earth to create volcanoes? Why do volcanoes occur in some places and not in others? Deep inside Earth, heat and pressure melt rocks. The liquid rock is called magma. Some rocks deep in Earth already are melted. Other rocks are so hot, the smallest rise in temperature or drop in pressure melts them into magma. What makes magma come to the surface? ☑

**Reading Check**

2. **Identify** What two factors work together to melt rock into magma?

______________________

______________________

## Why is magma forced upward?

Magma is not as dense as the rock around it. This difference in density forces the magma to rise toward Earth's surface. You can see this process if you turn a bottle of cold syrup upside down. The dense syrup will force the less dense air bubbles to slowly rise.

After many thousands or even millions of years, magma reaches Earth's surface. Magma flows out through an opening called a **vent.**

Once magma reaches Earth's surface, it is called lava. As lava flows out, it cools and becomes solid, forming layers of igneous rock around the vent. Often the area around the vent is bowl-shaped. The steep, bowl-shaped area around a volcano's vent is the **crater.** The figure below shows magma inside Earth being forced to the surface.

### Picture This

**3. Determine** Draw arrows along the path the magma travels from the magma chamber, through the vent, and out of the crater.

**How a Volcano Forms**

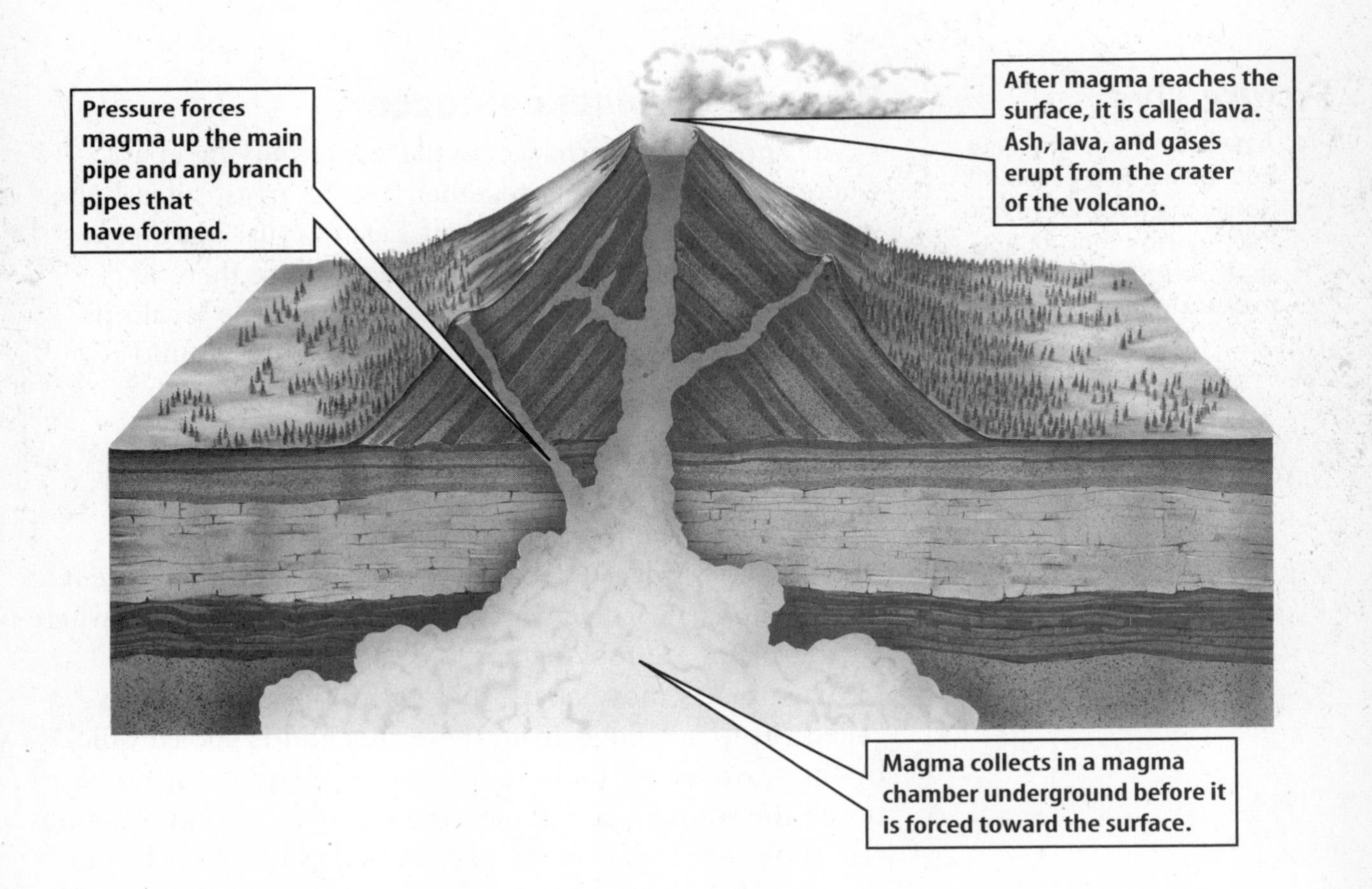

**Volcanoes, Hot Spots, and Plate Boundaries**

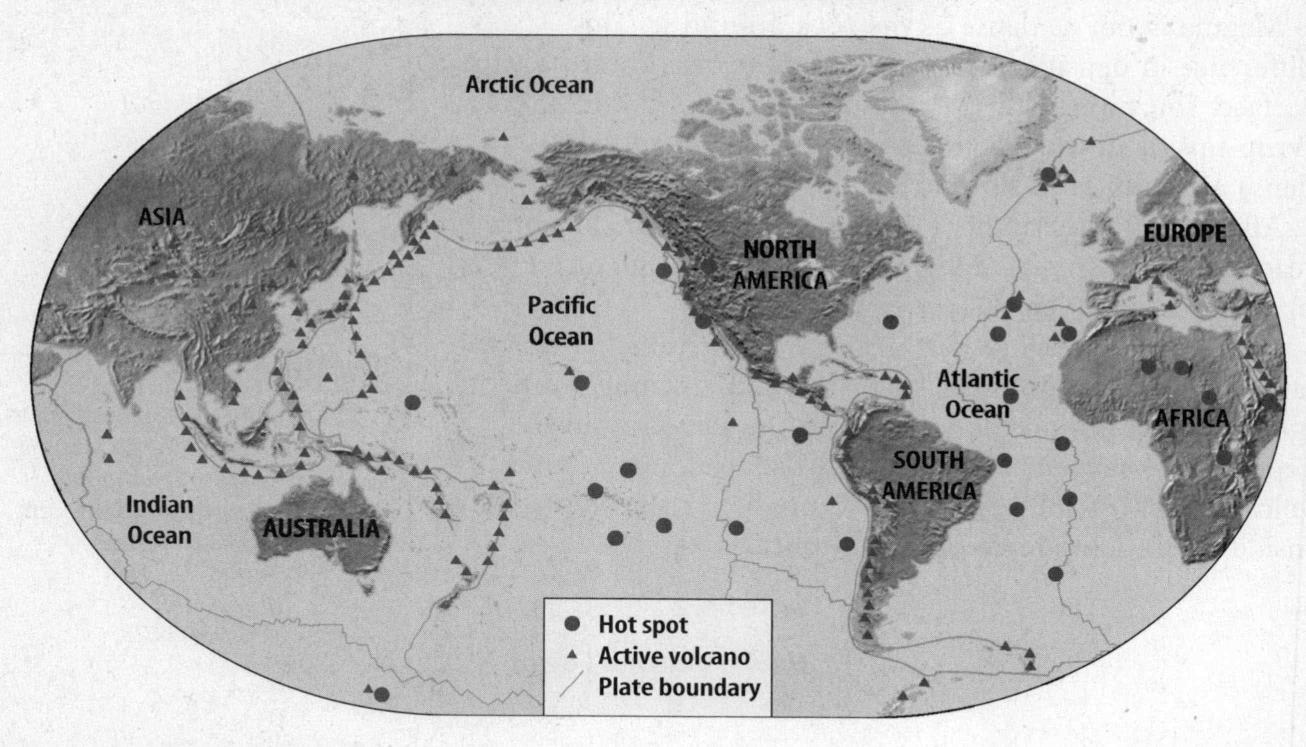


**Picture This**

4. **Analyze** Highlight plate boundaries on the map. Notice the volcanoes along the way. Now look for hot spots. Describe where they are found.

______________________

______________________

______________________

______________________

## Where do volcanoes occur?

Volcanoes often form where plates are moving apart, where plates are moving together, and in areas called hot spots. Plates are large sections of Earth's crust and upper mantle. Plate boundaries are the areas where there is movement of plates. The map above shows the locations of volcanoes, hot spots, and plate boundaries around the world.

## What are divergent plate boundaries?

Iceland is a large island in the North Atlantic Ocean. Iceland has volcanic activity because it is part of the Mid-Atlantic Ridge. The Mid-Atlantic Ridge is a divergent plate boundary. A divergent plate boundary is an area where Earth's plates are moving apart.

When plates move apart, they form long, deep cracks called rifts. Lava flows from these rifts and is cooled quickly by the seawater. As more lava flows and hardens, it builds up on the seafloor. Sometimes, the volcanos and rift eruptions rise above sea level. Islands such as Iceland formed in this way. In 1963, another island, Surtsey, formed nearby.

## Convergent Plate Boundaries

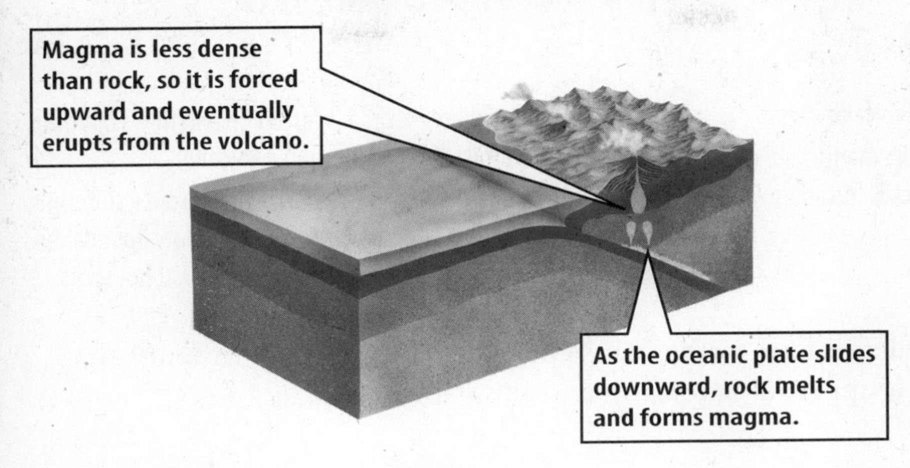


**Picture This**

5. **Interpret** Circle the plate that is going underneath the other plate.

## What occurs at convergent plate boundaries?

Areas where Earth's plates are pushing together are convergent plate boundaries. Sometimes an oceanic plate slides under a continental plate. Other times, an oceanic plate slides under another oceanic plate. The figure above shows how volcanoes can form where plates collide and one plate slides below the other. The Andes mountain range in South America began forming when an oceanic plate started sliding under a continental plate. Volcanoes that form on convergent plate boundaries usually have more explosive eruptions than other volcanoes. Magma forms when the plate sliding below another plate gets deep enough and hot enough to melt partially. The magma is forced to rise slowly to Earth's surface, forming volcanoes like Soufrière Hills on the island of Montserrat.

## What are hot spots?

The Hawaiian Islands are forming as a result of volcanic activity. But they haven't formed at a plate boundary. They are in the middle of the Pacific Plate, far from its edges.

Scientists think there are areas between Earth's core and mantle that are unusually hot. Hot rock in these areas is forced toward the crust where it partly melts to form a **hot spot**. ☑

**Reading Check**

6. **Identify** What is an unusually hot area between Earth's mantle and core called?

______________________

## How were the Hawaiian Islands formed?

The Hawaiian Islands sit on top of a hot spot under the Pacific Plate. Magma has broken through the crust to form several volcanoes. The volcanoes that rise above the ocean form the Hawaiian Islands.

# After You Read

## Mini Glossary

**crater:** steep, bowl-shaped area around a volcano's vent

**hot spot:** unusually hot area between Earth's mantle and core that forms volcanoes when melted rock is forced upward and breaks through the crust

**vent:** opening where magma is forced up and flows out onto Earth's surface as lava, forming a volcano

**volcano:** opening in Earth's surface that erupts sulfurous gases, ash, and lava; can form at Earth's plate boundaries, where plates move apart or together, and at hot spots

**1.** Review the terms and their definitions in the Mini Glossary. Then write a sentence that describes the materials that erupt from a volcano and the path they take.

______________________________________________

______________________________________________

______________________________________________

**2.** Write the following events in the order in which they occur.

**Magma slowly rises to Earth's surface.**
**Lava hardens to rock.**
**Heat and pressure deep inside Earth cause rock to melt.**
**Magma erupts through a vent.**

**First**

| |
|---|
| |

**Second**

| |
|---|
| |

**Third**

| |
|---|
| |

**Fourth**

| |
|---|
| |

**Science Online** Visit **earth.msscience.com** to access your textbook, interactive games, and projects to help you learn more about volcanoes and Earth's moving plates.

# Volcanoes

## section 2 Types of Volcanoes

## Before You Read

Imagine you have two balloons. One balloon is twice the size of the other. If you popped both balloons, which one would you expect to have the loudest bang?

_______________________________________________

_______________________________________________

### What You'll Learn

- what determines the explosiveness of volcanoes
- about three types of volcanoes

## Read to Learn

### What controls eruptions?

Some volcanic eruptions are explosive. These eruptions are rapid, powerful, and destructive, like the eruption that took place at the Soufrière Hills volcano. Other volcanic eruptions are quiet. The lava flows slowly from the vent, like the eruptions at the Kilauea volcano. What causes these differences?

Two factors determine whether an eruption will be explosive or quiet. One factor is the amount of water vapor and other gases trapped in the magma. The other factor is how much silica is in the magma. Silica is a compound made of the elements silicon and oxygen. ☑

### How do trapped gases affect eruptions?

What happens if a can of soda is shaken up and then opened quickly? Pressure from the gas in the drink is released suddenly, and the drink sprays everywhere. Something similar occurs with volcanoes. Gases, like water vapor and carbon dioxide, are trapped in magma by the pressure of the surrounding magma and rock. As magma nears the surface, it is under less pressure. This allows the gas to escape from the magma. Gas escapes easily from some magma during quiet eruptions. However, gas that builds up to high pressures causes explosive eruptions.

Study Coach

**Two-Column Notes** As you read, organize your notes in two columns. In the left column, write the main idea. Next to it, in the right column, write information to support the main idea.

**Reading Check**

1. **Explain** What determines how a volcano will explode?

_______________________

_______________________

_______________________

## Think it Over

2. **Explain** Why does some magma contain a lot of water vapor?

### How does water vapor affect eruptions?

The magma at some convergent plate boundaries contains a lot of water vapor. This happens because an oceanic plate and some of its water slides under other plate material at some convergent plate boundaries. The trapped water vapor in the magma can cause explosive eruptions.

## Composition of Magma

The second major factor that affects the nature of eruptions is the composition of magma. Magma can be divided into two major types—silica poor and silica rich.

### What causes quiet eruptions?

Magma that is low in silica is called basaltic (buh SAWL tihk) magma. It is fluid and produces quiet eruptions like those at Kilauea. This type of lava pours from the volcanic vents and runs down the sides of a volcano. This is called *pahoehoe* (pa HOY hoy) lava. When *pahoehoe* lava cools, it forms a ropelike structure. If the same lava flows at a lower temperature, it forms stiffer, slower moving lava called *aa* (AH ah) lava. ☑

**Reading Check**

3. **Determine** Is basaltic magma low or high in silica?

Quiet eruptions form volcanoes over hot spots, which is how the Hawaiian volcanoes formed. Basaltic magma also flows from rift zones, which are long, deep cracks in Earth's surface. Many lava flows in Iceland are from rift zones.

Basaltic magma is so fluid that when it is forced upward in a vent, the trapped gases can escape easily. As a result, the explosion is quieter, sometimes forming lava fountains. Lava that flows underwater forms pillow lava formations. Just as their name suggests, they are shaped like pillows.

### What causes explosive eruptions?

Magma that contains a lot of silica, or granitic magma, produces explosive eruptions, like those at Soufrière Hills volcano. This magma sometimes forms in areas where Earth's plates are moving together and one plate slides under the other. As the sinking plate goes deeper, some rock melts. The magma is forced upward because it is less dense than the rock around it. As it moves up, it comes in contact with Earth's crust, and becomes enriched in silica. Silica-rich granitic magma is thick. As a result, gas gets trapped inside, causing pressure to build up. When an explosive eruption occurs, the gases expand quickly, often carrying pieces of lava in the explosion.

## Think it Over

4. **Cause and Effect** Why does granitic magma cause explosive eruptions?

## What is andesitic magma?

There is another type of magma—andesitic magma. Andesitic magma contains more silica than basaltic magma, but less than granitic magma. It often forms at convergent plate boundaries where one plate slides under the other. Because of the higher silica content, andesitic magmas erupt more violently than basaltic magmas.

The word *andesitic* comes from the Andes mountain range located in South America. These mountains contain many andesite rocks. Many of the volcanoes that circle the Pacific Ocean also are made up of andesite.

### Think it Over

5. **Compare and Contrast** Of the three types of magmas, which causes the least explosive eruptions?

______________________

# Forms of Volcanoes

All volcanoes do not look alike. The shape of a volcano depends on whether it was formed by a quiet or explosive eruption. The shape also depends on what type of lava it is made of—basaltic, granitic, or andesitic. The three basic types of volcanoes are shield volcanoes, cinder cone volcanoes, and composite volcanoes.

### FOLDABLES™

**B Take Notes** Make a layered book Foldable using two sheets of notebook paper. As you read, take notes on the three kinds of volcanoes.

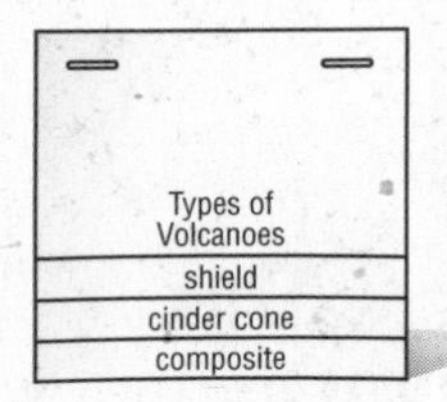


## What is a shield volcano?

Quiet eruptions of basaltic lava spread out in flat layers. These layers build up and form a broad volcano, as shown in the figure below. A broad, gently sloping volcano formed by quiet eruptions of basaltic lava is a **shield volcano.** The Hawaiian Islands are examples of shield volcanoes.

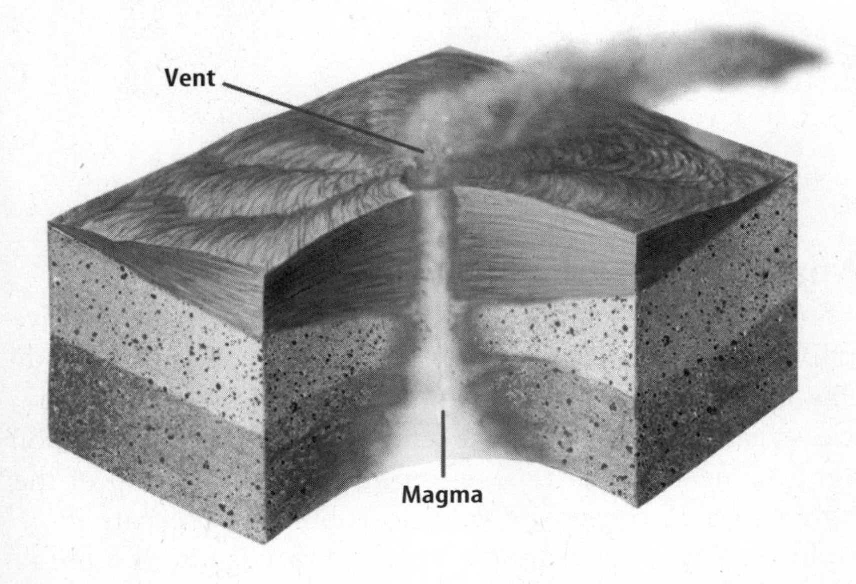


### Picture This

6. **Determine** Use your pen or pencil to draw how lava erupts out of the shield volcano in the figure.

**Flood Basalts** The same basaltic lava that forms shield volcanoes also forms flood basalts. Basaltic lava can flow out of Earth's surface through large cracks called fissures (FIH zhurz). This type of eruption does not form volcanoes. It forms flood basalts and accounts for the greatest amount of erupted volcanic material. The basaltic lava flows over Earth's surface, covering large areas with thick deposits of basaltic igneous rock. The Columbia Plateau, located in the northwestern United States, is a flood basalt. Much of the new seafloor that begins at mid-ocean ridges forms as underwater flood basalts. ☑

**Reading Check**

7. **Describe** What type of lava forms flood basalts?

______________________

______________________

## What is a cinder cone volcano?

Explosive eruptions throw lava and rock high into the air. Bits of rock and solidified lava dropped from the air during an explosive volcanic eruption are called **tephra** (TEH fruh). Tephra comes in different sizes from small pieces of volcanic ash to large rocks called bombs and blocks. **Cinder cone volcanoes** are steep-sided, loosely packed volcanoes formed when tephra falls to the ground. The figure below shows the tephra layers and steep sides of a cinder cone volcano.

**Picture This**

8. **Illustrate** Use your pen or pencil to draw how lava erupts from the cinder cone volcano in the figure.

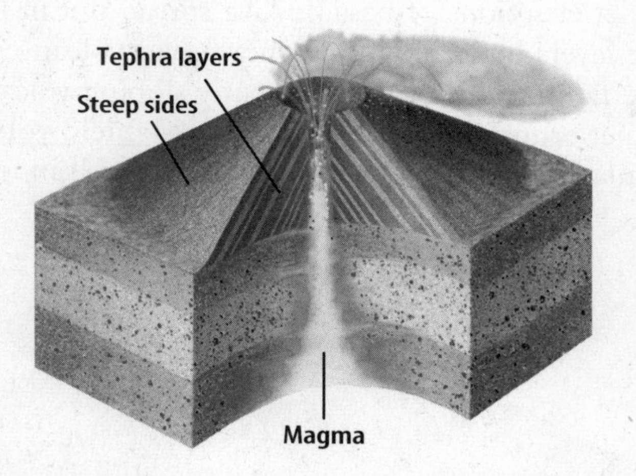


## What is a composite volcano?

Sometimes the same volcano has both quiet and explosive eruptions. How it erupts depends on the trapped gases and how much silica is in the magma. An explosive period can release gas and ash, forming a tephra layer. Then, the eruption can become a quieter type, erupting lava over the top of the tephra layer. A **composite volcano** is built by alternating explosive and quiet eruptions that produce layers of tephra and lava. Composite volcanoes are found mostly where Earth's plates come together and one plate sinks beneath the other.

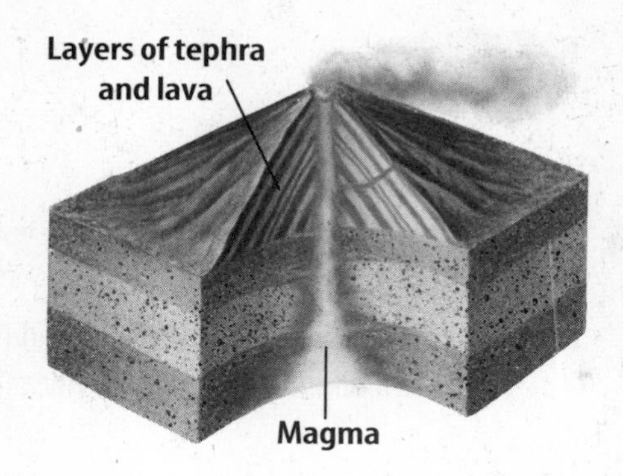


**Forming a Composite Volcano** Soufrière Hills volcano is a composite volcano like the one shown in the figure above. This volcano formed when the ocean floor of the North American Plate and South American Plate slid under the Caribbean Plate. Magma was formed. The alternating eruptions of lava and tephra produced the composite volcano.

## Where were other volcanic eruptions?

Soufrière Hills volcano is listed and described in the table below. As you read about some well-known volcanic eruptions, compare the types of volcanoes with their erupting force.

**Thirteen Selected Eruptions**

| Volcano and Location | Year | Type | Eruptive Force | Magma Content | | Ability of Magma to Flow | Products of Eruption |
|---|---|---|---|---|---|---|---|
| | | | | Silica | $H_2O$ | | |
| Mount Etna, Sicily | 1669 | composite | moderate | high | low | medium | lava, ash |
| Tambora, Indonesia | 1815 | cinder cone | high | high | high | low | cinders, ash |
| Krakatau, Indonesia | 1883 | composite | high | high | high | low | cinders, ash |
| Mount Pelée, Martinique | 1902 | cinder cone | high | high | high | low | gas, ash |
| Vesuvius, Italy | 1906 | composite | moderate | high | low | medium | lava, ash |
| Mount Katmai, Alaska | 1912 | composite | high | high | high | low | lava, ash, gas |
| Paricutín, Mexico | 1943 | cinder cone | moderate | high | low | medium | ash, cinders |
| Surtsey, Iceland | 1963 | shield | moderate | low | low | high | lava, ash |
| Mount St. Helens, Washington | 1980 | composite | high | high | high | low | gas, ash |
| Kilauea, Hawaii | 1983 | shield | low | low | low | high | lava |
| Mount Pinatubo, Philippines | 1991 | composite | high | high | high | low | gas, ash |
| Soufrière Hills, Montserrat | 1995 | composite | high | high | high | low | gas, ash, rocks |
| Popocatépetl, Mexico | 2000 | composite | moderate | high | low | medium | gas, ash |

### Think it Over

**9. Identify** Name the two kinds of alternating layers that form a composite volcano.

_______________

_______________

### Think it Over

**10. Interpret** What kind of volcano is Krakatoa? What was its eruptive force?

_______________

_______________

## After You Read

### Mini Glossary

**cinder cone volcano:** steep-sided, loosely packed volcano formed when tephra falls to the ground

**composite volcano:** volcano built by alternating explosive and quiet eruptions that produce layers of tephra and lava

**shield volcano:** broad, gently sloping volcano formed by quiet eruptions of basaltic lava

**tephra (TEH fruh):** bits of rock and solidified lava dropped from the air during an explosive volcanic eruption

**1.** Review the terms and their definitions in the Mini Glossary. Then write a sentence explaining which two kinds of volcanoes are likely to erupt tephra.

______________________________________________

______________________________________________

______________________________________________

**2.** Read the sentence in the box below labeled *cause.* Think of what happens as a result of this. Choose a sentence listed below that tells what will probably occur next and write it in the box labeled *effect.*

**CAUSE:**

| Magma is thick and traps gases. |
|---|

**EFFECT:**

| |
|---|

**At a convergent boundary, plates collide.**
**At a divergent boundary, plates pull apart.**
**There will be a violent eruption.**
**There will be a quiet eruption.**

**3.** Did this study strategy of writing your notes in two columns help you learn the concepts in this section? Why or why not?

______________________________________________

______________________________________________

______________________________________________

Visit **earth.msscience.com** to access your textbook, interactive games, and projects to help you learn more about the types of volcanoes.

# Volcanoes

## section 3 Igneous Rock Features

### Before You Read

You have learned about magma that is forced up and out of a volcano. Think about what happens to magma that remains under Earth's surface. On the lines below, describe how magma that cools underground might look.

_______________________________________________

_______________________________________________

_______________________________________________

**What You'll Learn**

- how igneous rock features form
- how volcanic necks and calderas form

### Read to Learn

## Intrusive Features

You can see volcanic eruptions because they occur at Earth's surface. However, a lot more volcanic activity occurs underground. In fact, most magma never reaches Earth's surface to form volcanoes or to flow as flood basalts. Most magma cools underground and produces underground rock bodies. Over time, these rock bodies may be seen at Earth's surface if erosion exposes them. These underground rock bodies are called intrusive igneous rock features. There are several different types of intrusive features. The most common types are batholiths (BATH uh lihths), sills, dikes, and volcanic necks.

**Study Coach**

**Think-Pair-Share** Work with a partner. As you read the text, discuss what you already know about the topic and what you learn from the text.

### What are batholiths?

**Batholiths** are the largest intrusive igneous rock bodies. They can be many hundreds of kilometers wide and long. They can be several kilometers thick. Batholiths form when magma bodies cool slowly and solidify before reaching Earth's surface. Some batholiths have been exposed at Earth's surface after millions of years of erosion. The remains of a huge batholith can be seen in Yosemite National Park.

**FOLDABLES™**

**C Write Definitions** Make the Foldable shown below. Write the definition of each word under the tab.

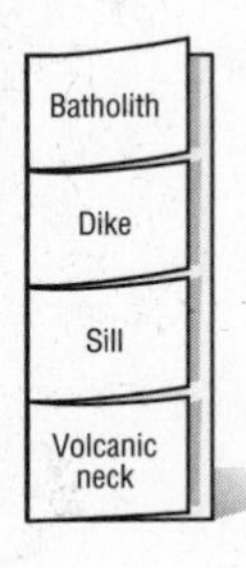

## What are dikes and sills?

Magma sometimes squeezes into cracks in rock below the surface. This is like squeezing toothpaste into the spaces between your teeth. Magma that is squeezed into a crack that cuts across rock layers and hardens underground is a **dike.** Magma that is squeezed into a crack parallel to rock layers and hardens underground is a **sill.** These two igneous rock features are shown in the figure below. Most dikes and sills run from a few meters to hundreds of meters long.

## Other Features

When a volcano stops erupting, the magma hardens inside the vent. Erosion begins to wear away the volcano. Because the cone is softer than the igneous rock in the vent, it erodes first, leaving behind a volcanic neck. A **volcanic neck** is the solid igneous core of a volcano that is left behind after the softer cone erodes.

### Picture This

1. **Determine** Color red any features that form above ground. Color blue any features that form underground.

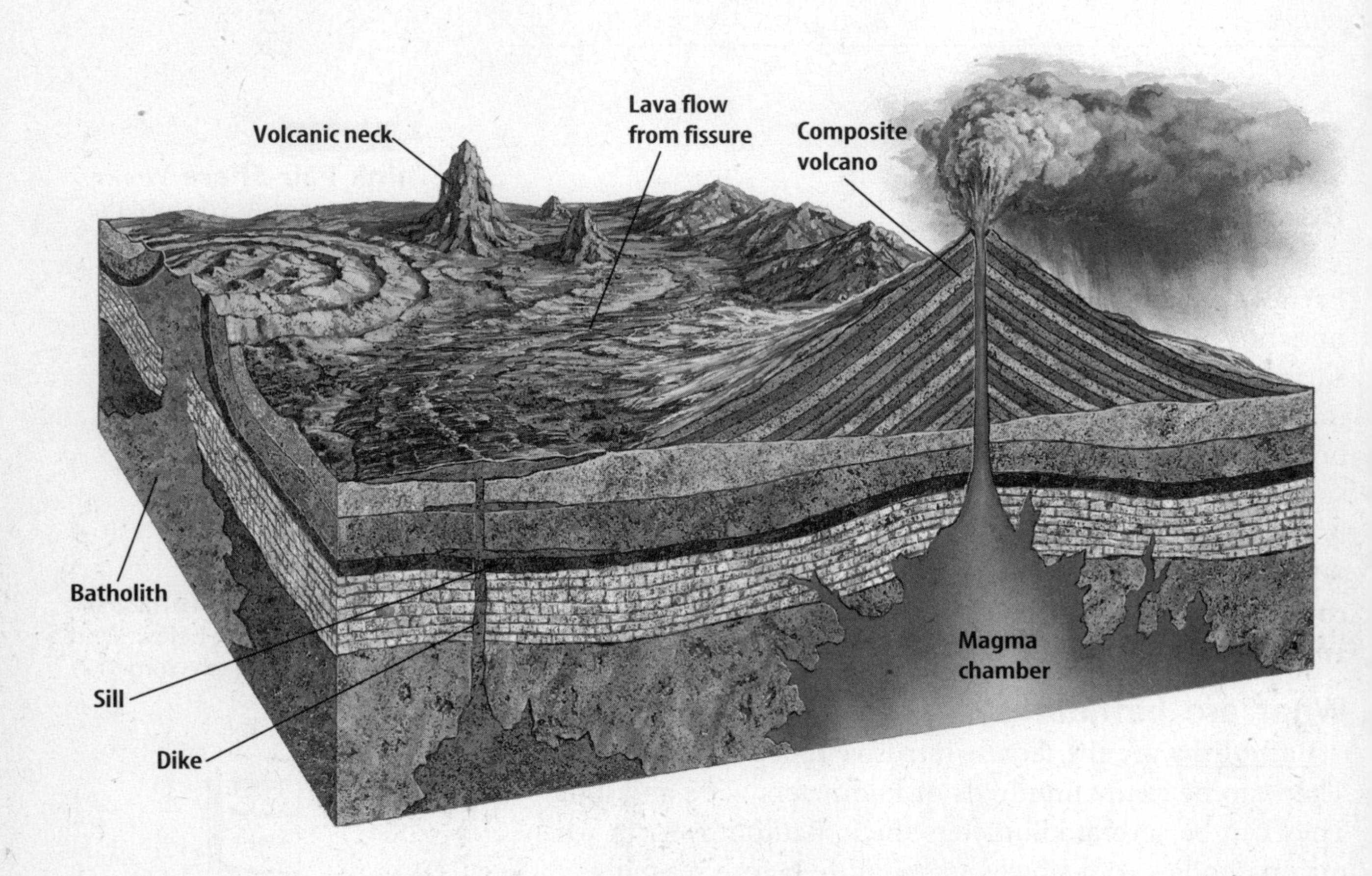

## What are calderas?

Sometimes after an eruption, the top of a volcano collapses. A **caldera** (kal DUR uh) is a large depression, or bowl shape, created when a volcano collapses. The figure below shows the process that forms a caldera. Crater Lake in Oregon is a caldera that filled with water and is now a lake. Crater Lake formed about 7,000 years ago when Mount Mazama erupted violently and then collapsed.

**How Calderas Form**

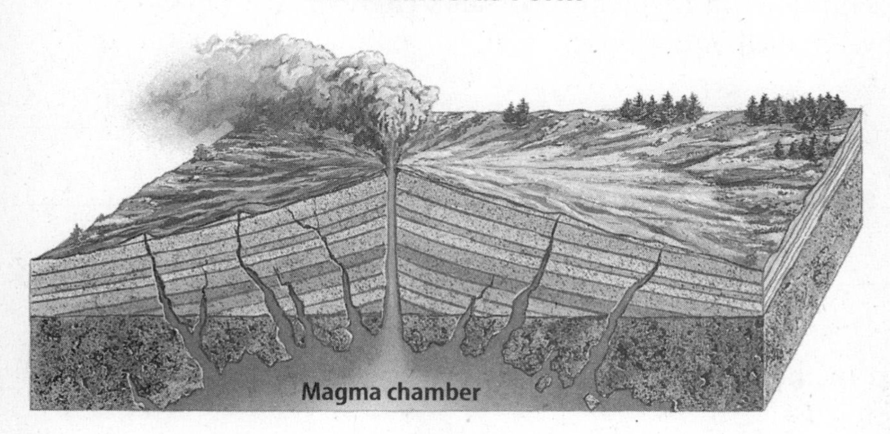


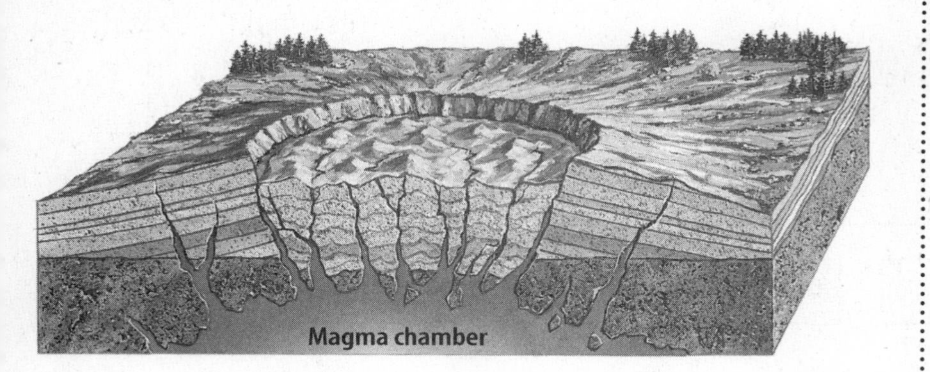


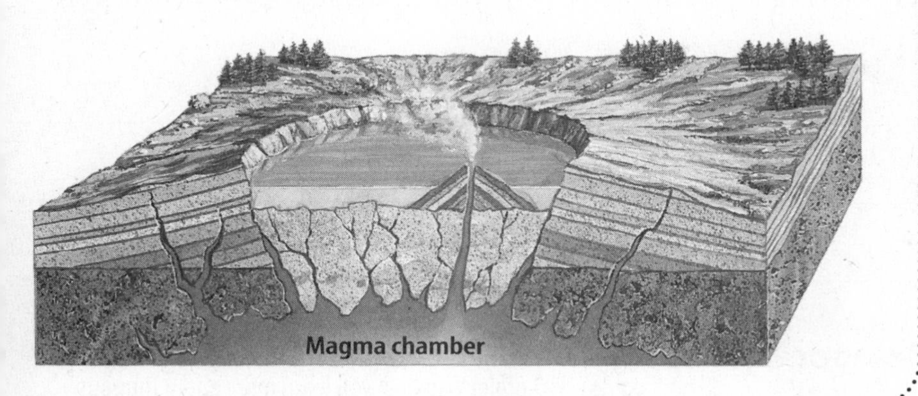


**Picture This**

2. **Identify** In the first figure, circle the area where the caldera later formed.

# After You Read

## Mini Glossary

**batholith (BATH uh lihth):** large intrusive igneous rock body that forms when magma moving upward cools slowly and hardens underground

**caldera (kal DUR uh):** large depression, or bowl shape, formed when a volcano collapses

**dike:** magma squeezed into cracks that cut across rock layers and hardens underground

**sill:** magma that is squeezed into a crack parallel to rock layers and hardens underground

**volcanic neck:** solid igneous core of a volcano left behind after the softer cone has been eroded

**1.** Review the terms and their definitions in the Mini Glossary. Then write a sentence explaining the difference between a sill and a dike.

______________________________________________

______________________________________________

______________________________________________

**2.** Details are listed in each box. In the first box, write a sentence telling the main idea related to all the details given.

| Main Idea: |
| --- |
| |

| Detail 1: sill and dike | Detail 2: batholiths | Detail 3: volcanic necks |
| --- | --- | --- |

**3.** Did the Think-Pair-Share partnering study strategy help you understand what you were reading? Why or Why not?

______________________________________________

______________________________________________

______________________________________________

End of Section

Science Online Visit **earth.msscience.com** to access your textbook, interactive games, and projects to help you learn more about igneous rock features.

# Clues to Earth's Past

## section 1 Fossils

## Before You Read

Every person's history may be revealed by photos, letters, and other things they own. Earth has a history, too. What things might be used to reveal Earth's history?

______________________________________________

______________________________________________

______________________________________________

**What You'll Learn**

- how fossils form
- how fossils are used to tell rock ages
- how fossils explain changes in Earth's surface and life forms

## Read to Learn

### Traces of the Distant Past

It's likely that you've read about dinosaurs and other animals that lived on Earth in the past. But how do you know that they were real? How do you know what they were like? The answer is fossils. Paleontologists, scientists who study fossils, can learn about extinct animals from their fossil remains.

Study Coach

**Two-Column Notes** As you read, organize your notes in two columns. In the left column, write the main idea of each paragraph. Next to it, in the right column, write details about it.

### Formation of Fossils

**Fossils** are the remains, imprints, or traces of animals or plants that died long ago. Scientists have used fossils to determine when life first appeared, when plants and animals first lived on land, and when organisms became extinct. Fossils can tell a lot about the past.

Most animals and plants decay soon after they die. Some animals may eat and scatter the remains of dead organisms. Fungi and bacteria may cause the remains to rot. In time, no trace is left. But some dead animals and plants do become fossils. Sometimes conditions are just right for fossils to form.

FOLDABLES™

**A Compare and Contrast** Make a Foldable as shown below to compare and contrast the ways that fossils are preserved.

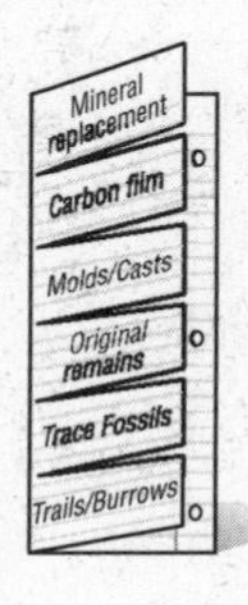

## How do fossils form?

Dead organisms that are protected from scavengers and other harmful forces may leave fossils. One way a dead organism can be protected is for sediment to bury the body quickly. For example, a dead fish might sink to the bottom of a lake. If it is quickly buried by sediment dropped from a stream, it would be protected from scavengers. Over time, the fish may become a fossil in a layer of rock. But quick burial alone isn't always enough to make a fossil.

Organisms with hard parts, such as bones, shells, or teeth, have a better chance of becoming fossils. These hard parts are less likely to be eaten by scavengers than soft parts. Hard parts also decay more slowly than soft parts do. Most fossils are the hard parts of organisms.

**Think it Over**

1. **Explain** What is one reason that hard parts of organisms have a better chance of becoming fossils than soft parts do?

_______________

_______________

_______________

## Types of Preservation

You may have seen the bones of a dinosaur in a museum. You may have seen drawings of dinosaurs in books. Artists who draw dinosaurs study their fossil bones. What preserves fossil bones?

### How do minerals help form fossils?

The hard parts of living things, such as bones, teeth, and shells, have tiny spaces in them. When organisms are alive, the spaces can be filled with cells, blood vessels, nerves, or air. When organisms die, the soft parts decay and leave empty spaces. If the hard part is buried, groundwater can seep into these spaces and deposit minerals. The result is a type of fossil. **Permineralized remains** are fossils in which the spaces are filled with minerals from groundwater. Sometimes minerals replace all of the original hard parts of an organism. ☑

Scientists learn about past forms of life from remains that are permineralized. Other types of fossils can be found as well.

**Reading Check**

2. **Identify** What is the source of the minerals that form permineralized remains?

_______________

### What are carbon films?

The tissues of living organisms contain carbon. Some fossils are made only of carbon. Fossils usually form when sediments bury a dead organism. As sediment builds up, heat and pressure force all the gases and liquids out of the organism. Then, just a thin layer, or film, of carbon is left. It looks like a shadow of the organism's body. A **carbon film** is a thin film of carbon left from an organism and preserved as a fossil.

## What is coal?

Large amounts of dead plants may build up in swamps. Over millions of years, heat and pressure change the plant material into coal which contains large amounts of carbon. Coal is another kind of fossil, but it doesn't reveal much about the past. As the coal forms, the structure of the original plant is usually lost.

## What are molds and casts?

Sometimes the hard parts of a dead organism fall into a soft sediment, such as mud. Then more sediment buries the object. In time, pressure and cementation turn the sediment into rock. Cementation is when minerals from water are deposited in the spaces between sediment particles. Then water and air flow through open spaces in the rock and dissolve the organism's hard parts. This leaves a hole, or cavity, in the rock called a mold. A **mold** is a body fossil that forms when an organism decays, or dissolves, and leaves a cavity in rock. ☑

Later, water and minerals may enter the mold and form new rock. This produces a copy, or cast, of the original object. A **cast** is a type of body fossil that forms when minerals fill a mold and harden into rock. The figure below shows a cast resulting when a mold fills with minerals.

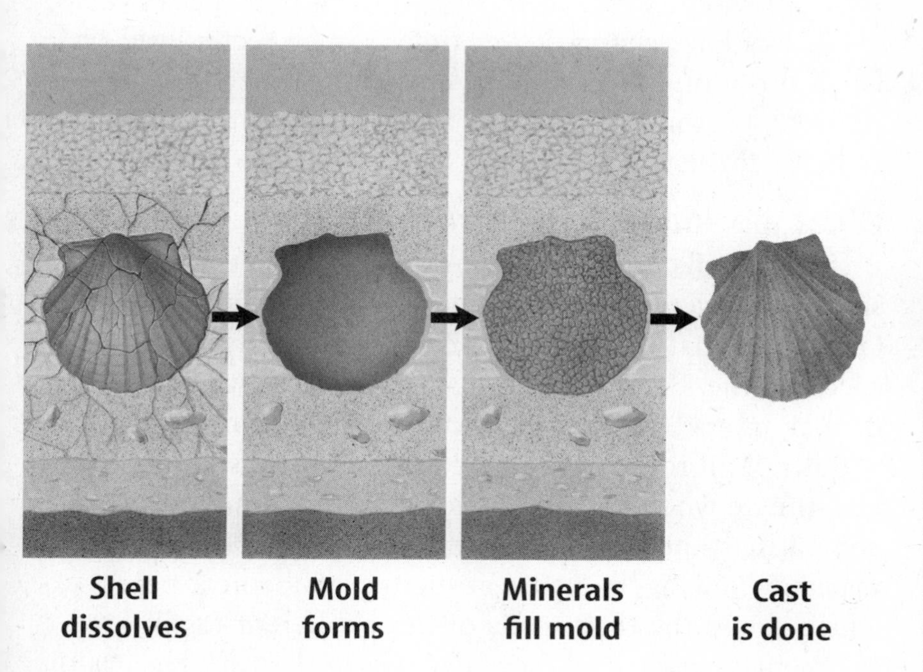


**Reading Check**

3. **Describe** What does a mold in a rock look like?

______________________

______________________

**Picture This**

4. **Explain** What is the key difference between the shell on the left and the cast fossil on the right?

______________________

______________________

______________________

### How are original remains preserved?

From time to time, conditions are just right for the soft parts of an organism to be preserved. Insects can be trapped in amber, a tree sap that turns into a solid. The amber surrounds and protects the insect's original parts. Some organisms, such as mammoths, have been found preserved in frozen ground in Siberia. Some original remains have been found in natural tar deposits.

### What are trace fossils?

Animals that walked on Earth long ago left tracks in soft mud. Some of those tracks have been preserved. They are trace fossils. Trace fossils may be footprints, trails, burrows, or any marks that tell something about how an animal moved and lived. In some cases, tracks can tell more about how an organism lived than any other type of fossil. ☑

Some trace fossils are burrows and trails left by worms and other animals. These fossils give scientists clues about an animal's way of life. For example, examining fossil burrows can reveal how firm the sediment was that the animal lived in. From this information, scientists can learn more about the animal who dug the burrow.

**Reading Check**

**5. Identify** List three different trace fossils.

______

______

______

## Index Fossils

By studying fossils, scientists can learn that species of organisms have changed over time. Some species lived on Earth for a long time without changing. Other species changed a lot in a short time. Scientists use these organisms as index fossils.

### What do index fossils reveal?

**Index fossils** are the remains of species that lived for a short time, were numerous, and were found in many places. Organisms that became index fossils lived only during a specific time. Because of this, scientists can estimate the ages of rock layers based on the index fossils they contain. ☑

But not all rocks have index fossils. There is another way to estimate when these rocks formed. It is done by comparing time spans in which more than one fossil appears. Look at the figure at the top of the next page which shows the time spans of three different fossils in sedimentary rock. The estimated age of the rock layer is the time period where the different fossil ranges overlap.

**Reading Check**

**6. Define** What are index fossils?

______

______

______

**Fossils in a Sequence of Sedimentary Rock**

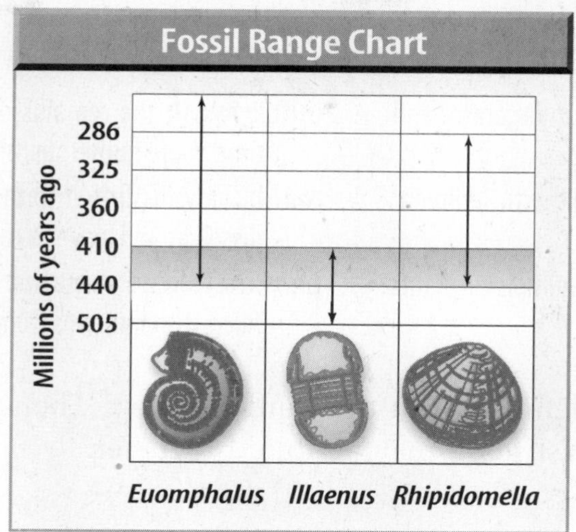


## Fossils and Ancient Environments

Scientists can use fossils to learn what an area was like long ago. Using fossils, they can determine whether an area was land or covered by ocean. If the area was covered by ocean, it might even be possible to learn how deep the water was.

Fossils also can give clues about the past climate of an area. For example, rocks in parts of the eastern United States contain fossils of tropical plants. But today the environment of this area isn't tropical. Because of these fossils, scientists know that the climate was tropical when these plants were living.

### Why might fossils of sea animals be found in a desert?

Crinoids are animals with many arms that usually live in warm, shallow waters. But fossils of crinoids have been found in deserts in parts of western and central North America. What do scientists learn from these fossils? When the fossil crinoids were living, a shallow sea must have covered this area of North America.

Fossils give clues about past life on Earth. Fossils give information about plants and animals that are now extinct. They provide information about the rock layers that contain them and about the ages of the rock layers. By studying fossils, scientists learn about the climate and environment that existed when the rocks formed.

**Picture This**

7. **Interpret** According to the chart, during what span of time did all three fossils appear together on Earth?

_______________

**Think it Over**

8. **Infer** If fossil sea shells are found on the tops of mountains today, what does that tell you about the location of the mountains long ago?

_______________

_______________

_______________

# After You Read

## Mini Glossary

**carbon film:** thin film of carbon left from an organism and preserved as a fossil

**cast:** body fossil that forms when a mold fills with sediment or minerals and then hardens into rock

**fossil:** the remains, imprints, or traces of animals or plants that died long ago

**index fossil:** the remains of species that lived for a short time, were numerous, and were found in many places

**mold:** body fossil that forms when an organism decays or dissolves and leaves a cavity in the rock

**permineralized remains:** fossils in which the spaces are filled with minerals from groundwater

1. Review the terms and their definitions in the Mini Glossary. Then write a sentence that explains why a mold and a cast fossil may be found together.

______________________________________________________________________

______________________________________________________________________

______________________________________________________________________

2. Complete the concept map below.

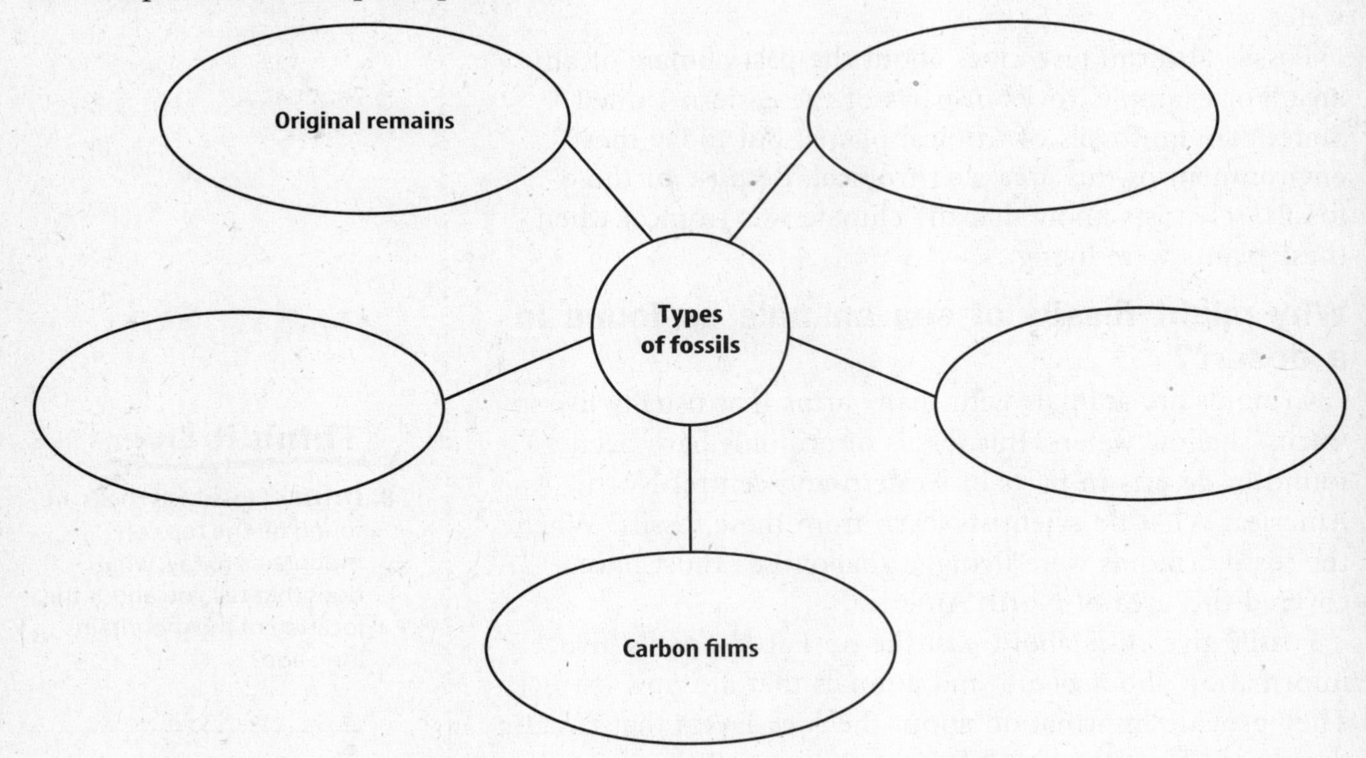

Science Online Visit **earth.msscience.com** to access your textbook, interactive games, and projects to help you learn more about fossils.

# Clues to Earth's Past

## section 2 Relative Ages of Rocks

## Before You Read

Think of two friends. You want to know who is older. What information do you need to figure out who's older?

______________________________________________

______________________________________________

### What You'll Learn

- how to tell the relative ages of rock layers
- how to interpret gaps in the rock record

## Read to Learn

### Superposition

Imagine that you see an interesting car drive by. Then you remember seeing a picture of the car in the January edition of a magazine you have at home. In your room is a pile of magazines from the past year. As you dig down through the pile, you find magazines from March, then February. January must be next. How did you know that the January issue would be at the bottom?

To find the older magazine under newer ones, you used the principle of superposition. How does this principle apply to rocks? The **principle of superposition** states that in layers of rock that have not been disturbed, the oldest rocks are on the bottom and the rocks become younger and younger toward the top.

### Why are rocks in layers?

Sediments build up, forming layers of sedimentary rocks. The first layer to form is on the bottom. A new layer forms on top of the first one. A third layer forms on top of the second layer. The bottom layer is the oldest, because it was formed first. Sometimes, the layers of rock are disturbed. When layers have been turned upside down, other clues are needed to tell which rock layer is oldest.

Study Coach

**Make Flash Cards** As you read this section, make flash cards for each main topic. On one side of the card, write the topic. On the other side, write key information.

FOLDABLES™

**B Compare** Make a three-tab Foldable to compare the concept of relative age and the principle of superposition.

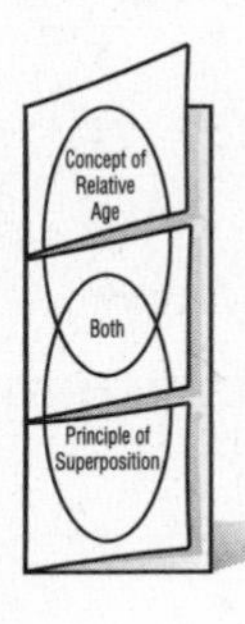

## Relative Ages

Remember the old magazine you were trying to find? What if you want to find another magazine? You don't know how old it is, but you know it came after the January issue. You can find it in the stack by using the principal of relative age. **Relative age** is the age of something compared with the ages of other things. ☑

Scientists figure out the relative ages of rocks by studying their places in a sequence. For example, if layers of sedimentary rock have been moved by a fault, or a break in Earth's surface, the rock layers had to be there before the fault cut through them. So, the relative age of the rocks is older than the relative age of the fault. Relative age doesn't tell you how old the rock is in actual years. The rock layer could be 10,000 years old or one million years old. The relative age only tells you that the rock layer is younger than the layers below it and older than the fault cutting through it.

**Reading Check**

1. **Identify** What term describes the age of something compared with the age of other things?

______________________

### How do other clues help?

It's easy to figure out relative age if the rocks haven't been moved. Look at the figure below on the left showing rock layers that haven't been disturbed. Which layer is the oldest? According to the principle of superposition, the bottom layer is oldest.

Now look at the figure on the right where the rock layers have been disturbed. If a fossil is found in the top layer that is older than a fossil in the lower layer, it shows that the layers have been turned upside down. This could have been caused by folding during mountain building.

**Picture This**

2. **Interpret** Highlight the layer of limestone in both figures. In which rock layer might the oldest fossils be found?

______________________

______________________

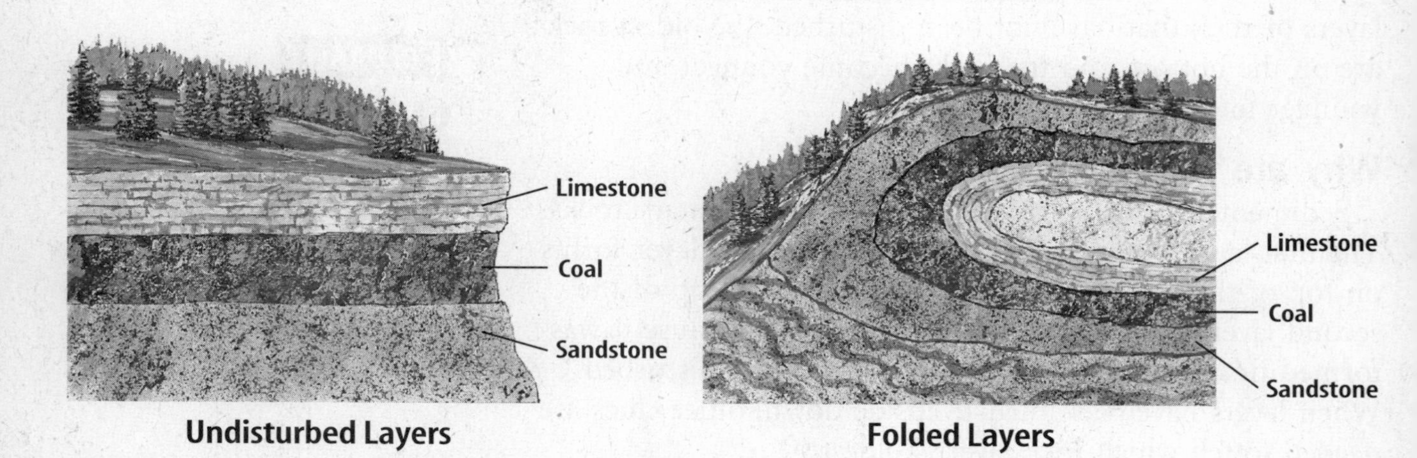


**Undisturbed Layers** **Folded Layers**

## Unconformities

Layers of rock form a record of the past. But the record may not be whole. Layers or parts of layers might be missing. These gaps in the rock layers are called **unconformities** (un kun FOR muh teez). Unconformities develop when erosion removes rock layers by washing or scraping them away. There are three types of unconformities. ☑

### What are angular unconformities?

Forces below Earth's surface can lift and tilt layers of sedimentary rock as shown in the figure below. Over time, erosion and weathering wear down the tilted rock layers. Later, new layers of sedimentary rock are deposited on top of the tilted and eroded layers. The unconformity that results when new layers form on tilted layers is called an angular unconformity.

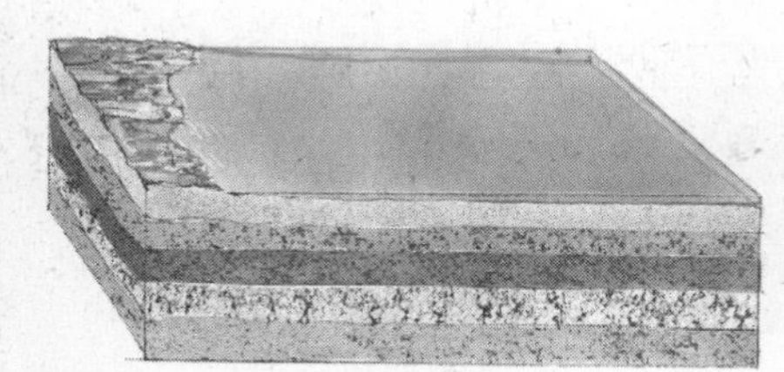

**Sedimentary rocks are deposited originally as horizontal layers.**

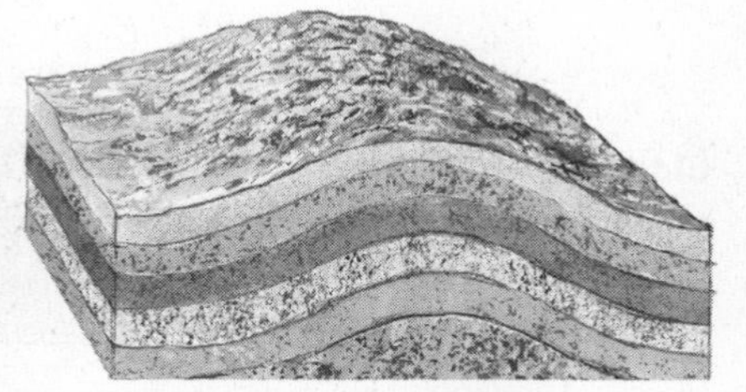

**The horizontal rock layers are tilted as forces within Earth deform them.**

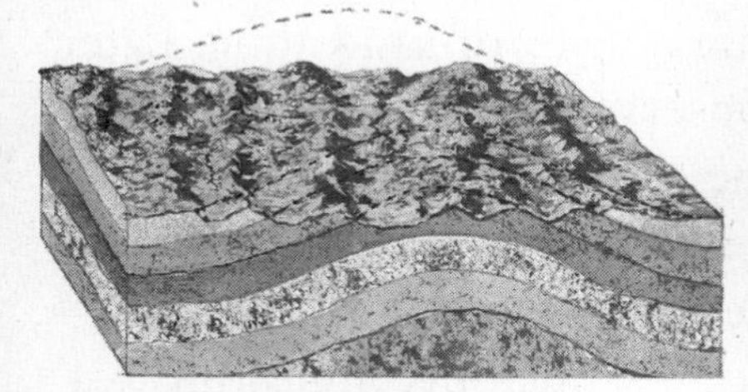

**The tilted layers erode.**

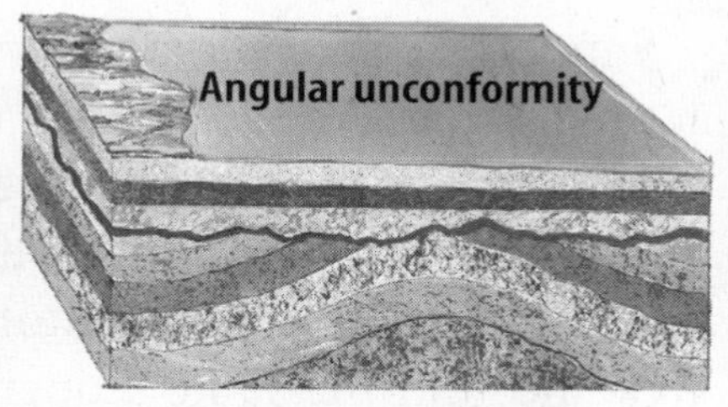


**An angular unconformity results when new layers are deposited on the tilted layers.**

### Why would a layer of rock be missing?

Now and then, a layer of rock is missing from a stack of sedimentary rock layers. Careful study reveals an old surface of erosion. At one time the rocks were exposed and eroded. Later, younger rocks formed above the erosion surface when sediments were deposited again. Even though all the layers are parallel, the rock record still has a gap.

**Reading Check**

3. **Identify** What are gaps in the rock layers called?

_______________

**Picture This**

4. **Identify** Highlight the angular unconformity in the last figure.

## Picture This

5. **Identify** Highlight the surface where rocks were exposed and eroded before new sediments were deposited over them.

**Disconformity** When a rock layer is missing, this type of unconformity is called a disconformity, shown in the figure below. A disconformity also forms when a long period of time passes without any new layers of rock forming.

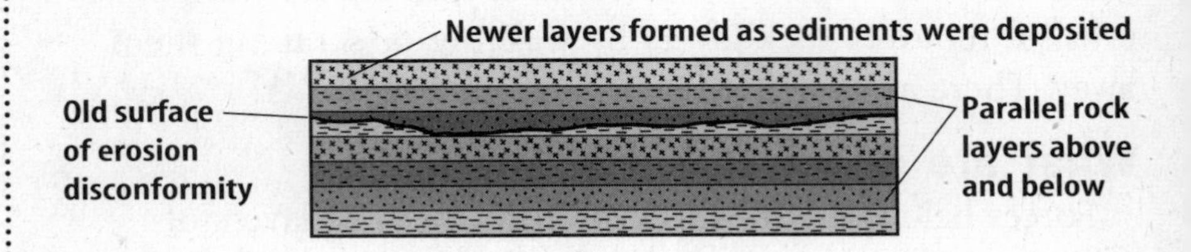


## Picture This

6. **Determine** In the figure color the rock being uplifted red. Color the sedimentary rock being deposited blue.

### What are nonconformities?

Another type of unconformity is shown in the figure below. A nonconformity occurs when metamorphic or igneous rocks are uplifted and eroded. Sedimentary rocks are then deposited on top of the erosion surface. The surface between the two rock types is a nonconformity.

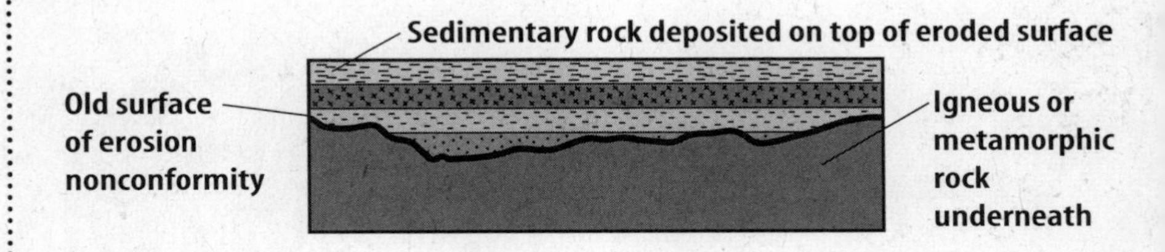


## Matching Up Rock Layers

Suppose scientists are studying a layer of sandstone. Later, at an area 250 km away, they observe a layer of sandstone that looks like the sandstone they studied in the first location. Above the sandstone is a layer of limestone and then another layer of sandstone. They return to the first area and find the same sequence—sandstone, limestone, sandstone. Based on their observations, they theorize that the same layers of rock are in both locations. Often, layers of rocks that are far apart can be matched up, or correlated.

## Think it Over

7. **Explain** What are two ways to correlate rock layers?

______________________

______________________

______________________

### What evidence can correlate rock layers?

One way to correlate exposed rock layer from two places that are far apart is to walk along the layer from one place to the next. Walking along a layer can prove it is unbroken. Layers can also be matched using fossil evidence. If the same types of fossils are found in the same rock layer in both places, it shows that the rock layer in each place is the same age and also that it is from the same deposit.

## After You Read

### Mini Glossary

**principle of superposition:** states that in undisturbed rock layers, the oldest rock is at the bottom and the rocks become younger and younger toward the top

**relative age:** age of something compared to the age of other things

**unconformity:** a gap in the rock layers due to erosion or a period without rock deposit

1. Review the terms and their definitions in the Mini Glossary. Choose one term and explain in your own words what it means.

_______________________________________________

_______________________________________________

_______________________________________________

2. Complete the table about unconformities.

| Unconformities | | |
|---|---|---|
| **Type** | **Description** | **Causes** |
| Angular unconformity | | |
| | | Erosion of whole layers or no new deposition |
| | Sedimentary rock layers over igneous or metamorphic rock | |

3. As you read this section, you made flash cards to help you learn. How did the flash cards help you learn about how layers can be correlated?

_______________________________________________

_______________________________________________

Visit **earth.msscience.com** to access your textbook, interactive games, and projects to help you learn more about the relative ages of rocks.

# Clues to Earth's Past

## section 3 Absolute Ages of Rocks

**What You'll Learn**
- how absolute age differs from relative age
- how the half-lives of isotopes are used to tell a rock's age

## Before You Read

How old are you? How do you know what your exact age is? On the lines below, tell different ways you could verify your exact age.

______________________________

______________________________

______________________________

**Mark the Text**

**Highlight** As you read this section, highlight each of the vocabulary terms and their definitions.

## Read to Learn

### Absolute Ages

After you sort through your stack of magazines looking for that article about the car you saw, you decide that you need to get your magazines back into a neat pile. By now, they are all in a jumble. They are no longer in order according to their relative age. How can you stack them so the oldest are on the bottom and the newest are on the top? Luckily, all the magazines have dates on their covers. The dates make your job easy. By using the dates as your guide, you can put the magazines back in order easily.

**FOLDABLES™**

**C Explain** Use quarter sheets of notebook paper to explain absolute age, radioactive decay, half-life, and radiometric dating.

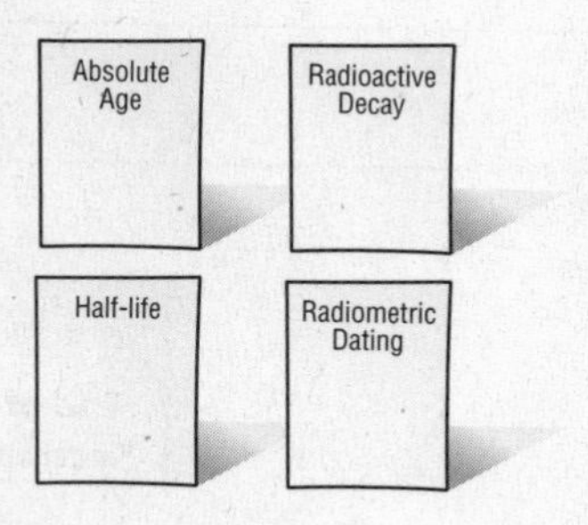


### What is absolute age?

Rocks don't have dates stamped on them. Or do they? **Absolute age** is the age, in years, of a rock or other object. Scientists who study rocks, or geologists, are able to figure out the absolute age of rocks. Geologists use the properties of atoms in rock material to determine absolute age. Knowing the absolute age of rocks leads to a better understanding of events in Earth's history.

## Radioactive Decay

Each atom has a dense center called the nucleus, which is surrounded by particles with a negative charge called electrons. Inside the nucleus are protons, which have a positive charge, and neutrons, which have no electric charge. The number of protons determines the identity of the element. The number of neutrons determines the form of the element, or isotope. For example, every atom with just one proton is a hydrogen atom. Hydrogen atoms can have no neutrons, one neutron, or two neutrons. This means that there are three isotopes of hydrogen. Some isotopes break down into other isotopes, giving off a lot of energy. **Radioactive decay** is the process in which the nucleus of an atom breaks down. ☑

### What are alpha and beta decay?

In some isotopes, a neutron breaks down into a proton and an electron. This type of radioactive decay is called beta decay, because the electron leaves as a beta particle. The nucleus loses a neutron but gains a proton. Other isotopes give off two protons and two neutrons in the form of an alpha particle. This is called alpha decay. Alpha and beta decay are shown in the figure below.

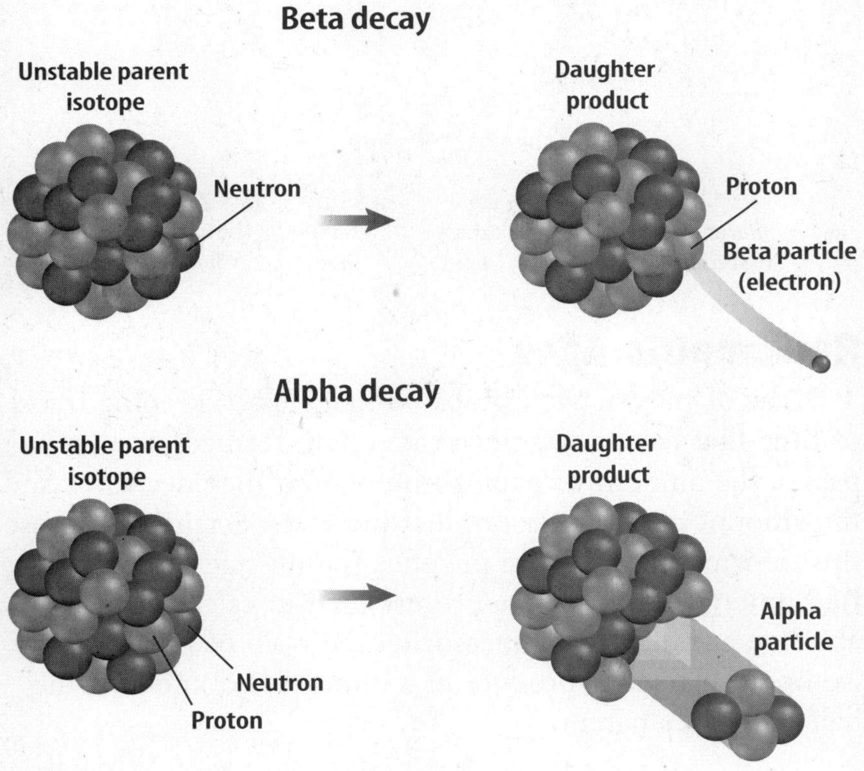

**Reading Check**

1. **Identify** What is the process in which the nucleus of an atom breaks down called?

______________________

**Picture This**

2. **Determine** the beta particle that is given off during beta decay and the alpha particle given off during alpha decay.

**Think it Over**

3. **Explain** What has to happen to the parent isotope before the daughter product can form?

______________________________

______________________________

## What is a half-life?

In radioactive decay, the parent isotope breaks down. The daughter product is formed. Each parent isotope decays to its daughter product at a certain rate. Based on its decay rate, it takes a certain period of time for one half of the parent isotope to decay to its daughter product. The **half-life** of an isotope is the time it takes for half of the atoms in the isotope to decay.

The figure below shows how during each half-life, one half of the parent material decays to the daughter product. For example, the half life of carbon-14 is 5,730 years. So, it will take 5,730 years for half of the carbon-14 atoms to change into nitrogen-14 atoms. You might think that in another 5,730 years, all the remaining carbon-14 atoms will decay into nitrogen-14 atoms. But they don't. Only half the remaining atoms will decay during the next 5,730 years. So, after two half-lives, one fourth of the original carbon-14 atoms will remain. After many half-lives, such a small amount of isotope remains that it is not measurable.

**Picture This**

4. **Determine** the fraction that shows what remains of the parent material after 4 half-lives. Write the fraction below.

______________________________

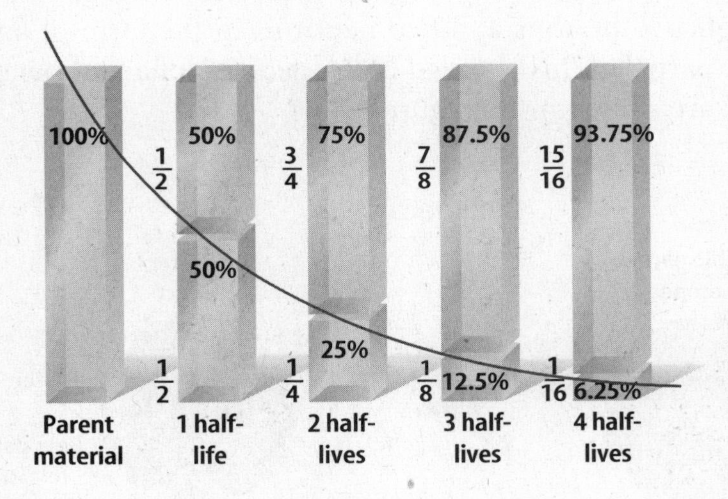


## Radiometric Ages

Decay of radioactive isotopes is like a clock keeping track of time that has passed since rocks have formed. As time passes, the amount of parent isotope in a rock decreases and the amount of daughter product increases. Scientists can use this information to figure out the absolute age of the rock. **Radiometric dating** is the process used to calculate the absolute age of rock by measuring the ratio of parent isotope to daughter product in a mineral and knowing the half-life of the parent.

## What does radiocarbon dating show?

Carbon-14 is useful for dating bones, wood, and charcoal up to 75,000 years old. Living organisms take in carbon from the environment to build their bodies. Most of the carbon is carbon-12, but some is carbon-14. The ratio of these two isotopes in the environment is always the same. After the organism dies, the carbon-14 slowly decays. Scientists can compare the isotope ratio in the sample to the isotope ratio in the environment. Once scientists know the amount of carbon-14 in a sample, they can determine the age of bones, wood, or charcoal.

## Can radiometric dating be used on all rocks?

Aside from carbon-14 dating, rocks that can be radiometrically dated are usually igneous and metamorphic rocks. Most sedimentary rocks can't be dated this way. Why? Many sedimentary rocks are made up of particles that eroded from older rocks. Dating these pieces only gives the age of the original rocks they came from.

**Think it Over**

5. **Explain** Why doesn't radiometric dating work on sedimentary rock?

______________________

______________________

______________________

## What are the oldest known rocks?

Radiometric dating has been used to date the oldest rocks on Earth. These rocks are about 3.96 billion years old. Scientists estimate Earth is about 4.5 billion years old. Rocks older than 3.96 billion years probably were eroded or changed by heat and pressure.

# Uniformitarianism

Before radiometric dating was used, many people thought Earth was only a few thousand years old. But in the 1700s, Scottish scientist James Hutton estimated the Earth to be much older. He used the principle of uniformitarianism. **Uniformitarianism** states that Earth processes occurring today are similar to those that occurred in the past.

Hutton observed that the processes that changed the landscape around him were slow. He inferred that they were just as slow all through Earth's history. Hutton hypothesized that it took much longer than a few thousand years to form rock layers and erode mountains.

Today, scientists agree that Earth has been shaped by two types of change. There are slow, everyday processes that take place over millions of years. There are also sudden, violent events such as the collision of a comet that might have caused the dinosaurs to become extinct. ☑

**Reading Check**

6. **Describe** What are the two types of change that have changed Earth?

______________________

______________________

## After You Read

### Mini Glossary

**absolute age:** age, in years, of a rock or other object

**half-life:** time it takes for half of the atoms in an isotope to decay

**radioactive decay:** process in which the nucleus of an atom breaks down

**radiometric dating:** process used to calculate the absolute age of rock by measuring the ratio of parent isotope to daughter product in a mineral and knowing the half-life of the parent

**uniformitarianism:** principle stating that Earth processes occurring today are similar to those that occurred in the past

**1.** Review the terms and their definitions in the Mini Glossary. Then explain the difference between absolute age and relative age.

______________________________________________

______________________________________________

______________________________________________

**2.** Fill in the half-life chart to show the decay of carbon-14 over time.

**Half-Life of Carbon-14**

| Percent Carbon-14 | Years Passed |
|---|---|
| 100 | 0 |
| | |
| | |
| 12.5 | |
| 6.25 | |
| 3.125 | |

**3.** In this section you highlighted vocabulary terms. Was this strategy helpful? Explain why or why not.

______________________________________________

______________________________________________

Visit **earth.msscience.com** to access your textbook, interactive games, and projects to help you learn more about the absolute ages of rocks.

# chapter 14 Geologic Time

## section 1 Life and Geologic Time

### Before You Read

Think about a giraffe you have seen. On the lines below, describe the giraffe and tell why you think it has a long neck.

_______________________________________________

_______________________________________________

_______________________________________________

**What You'll Learn**

- how geologic time is divided
- how plate tectonics and other changes on Earth affect species

### Read to Learn

#### Geologic Time

A group of students is searching for fossils. By looking in rocks that are hundreds of millions of years old, they hope to find fossils of organisms called trilobites (TRI loh bites). Trilobites are small, hard-shelled animals that lived in ancient seas. Trilobites are considered to be index fossils. Index fossils lived over vast regions of the world during specific periods of geologic time. The students hope that by studying trilobite fossils, they can help piece together a puzzle. They want to know what caused the trilobites to disappear from Earth millions of years ago.

**Study Coach**

**Read-and-Say Something** Work with a partner. As you read each paragraph of the text, take turns saying something about the main idea of the paragraph. Help each other understand the information in the text.

#### What is the geologic time scale?

The appearance or disappearance of types of organisms throughout Earth's history marks important events in geologic time. Paleontologists, scientists who study the prehistoric world, divide Earth's history into time units based on life-forms that existed only during certain periods. This division of Earth's history is known as the **geologic time scale**. Sometimes few fossils remain from a period. Then paleontologists use other methods to define a division of geologic time.

**FOLDABLES™**

**A Classify** Make the following Foldable to help you organize geologic time periods and events into groups based on their characteristics.

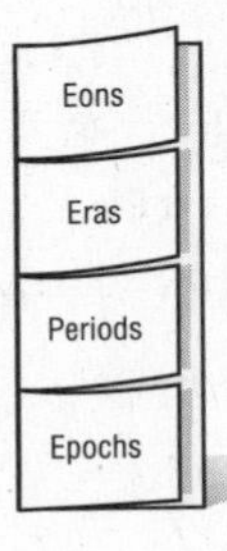

## What are major subdivisions of geologic time?

The fossil record is used to divide Earth's history into geologic time periods. The figure below shows the four major subdivisions of geologic time—eons, eras, periods, and epochs. **Eons** are the longest subdivision and are based on the abundance of certain fossils. ☑

Eons are divided into smaller time periods called eras. An **era** is marked by major worldwide changes in the types of fossils present. For example, at the end of the Mesozoic Era, many kinds of invertebrates, birds, mammals and reptiles became extinct.

Eras are subdivided into periods. A **period** is a unit of geologic time during which certain types of life-forms existed all over the world.

Geologic periods are divided into epochs. An **epoch** is also characterized by differences in life-forms, but these may vary from continent to continent. Epochs may be given names, like those in the Cenozoic Era or may be called simply early, middle, or late.

**Reading Check**

1. **Identify** What is the longest subdivision of geologic time?

_______________

## What limits the divisions of geologic time?

There is a limit to how finely geologic time can be subdivided. It depends on the kind of rock record that is being studied. Sometimes it is possible to distinguish different layers of rock that formed during a single year. In other cases, there is little information to help scientists subdivide geologic time.

**Think it Over**

2. **Interpret** What major event occurred at the end of the Paleozoic Era?

_______________

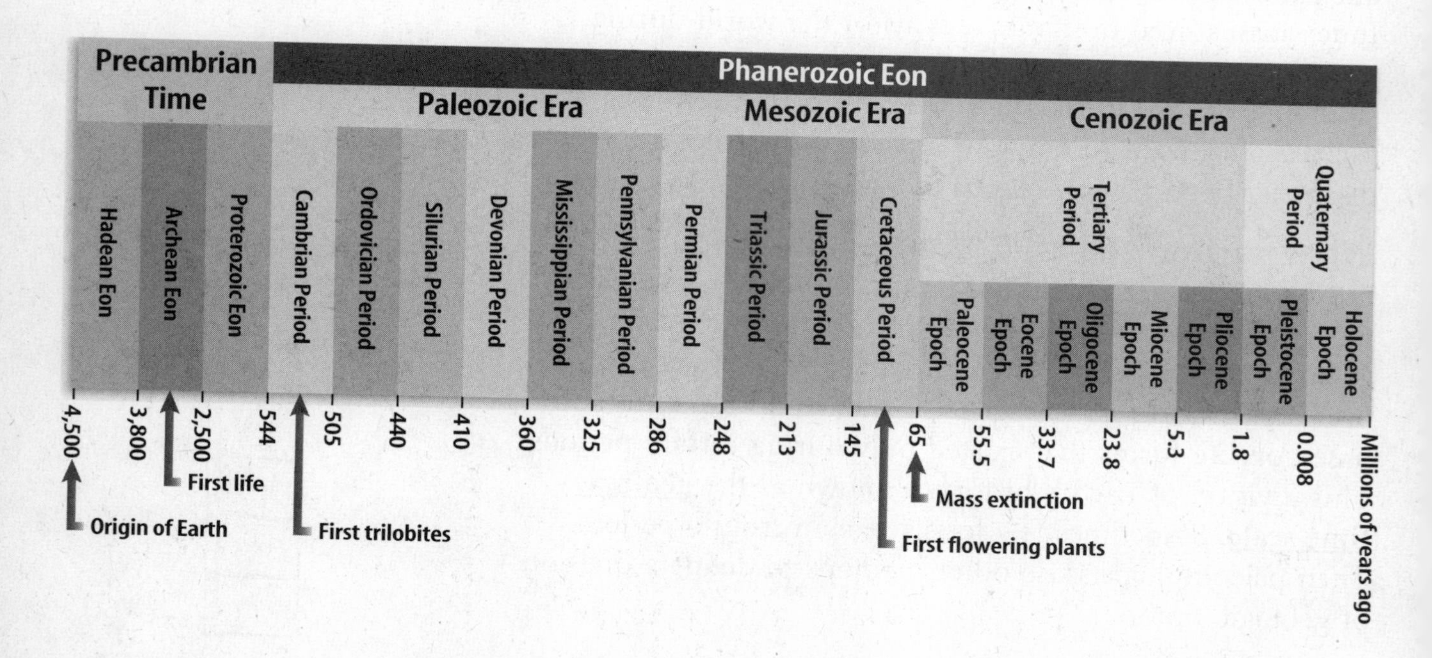

## Organic Evolution

The fossil record shows that species (SPEE sheez), or types of life-forms, have changed over geologic time. **Organic evolution** is the change in species over geologic time. Most scientists believe that changes in the environment affect an organism's survival. Organisms that do not adapt to the changes are less likely to survive or reproduce. Over time, the disappearance of individuals that are not adapted to the new conditions can cause changes to species of organisms.

### What is a species?

There are many ways to define the term species. Life scientists often define a **species** as a group of organisms that normally reproduce only with other members of their group. For example, horses generally reproduce only with other horses. Sometimes, members of different species can mate and produce offspring. Horses sometimes mate with donkeys and produce mules. However, the offspring of two different species are often sterile. ☑

**Reading Check**

3. **Identify** What is the name for a group of organisms that normally reproduce only with other members of their group?

_______________

### What is natural selection?

Charles Darwin was a naturalist—someone who studies the natural world. Between 1831 and 1836, Darwin sailed around the world, carefully observing plants and animals. He collected samples of life-forms and studied them to learn how they were related.

After returning home to England, Darwin explained his theory of natural selection. Darwin defined **natural selection** as the process by which organisms with characteristics suited to a certain environment have a better chance of surviving and reproducing than organisms that do not have these characteristics. Darwin understood that all organisms compete for resources, such as food and living space. He also knew that individuals within a species could be different, or show variation. An individual's differences might help or hurt its chances of surviving in a changing environment. ☑

**Reading Check**

4. **Identify** Who proposed the theory of natural selection?

_______________

Some organisms that were well suited to their environment lived longer and had a better chance of producing offspring. Organisms that were poorly adapted to their environment produced few or no offspring. Because many characteristics are inherited, the characteristics of organisms that are better adapted to the environment get passed on to offspring more often. According to Darwin, this can cause a species to change over time.

## What is natural selection within a species?

Suppose that an animal species exists in which a few of the individuals have long necks, but most have short necks. The main food for the animal is the leaves on trees in the area. Then suppose the climate changes and the area becomes dry. The lower branches of the trees might not have any leaves. Now which of the animals will be better suited to survive? In this case, the animals with the longer necks will be better able to eat the leaves. Clearly, the long-necked animals have a better chance of surviving and reproducing. Their offspring will have a greater chance of inheriting the important characteristic. Gradually, as the number of long-necked animals becomes greater, the number of short-necked animals decreases. Over time, the species might change so that nearly all of its members have long necks—just like the giraffe.

It is important to notice that individual, short-necked animals did not change into long-necked animals. A new characteristic becomes common in a species only under two conditions. First, some members must already have that characteristic. Second, the trait must increase the animal's chance of survival. If no animal in the species had a long neck in the first place, a long-necked species could not have evolved by means of natural selection.

**Think it Over**

**5. Predict** What might have happened to the animals in their new environment if none of them had had a longer neck?

______________________

______________________

## What is artificial selection?

Humans have long used artificial selection to breed domestic animals. Animal breeders carefully choose individuals animals with desired characteristics to mate. Their offspring also have the desired characteristics. In this way, animal breeders have created many different breeds of cats, dogs, cattle, and chickens.

## How do new species evolve?

Natural selection explains how characteristics change and how new species arise. Remember the animals with short necks? If they had moved, or migrated, to a different area, they may have survived. Then they may have reproduced in the new area, developing different characteristics from the long-necked animals. If the short-necked animals were different enough from the long-necked animals that they could no longer breed, then a new species would have evolved. ☑

**Reading Check**

**6. Explain** How could the animals with short necks survive without evolving?

______________________

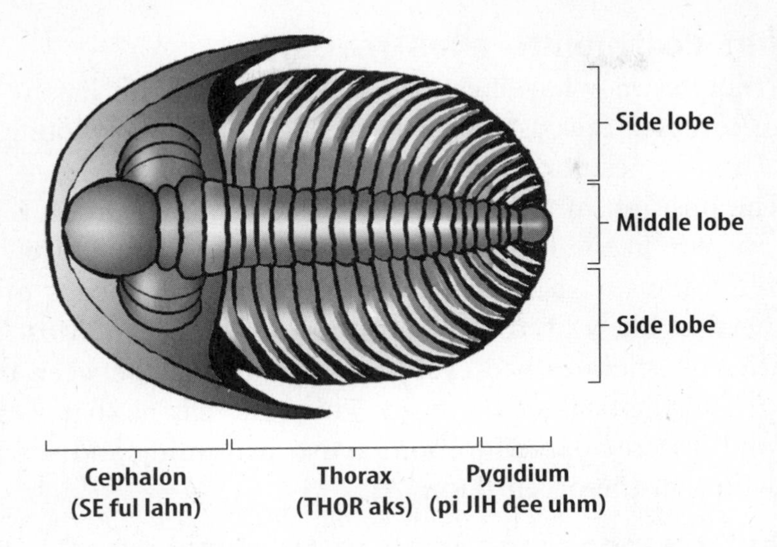


## Trilobites

Remember the trilobites you read about earlier? A **trilobite** is an ancient organism with a three-lobed exoskeleton. An exoskeleton is a hard outer skeleton. These ancient animals were given the name trilobite because they have a three-part shell. The figure above shows the three parts of the trilobite shell. The three parts, called lobes, run the length of its body. Trilobites also have a head (cephalon), a middle that is segmented or divided into sections (thorax), and a tail (pygidium).

### What do the changing characteristics of trilobites tell scientists?

Trilobites lived in Earth's oceans for more than 200 million years. All through the Paleozoic Era, some species of trilobites became extinct and other species of trilobites evolved. Different periods of the Paleozoic Era had different species of trilobites. Each species of trilobites had its own particular characteristics that were different from all other species. ☑

Paleontologists use the differences in trilobite species to explain how trilobites evolved over geologic time. These changes tell how different trilobites from different periods lived. The changes also tell how trilobites responded to changes in their environment.

**Picture This**

**7. Identify** Look carefully at the picture of a trilobite. With a red pencil, outline the trilobite's lobes. With a blue pencil, circle its head, middle, and tail.

**Reading Check**

**8. Identify** What were the two things that happened to trilobite species during the Paleozoic Era?

_______________

_______________

## What do trilobite eyes reveal?

Trilobites may have been the first organisms on Earth with complex eyes as shown in the figure above. Trilobite eyes are the result of natural selection.

The position of an organism's eyes tells how it lived. If its eyes are in the front of its head, it likely swam actively through the sea. If its eyes are located toward the back of its head, it likely lived on the bottom of the ocean. Most species of trilobites had eyes that were midway between the front and the back of the head. This clue tells us that trilobites were adapted to both active swimming and crawling on the ocean floor. ☑

**Reading Check**

**9. Identify** What characteristic helps scientists figure out where different trilobite species lived?

______________________

______________________

## What changes occurred in trilobite eyes?

Over time, the eyes in some trilobites changed. Gradually, the eyes of many trilobite species became smaller and smaller. Eventually, their eyes disappeared completely. These blind trilobites, shown above, might have burrowed into sediments on the ocean floor. Or they lived in a part of the deep ocean where there was no light.

Not all trilobite species lost their eyes. Some trilobite species developed highly complex eyes. One species of trilobite had compound eyes—eyes with many individual lenses. These trilobites had excellent vision. Still other trilobite species developed complex eyes on stalks that extended from their head. They also could see their world very well.

## What changes occurred in trilobite bodies?

The trilobite body also changed over geologic time, as shown in the figure below. Some early trilobite species had many segments in the middle part of the body. Later trilobites had fewer segments.

**Picture This**

**10. Identify** Highlight the body segments of the trilobites in the figure.

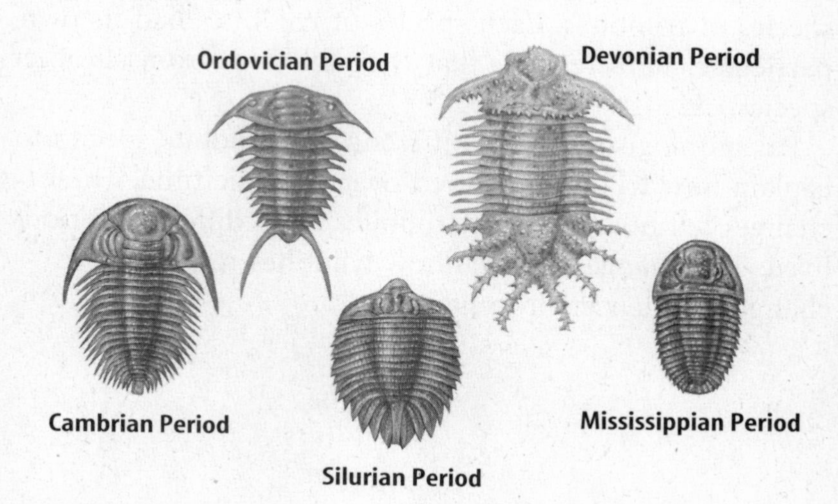

## What evidence do fossils provide?

The exoskeletons of trilobites changed as their environment changed. Each change in the trilobite body shows how different species of trilobite adapted to new conditions. Some species of trilobite could not adapt. These species disappeared, or became extinct. ☑

**Reading Check**

11. **Explain** What did each change in the trilobite body show?

______________________

______________________

## Plate Tectonics and Earth History

Earth's crust is made up of several plates. These plates are in slow but constant motion. This motion, called plate tectonics, caused continents to split apart or to collide. At the time the trilobites dominated Earth's seas, Earth's plates were moving together. When all the continents collided, they formed a single, enormous continent known as Pangaea (pan JEE uh), shown in the figure below. **Pangaea** was one giant landmass, or supercontinent. When Pangaea was forming, sea levels were dropping. Because trilobites lived in the ocean, they could not survive in the changed environment. At the end of the Paleozoic Era, trilobites became extinct.

Some scientists do not accept that the formation of Pangaea caused the extinctions at the end of the Paleozoic Era. Changes in the climate or other conditions may have led to the Paleozoic extinctions. As in all scientific debates, evidence must be considered carefully, and conclusions must be drawn based on the evidence.

**Picture This**

12. **Outline** In the figure, outline the borders of the continents that crashed together to form the supercontinent, Pangaea.

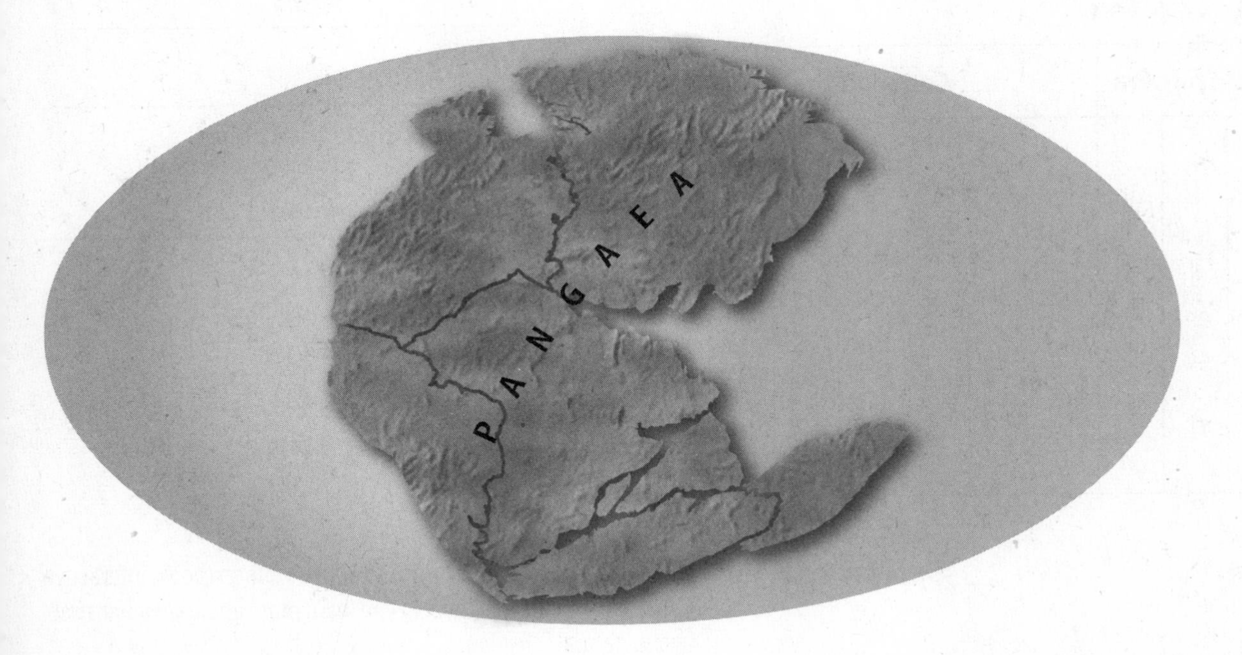

# After You Read

## Mini Glossary

**eon:** largest geologic time division that is based on the abundance of certain fossils

**epoch:** geologic time division characterized by differences in life-forms, which may vary from continent to continent

**era:** geologic time division marked by major worldwide changes in the types of fossils present

**geologic time scale:** divisions of time in Earth's history

**natural selection:** process by which organisms with characteristics suited to a certain environment have a better chance of surviving and reproducing than organisms that do not have these characteristics

**organic evolution:** change in species over geologic time

**Pangaea:** one giant landmass, or supercontinent, that formed at the end of the Paleozoic Era

**period:** subdivision of geologic time during which certain types of life-forms existed all over the world

**species:** group of organisms that normally reproduce only with other members of their group

**trilobite (TRI loh bite) :** small, hard-shelled animal that lived in ancient seas

1. Review the terms and their definitions in the Mini Glossary. Then write two sentences about geologic time and natural selection. Use at least four vocabulary words in your sentences.

_______________________________________________

_______________________________________________

_______________________________________________

2. Fill in the correct term to show how the Geologic Time Scale is divided.

<table>
<tr><td colspan="10">Phanerozoic Eon</td><td>a. ______________</td></tr>
<tr><td colspan="3">Mesozoic Era</td><td colspan="7">Cenozoic Era</td><td>b. ______________</td></tr>
<tr><td rowspan="2">Triassic Period</td><td rowspan="2">Jurassic Period</td><td rowspan="2">Cretaceous Period</td><td colspan="5">Tertiary Period</td><td colspan="2">Quaternary Period</td><td>c. ______________</td></tr>
<tr><td>Paleocene Epoch</td><td>Eocene Epoch</td><td>Oligocene Epoch</td><td>Miocene Epoch</td><td>Pliocene Epoch</td><td>Pleistocene Epoch</td><td>Holocene Epoch</td><td>d. ______________</td></tr>
</table>

Science Online Visit earth.msscience.com to access your textbook, interactive games, and projects to help you learn more about geologic time and natural selection.

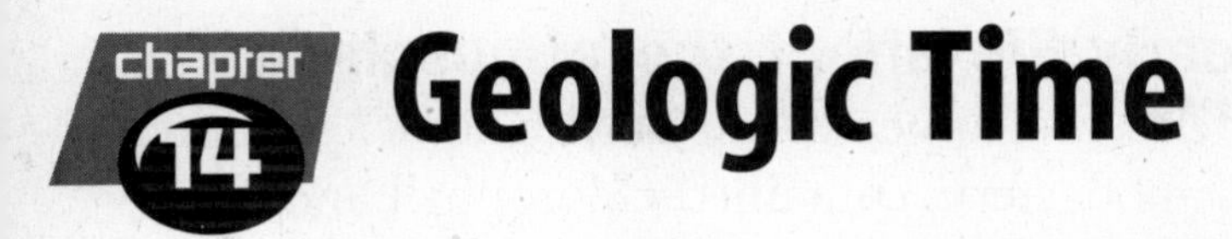

# Chapter 14 Geologic Time

## section 2 Early Earth History

### Before You Read

Think of a picture of a volcano you have seen. Describe what Earth would be like if the land were almost completely covered with volcanoes.

______________________________________________

______________________________________________

______________________________________________

**What You'll Learn**

- what Earth was like during its early history
- how species adapted to changes on Earth

### Read to Learn

#### Precambrian Time

It may seem strange, but during the first billion years of Earth's history, the land was covered with volcanoes.

Over the next 3 billion years, simple life-forms began to live in the oceans. Precambrian (pree KAM bree un) time is the longest part of Earth's history. **Precambrian time** lasted from about 4.5 billion years ago to about 544 million years ago.

**Mark the Text**

**Highlight** As you read, highlight the key terms and main ideas in each paragraph.

#### What is known about early life forms?

Little is known about the organisms that lived during Precambrian time. Most Precambrian rocks are buried deep within Earth where they have been changed by heat and pressure. Few fossils can survive these conditions. Most Precambrian organisms had soft bodies. These organisms did not have hard body parts that leave fossil imprints in rock.

One clue to early history of life is found in ancient stromatolites (stroh MA tuh lites). Stromatolites are layered mats formed by colonies, or groups, of cyanobacteria. **Cyanobacteria** are blue-green algae thought to be one of the earliest life-forms on Earth.

**FOLDABLES™**

**B Compare and Contrast** Make the following Foldable to compare the Precambrian and the Paleozoic Eras.

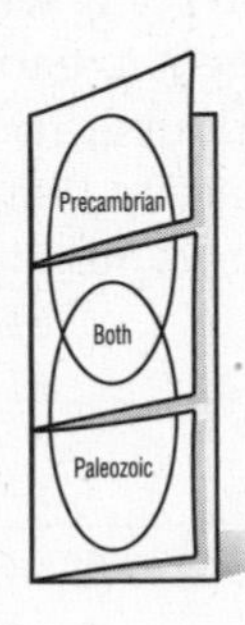

## How did early life affect the atmosphere?

Cyanobacteria first appeared on Earth about 3.5 billion years ago. Cyanobacteria contained chlorophyll and used photosynthesis. This is important because during photosynthesis, they produced oxygen, which helped change Earth's atmosphere. Following the appearance of cyanobacteria, oxygen became a major gas in the atmosphere.

Also of importance was that the ozone layer in the atmosphere began to develop, shielding Earth from ultraviolet rays. It is hypothesized that these changes allowed species of single-celled organisms to evolve into more complex organisms.

Near the end of Precambrian time, invertebrates (ihn VUR tuh brayts) appeared. Invertebrates are animals without backbones. Early invertebrates had soft bodies, so few were preserved as fossils. Because of this, many Precambrian fossils are trace fossils. Examples of trace fossils are tracks, trails, or burrows. Trace fossils provide information about how organisms lived and behaved.

### Think it Over

1. **Infer** How might Earth be different today if cyanobacteria had not lived in ancient seas?

______________________

______________________

______________________

## What were other unusual life-forms?

In the late Precambrian, a group of animals lived that were similar to some animals today. These soft-bodied animals looked like modern jellyfish and worms. The first fossils of these earliest invertebrates were found in the Ediacara Hills of Australia. This group of organisms became known as the Ediacaran (ee dee uh KAR un) animals. They have been found on every continent, except Antarctica.

Ediacaran animals lived on the bottom of Precambrian seas. Some scientists think these animals may have had tough outer coverings on their bodies. Trilobites may have outcompeted the Ediacaran animals and caused their extinction. However, no one knows for sure why the Ediacaran animals disappeared.

### Think it Over

2. **Explain** why scientists have been able to learn more about life in the Paleozoic Era than about life in Precambrian time.

______________________

______________________

# The Paleozoic Era

Beginning in the Paleozoic (pay lee uh ZOH ihk) Era, animals with shells and other hard body parts began to appear. Because hard body parts are well preserved in fossils, it is easier to find traces of life in this era. The **Paleozoic Era**, or era of ancient life, began about 544 million years ago and ended about 248 million years ago.

## What organisms lived during this era?

During most of the Paleozoic Era, warm, shallow seas covered much of the planet. Many life-forms lived in the oceans. Trilobites were common, especially early in the Paleozoic. Other organisms developed shells. As a result, the fossil record of this time contains many shells. However, invertebrates weren't the only animals in Paleozoic seas.

Animals with backbones, called vertebrates, evolved during this time. The earliest vertebrates were fishlike creatures without jaws. During the Devonian Period, fish with strong jaws evolved. These huge fish, armed with heavy protective coverings on their bodies, could eat large sharks. By the Devonian Period, forests began to grow on land. Some vertebrates adapted to the land environment. ☑

**Reading Check**

**3. Describe** What did the earliest vertebrates look like?

_______________

_______________

## How did early life forms move onto land?

Most fish, both ancient and modern, breathe through gills. But in the Devonian Period, many fish also had lungs. Because of their lungs, they could live in water that had low levels of oxygen and swim to the surface to breathe air.

One kind of ancient fish with lungs also developed fins that were like legs, as shown in the figure below. These leglike fins were used to swim and to crawl around on the ocean floor. Paleontologists hypothesize that today's amphibians might have evolved from these fish. Modern amphibians, such as frogs, live both in water and on land. All amphibians have one thing in common—they all lay their eggs in water or in a moist place.

**Picture This**

**4. Circle** Draw a circle around the leglike fins on the fish shown in the figure.

## What adaptations allowed reptiles to remain on land?

By the Pennsylvanian Period, some amphibians evolved eggs that were covered by a protective coating. The coating helped prevent the eggs from drying out. As a result, these animals, called reptiles, did not need to lay their eggs in water. Reptiles also have skin covered with hard scales. The scales prevent water loss from their bodies. These adaptations allow reptiles to live farther from water and in dry climates where many amphibians cannot live. ☑

**Reading Check**

5. **Identify** What was one adaptation that allowed reptiles to live out of water?

______________________

______________________

## How were mountains formed?

During the Paleozoic Era, there were great changes on Earth's surface. Several mountain ranges formed during this time. One example is the Appalachian Mountains in the Eastern United States. Mountain building occurred in several stages.

First, North America moved closer to Europe and Africa. This closed the ocean that had separated them. Several volcanic island chains that had been in the ocean collided with the North American Plate. The collision of the plate and the island chains created high mountains. ☑

**Reading Check**

6. **Identify** What collided with the North American Plate and created high mountains?

______________________

______________________

The next mountain-building event occurred when the African Plate crashed into the North American Plate. This collision formed mountains on both North America and Africa. Rock layers were folded and faulted. Some rocks that were near the eastern coast of the North American Plate were pushed west along faults as far as 65 km. Sediments were uplifted to form an immense mountain belt. Then, about 200 million years ago, the North American and African plates began to separate. The ocean between them began to open up again.

## What caused the end of the Paleozoic Era?

At the end of the Paleozoic Era, more than 90 percent of all ocean species and 70 percent of all land species died. Perhaps as the supercontinent Pangaea formed, changes in the ocean and land caused species to die. Another hypothesis is that erupting volcanoes changed Earth so much that organisms could not survive. Perhaps an asteroid hit the planet and destroyed its environments.

# After You Read

## Mini Glossary

**cyanobacteria:** blue-green algae thought to be one of the earliest life-forms on Earth

**Paleozoic Era:** era that began about 544 million years ago and ended about 248 million years ago; has abundant fossils

**Precambrian time:** longest part of Earth's history, lasting from about 4.5 billion years ago to about 544 million years ago

1. Review the terms and their definitions in the Mini Glossary. Then write two sentences, one about each time period covered in this section. Use the three vocabulary words in your sentences.

_______________

_______________

_______________

2. Fill in the blanks in the boxes below.

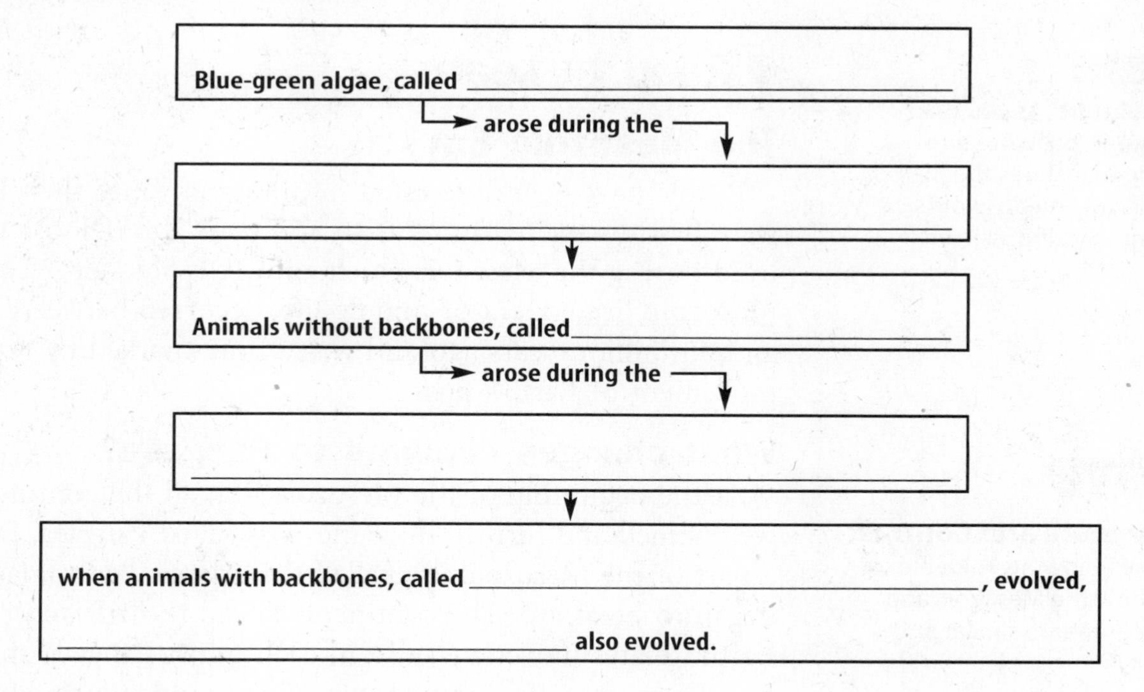


3. Did highlighting help you understand the information in this section?

_______________

Visit **earth.msscience.com** to access your textbook, interactive games, and projects to help you learn more about the early history of Earth.

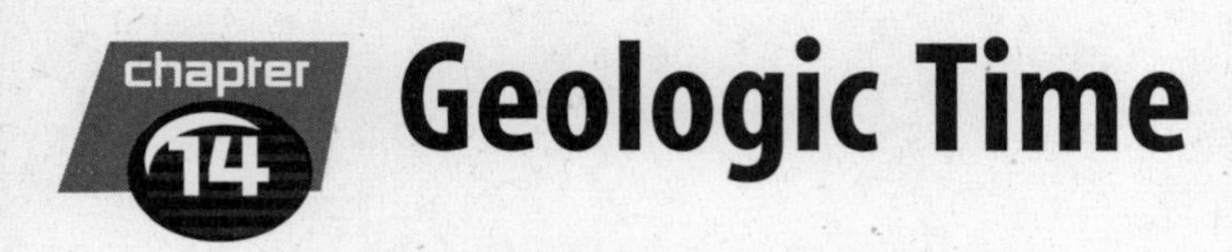


# Geologic Time

## section 3 Middle and Recent Earth History

### What You'll Learn

- about Mesozoic and Cenozoic life-forms
- how plate tectonics affected organisms during the Mesozoic Era
- when humans first appeared

## Before You Read

Think about a picture of a dinosaur you have seen. On the lines below, describe how dinosaurs are different from modern animals.

_______________________________________________

_______________________________________________

_______________________________________________

Study Coach

**Summarize** As you read each paragraph, write one sentence that states the main idea. Use vocabulary words in your sentence, if possible.

FOLDABLES™

**C Compare and Contrast** Make the following Foldable to record how the Mesozoic and Cenozoic Eras are similar and different.

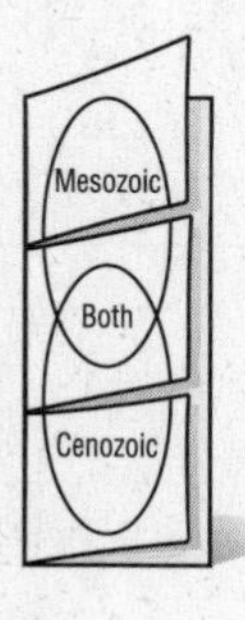


## Read to Learn

### The Mesozoic Era

People have been interested in dinosaurs since their bones were first discovered more than 150 years ago. Dinosaurs lived during the Mesozoic (meh zuh ZOH ihk) Era. The **Mesozoic Era**, or era of middle life, occurred between 248 and 65 million years ago and was a time marked by rapid movement of Earth's plates.

### What changes occurred to Pangaea?

At the beginning of the Mesozoic Era, all the continents were joined and formed one landmass called Pangaea. During a part of the Mesozoic Era called the Triassic Period, Pangaea began to break up. The continents began to drift apart as Earth's plates moved away from each other. Pangaea split in two. Over time, the continents we know today formed.

Some species, such as reptiles, survived the tremendous changes and mass extinction that occurred at the end of the Paleozoic Era. In the early Mesozoic Era, the climate became drier. The reptile's scaly skin kept in moisture, so reptiles could live in this drier climate. Reptile eggs are protected by a shell, so their young survived as well. Reptiles became the most abundant animals of the Mesozoic Era.

## What were the dinosaurs like?

Dinosaurs came in all shapes and sizes. Some were less than 1 m tall. Others, like *Apatosaurus* and *Tyrannosaurus*, were enormous. The first small dinosaurs appeared in the Triassic Period. Larger dinosaur species evolved during the Jurassic and Cretaceous Periods. Throughout the Mesozoic Era, some dinosaurs became extinct as others evolved.

## Were dinosaurs active?

Generally, the faster an animal runs, the farther apart its footprints are. Paleontologists have studied fossil footprints from dinosaurs. They have found that some dinosaur footprints were far apart, indicating that these species were fast runners. Scientists figured out that some dinosaurs could run up to 65 km/h—as fast as a modern racehorse. ☑

**Reading Check**

1. **Explain** How do scientists know that some dinosaurs were fast runners?

______________________________

______________________________

Studies also indicate that some dinosaurs might have been warm-blooded animals, not cold-blooded like reptiles. What evidence shows this? The bones of warm-blooded animals don't have growth rings, like the bones of cold-blooded animals. Dinosaur bones don't have growth rings either. They are similar to mammal bones as shown in the figures below.

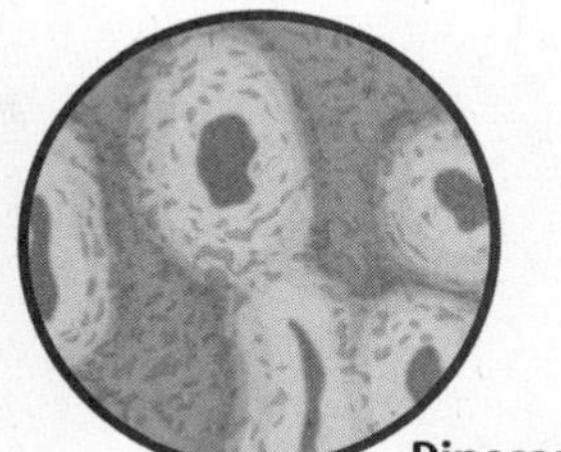

Dinosaur bone

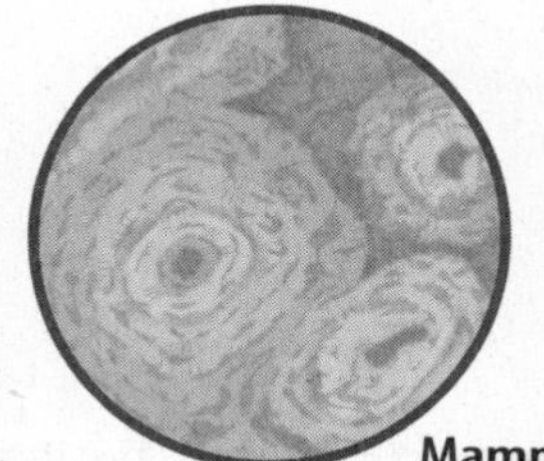
Mammal bone

**Picture This**

2. **Describe** How are the dinosaur bone and the mammal bone alike?

______________________________

______________________________

______________________________

These findings indicate that some dinosaurs were likely active warm-blooded, fast-running animals, similar to modern mammals and birds. In fact, dinosaurs may have been very different from present-day reptiles.

## Were dinosaurs good mothers?

The fossil record shows that some dinosaur species fed and took care of their young. These dinosaurs also traveled in herds, so the young were protected. *Maiasaura* acted in this way. *Maiasaura* mothers gathered together to build nests in colonies. The mothers laid their eggs and, after the young hatched, fed and cared for their offspring. Scientists have found fossils of hatchlings and adult dinosaurs in the same nest. *Maiasaura* mothers may have fed and tended their young until they were old enough to leave the nest.

## What were the first bird-like creatures?

Some paleontologists believe that modern birds evolved from small, meat-eating dinosaurs. The earliest bird-like dinosaur known, *Archaeopteryx*, had both wings and feathers. Since *Archaeopteryx* had some features different from modern birds, it is not a direct ancestor of today's birds.

## What were the first mammals like?

The first mammals appeared in the Triassic Period. These tiny, mouselike animals were warm-blooded and covered with fur, as shown in the figure below. The females produced milk to feed their young. Because of their furry coat and milk production, mammals where able to survive many changing environments.

### Picture This

**3. Describe** How does the early mammal resemble modern mammals, like mice or shrews?

## What are gymnosperms?

Gymnosperms (JIHM nuh spurmz) first appeared during the Paleozoic Era. By the Mesozoic Era, gymnosperms dominated the land. Gymnosperms are plants that produce seeds in cones, like pine cones. They do not produce flowers. There are many species of gymnosperms on Earth today, including pine trees and ginkgo trees.

**4. Compare** How are gymnosperms and angiosperms alike?

## What are angiosperms?

Angiosperms (AN jee uh spurmz) are flowering plants that evolved during the Cretaceous Period. Angiosperms produce seeds with hard outer coverings. Because their seeds are covered and protected, angiosperms can survive in many environments. Angiosperms are the most diverse and abundant plants on Earth today. Today's magnolia trees and oak trees first evolved during the Mesozoic Era.

## What ended the Mesozoic Era?

The Mesozoic Era ended about 65 million years ago, when most land and ocean species became extinct. The dinosaurs disappeared. Many paleontologists hypothesize that this mass extinction was caused by an asteroid that collided with Earth. The impact put a huge cloud of dust and smoke in the air, blocking sunlight. Without sunlight, plants died. As a result, the animals that fed on plants died. Some organisms managed to survive. They are the ancestors of the many species on Earth today.

# The Cenozoic Era

The **Cenozoic** (se nuh ZOH ihk) **Era** began about 65 million years ago and continues today. During this time, mountain ranges in North and South America formed. In the late Cenozoic, the climate cooled and ice ages occurred. The early Cenozoic Era is called the Tertiary Period. The present time is part of the Quaternary Period, which began about 1.8 million years ago. ☑

**Reading Check**

**5. Identify** What geologic era are we living in today?

_______________

## Which mountain ranges formed during this era?

Many mountain ranges formed during the Cenozoic Era as Earth's plates moved and collided. These include the Alps in Europe and the Andes in South America. Many people think the growth of these mountains helped create cooler worldwide climates.

## How have mammals evolved?

During the Cenozoic Era, grasslands expanded. As a result, grazing mammals like horses, deer, and elephants survived and grew larger. Horses evolved from small animals with many toes into the large, hoofed animals of today. Some mammals evolved to live in the sea, such as dolphins and whales.

As the continents continued to move apart, some species became isolated. For this reason, animals like kangaroos and koalas evolved in Australia and are not found anywhere else on Earth.

Modern humans, *Homo sapiens*, probably first appeared about 140,000 years ago. The appearance of early humans may have caused the extinction of many other mammals. Humans competed for food that other animals ate and also hunted animals. ☑

**Reading Check**

**6. Identify** What is the scientific name for the modern humans?

_______________

## After You Read

### Mini Glossary

**Cenozoic Era:** geologic time that began about 65 million years ago and is continuing today

**Mesozoic Era:** geologic time between about 248 and 65 million years ago, that was marked by rapid movement of Earth's plates and was when dinosaurs lived

1. Review the terms and their definitions in the Mini Glossary. Then write one sentence describing one of the eras.

2. Fill in the blanks in the boxes below by listing the plants, animals, and Earth changes for each era.

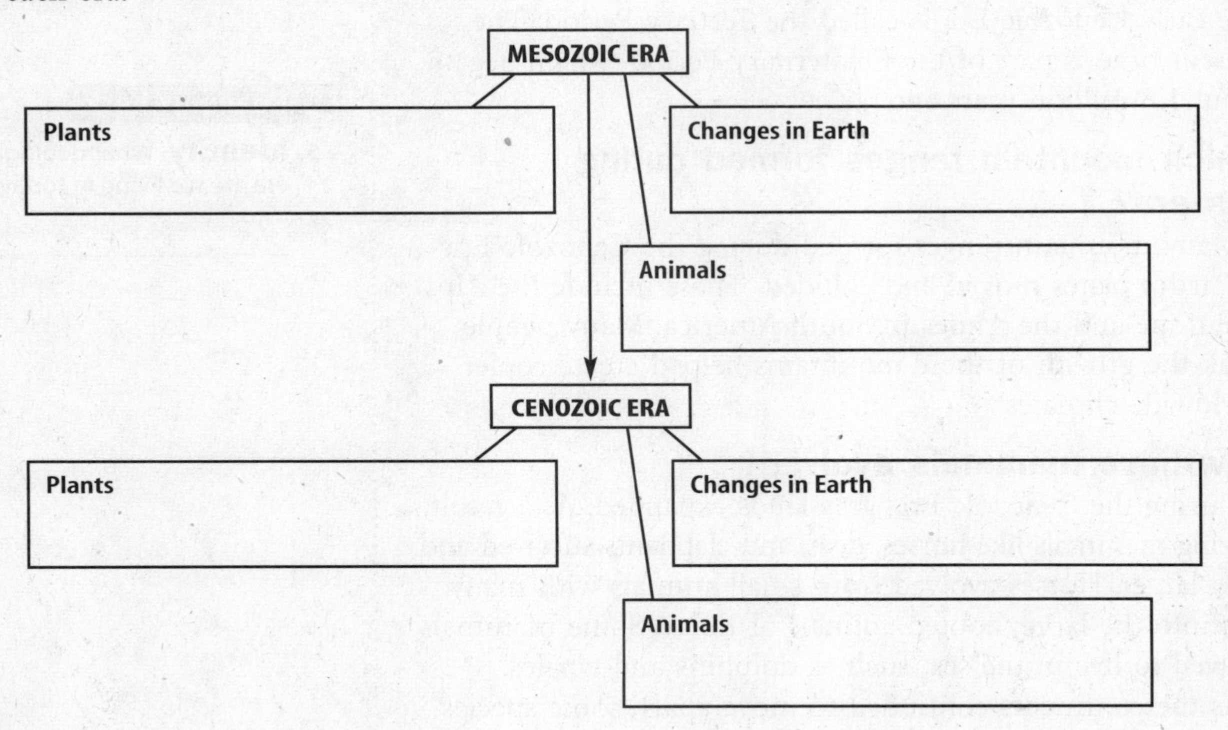


3. You summarized the main idea in each paragraph as you read this section. How did summarizing help you understand and remember the information in the text?

**Science Online** Visit **earth.msscience.com** to access your textbook, interactive games, and projects to help you learn more about the Mesozoic and Cenozoic eras.

# Atmosphere

## section 1 Earth's Atmosphere

### Before You Read

Imagine you are on a spaceship looking down at Earth. Would the view be perfectly clear? What do you think you might see surrounding Earth? Write your thoughts on the lines below.

______________________________________________

______________________________________________

______________________________________________

**What You'll Learn**
- the gases in Earth's atmosphere
- the structure of Earth's atmosphere
- what causes air pressure

### Read to Learn

#### Importance of the Atmosphere

Earth's **atmosphere** is a thin layer of gases, solids, and liquids that surround the planet forming a protective covering. The covering keeps Earth from getting too hot or too cold. The atmosphere keeps Earth from absorbing too much heat from the Sun. It also keeps too much heat from escaping into space. Without protection from the atmosphere, life on Earth could not exist.

#### Makeup of the Atmosphere

Viewed from space Earth's atmosphere today has a thin layer of gases. White clouds usually cover at least half the planet. Between gaps in the clouds, the blue color of the ocean waters shows through.

Earth's early atmosphere was very different from the atmosphere we know today. The early atmosphere was produced by erupting volcanoes. They released nitrogen and carbon dioxide, but little oxygen. Then, about 2 billion years ago, the amount of oxygen in the atmosphere began to increase.

Study Coach

**Flash Cards** Make flash cards to help you learn more about this section. Write the question on one side of the flash card and the answer on the other. Keep quizzing yourself until you know all the answers.

FOLDABLES™

**A Organize** Make a three-tab Foldable to help you learn about gases, solids, and liquids in Earth's atmosphere.

## What caused the atmosphere to change?

Early organisms that lived in the ocean used sunlight to make food. While making food, the organisms released oxygen into the atmosphere.

The oxygen formed a layer of ozone molecules around Earth. The ozone layer protects Earth from the Sun's harmful rays. Over time, this protective ozone layer enabled green plants to grow on land. The green plants released even more oxygen. Today, many living things on Earth, including humans, depend on oxygen to survive.

**Applying Math**

1. **Fractions** Which is the most accurate fraction for the amount of oxygen contained in Earth's atmosphere—about 1/5, about 2/3, about 3/4?

_______________

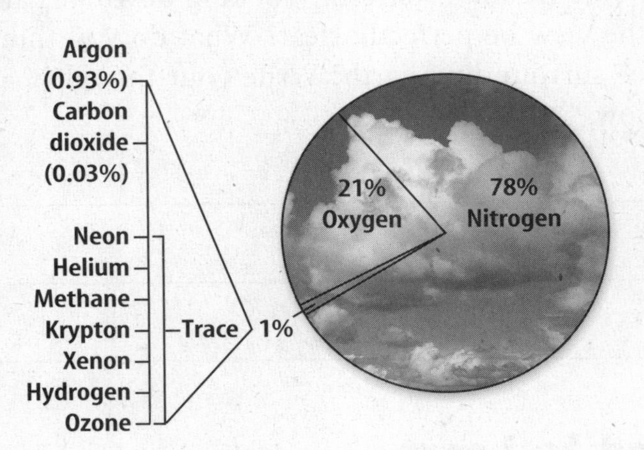


## What gases make up Earth's atmosphere?

As shown in the figure above, 78 percent of Earth's atmosphere is made up of nitrogen. Oxygen makes up 21 percent of the atmosphere. Small amounts of other gases make up the remaining 1 percent.

The composition of the atmosphere is changing in small but important ways. For example, humans burn fuel for energy. As fuel is burned, carbon dioxide is released as a by-product into Earth's atmosphere. Increasing energy use may increase the amount of carbon dioxide in the atmosphere.

## What solids and liquids are in Earth's atmosphere?

In addition to gases, Earth's atmosphere contains small, solid particles such as dust, salt, and pollen. Dust particles get into the atmosphere when wind picks them up off the ground. Salt is picked up from ocean spray. Plants give off pollen that becomes mixed throughout part of the atmosphere. ☑

The atmosphere also contains small liquid droplets, other than water droplets in clouds. The atmosphere moves these liquid droplets and solids from one area to another.

**Reading Check**

2. **List** three solids found in Earth's atmosphere.

_______________

_______________

_______________

## Layers of the Atmosphere

What would happen if a glass of chocolate milk was left untouched on a kitchen counter? Eventually, a lower layer, heavy with chocolate, would separate and fall to the bottom of the glass.

Like a glass of chocolate milk, Earth's atmosphere has layers. There are five layers in the atmosphere, as shown in the figure below. There are two lower layers: the troposphere (TRO puh sfihr) and the stratosphere (STRA tuh sfihr). The three upper layers of the atmosphere are: the mesosphere (MEZ uh sfihr), the thermosphere (THUR muh sfihr), and the exosphere (EK soh sfihr). Most of the air is contained in the troposphere and the stratosphere. ☑

**Reading Check**

3. **Identify** What are the two lower layers of the atmosphere?

______________________

______________________

### What are the lower layers of the atmosphere?

The **troposphere** is the layer of Earth's atmosphere that is closest to the ground. It contains 99 percent of the water vapor and 75 percent of the atmospheric gases. The troposphere is where clouds and weather occur. It extends up to about 10 km from Earth's surface.

The **stratosphere** is the layer of Earth's atmosphere directly above the troposphere. As the figure shows, the ozone layer is found within the stratosphere.

**Layers of the Atmosphere**

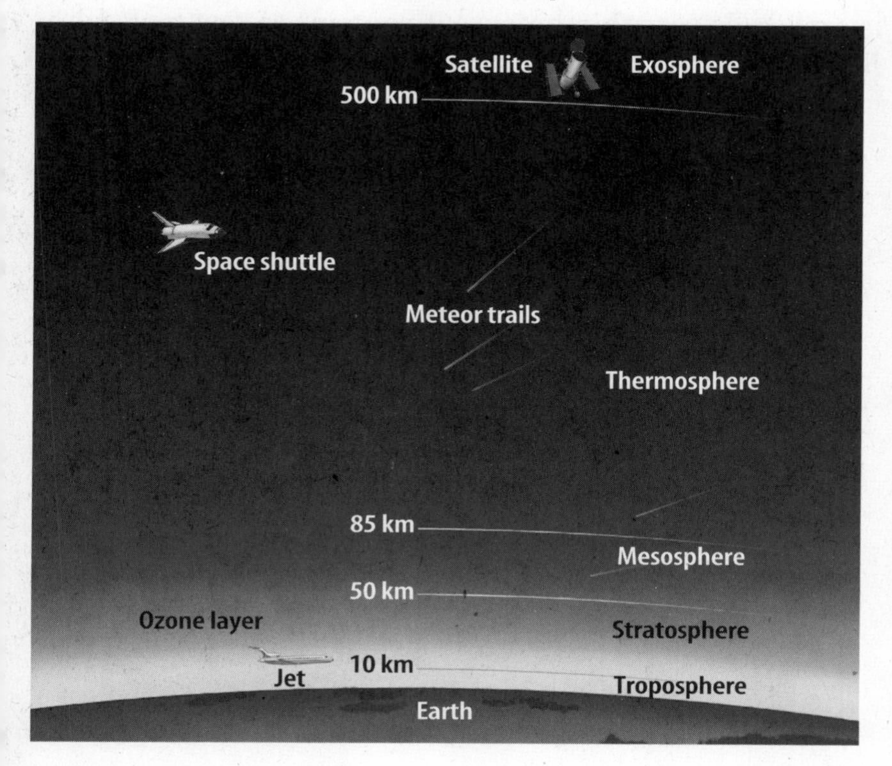


**Picture This**

4. **Interpret** In what layer of the atmosphere do satellites orbit?

______________________

## What are the upper layers of the atmosphere?

The **mesosphere** is the third layer of the atmosphere. It extends from the top of the stratosphere to about 85 km above Earth.

The thermosphere is named for its high temperatures. The **thermosphere** is the fourth layer of the atmosphere and is its hottest and thickest layer. It is found between 85 km and 500 km above Earth's surface. ☑

The ionosphere (I AH nuh sfihr) is within the mesosphere and thermosphere. The **ionosphere** is a layer of electrically charged particles that absorbs AM radio waves during the day and reflects them back at night. Because of this, daytime listeners cannot hear AM radio broadcasts from distant stations. When the ionosphere reflects radio waves at night, listeners can hear the distant stations they could not pick up during the day.

What causes this night and day difference between how far radio waves travel? During the day, energy from the Sun interacts with particles in the ionosphere. The interaction causes them to absorb AM radio frequencies. At night, the Sun's energy is not available and it does not interact with the particles in the ionosphere. This is why radio waves can travel greater distances at night as shown in the figure below.

The **exosphere** is the top layer of the atmosphere. The exosphere is very thin because it contains so few molecules. Beyond the exosphere is outer space.

**✔ Reading Check**

5. **Identify** Which of these is the hottest and thickest layer of the atmosphere: the mesosphere, the thermosphere, or the ionosphere?

_______________________

**Picture This**

6. **Determine** What is reflected by the ionosphere at night but not during the day?

_______________________

**Radio Waves in the Ionosphere**

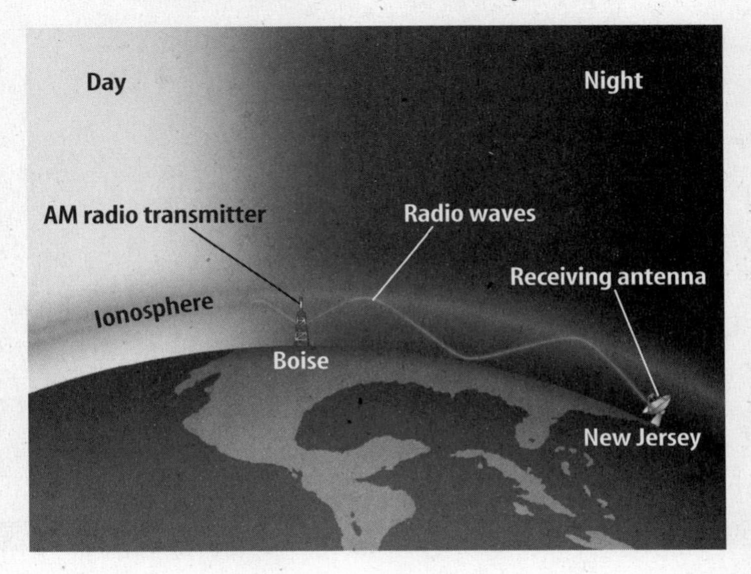

## Atmospheric Pressure

Imagine a football player running with the ball. Suddenly, six other players tackle him. They pile one on top of the other. Who feels the weight more—the player on the bottom holding the ball, or the one on top? The player on the bottom, of course. Why? A great mass of bodies is pressing down on him.

### What is pressure?

The molecules that make up human beings have mass. Atmospheric gases have mass, too. Atmospheric gases extend hundreds of kilometers above Earth's surface. Earth's gravity pulls these gases toward its surface. The weight of these gases presses down on the air below. As a result, the gas molecules nearer Earth's surface are closer together as shown in the figure to the right. This dense air close to the ground exerts more force than the less dense air near the top of the atmosphere. **Pressure** is the force exerted on an area.

**Air Molecules**

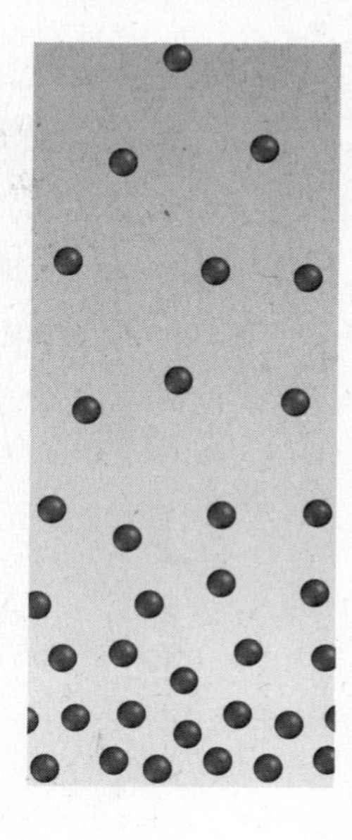

**Picture This**

7. **Interpret** Why do the air molecules at the bottom of the figure exert more pressure than those at the top?

______________________

______________________

______________________

### What affects air pressure?

Air pressure is greater near Earth's surface, where molecules are closer together. Air pressure decreases in air that is further from Earth's surface. In other words, air pressure decreases with altitude as shown in the graph below.

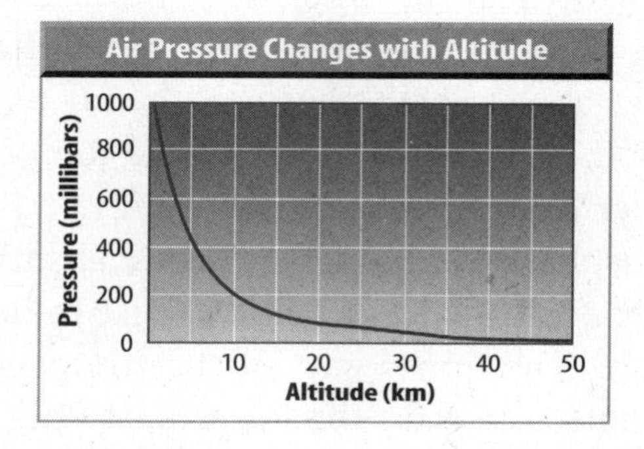


**Picture This**

8. **Estimate** What is the air pressure at an altitude of 5 km?

______________________

Because air pressure decreases with altitude, it is harder to breathe in the mountains than it is at the seashore. Jet airplanes maintain an inside air pressure that matches the air pressure on the ground. If the inside of the plane was not pressurized, people flying high above Earth's surface could not breathe.

## Think it Over

**9. Explain** Why do different layers of Earth's atmosphere have different temperatures?

## Picture This

**10. Interpret** Look at the graph of atmospheric temperatures. Does the temperature in the thermosphere increase or decrease with altitude?

**Reading Check**

**11. Compare** Are the air molecules in the troposphere warmed mainly by the Sun's heat or by the heat from Earth's surface?

## Temperature in Atmospheric Layers

Most of the energy on Earth comes from the Sun. This energy must pass through the atmosphere before it reaches Earth's surface. Some layers of the atmosphere contain gases that easily absorb the Sun's energy. Other layers do not contain these gases. As a result, different atmospheric layers have different temperatures as shown in the graph below.

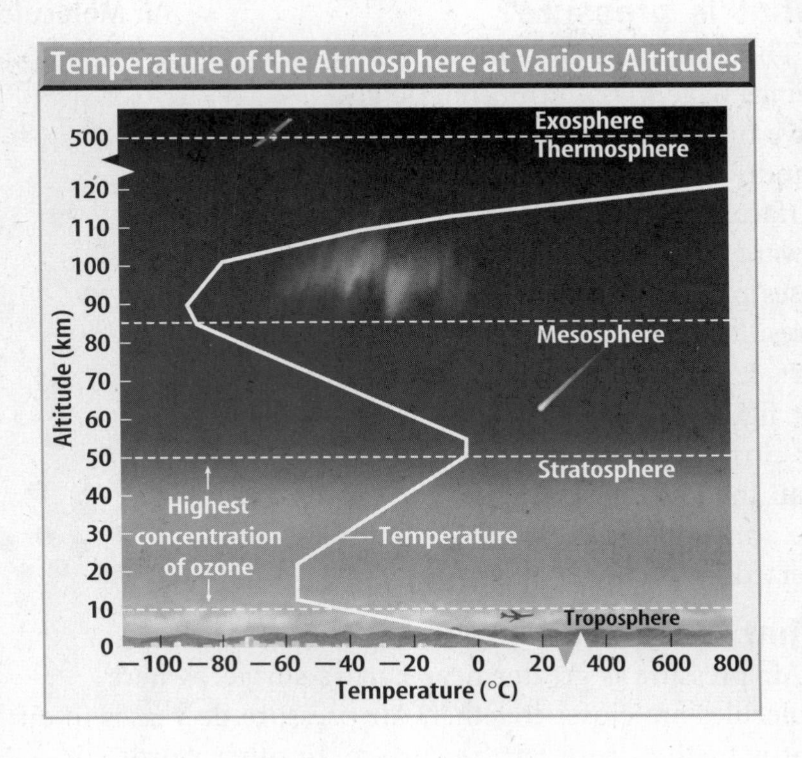


Molecules that make up the air in the troposphere are warmed mostly by heat from Earth's surface. The Sun warms Earth's surface, which then warms the air right above it. For every kilometer above Earth's surface, the air temperature falls about 6.5°C. As a result, the air at the top of a mountain usually is cooler than the air at the bottom. ☑

In the stratosphere, molecules of ozone absorb some of the Sun's energy. Energy absorbed by these molecules raises the temperature. The upper part of the stratosphere has more ozone molecules than the lower part does. Therefore, the temperature in this layer rises with increasing altitude.

Like the troposphere, the temperature in the mesosphere decreases with altitude. The thermosphere and the exosphere are closest to the Sun. These layers have fewer molecules, but each molecule has a lot of energy. Temperatures here are high.

## The Ozone Layer

The **ozone layer** lies within the stratosphere about 19 km to 48 km above the ground. Ozone is made of oxygen. All life depends on the ozone layer.

The oxygen you breathe has two atoms per molecule. An ozone molecule is made up of three oxygen atoms. The ozone layer contains a high concentration of ozone and shields you from the Sun's harmful energy. How? Ozone absorbs most of the ultraviolet radiation that enters the atmosphere. **Ultraviolet radiation** is a type of energy that comes to Earth from the Sun. Too much exposure to ultraviolet radiation can damage your skin and cause cancer. ☑

### What are CFCs?

**Chlorofluorocarbons** (CFCs) are chemical compounds used in some refrigerators, air conditioners, aerosol sprays, and production of foam packaging. Evidence exists that CFCs are one type of air pollutant destroying Earth's protective ozone layer.

CFCs can enter the atmosphere in different ways. CFCs can leak from appliances. Sometimes CFCs escape when products containing them are not disposed of properly. ☑

Molecules from CFCs can break apart ozone molecules. Each ozone molecule is made up of three oxygen atoms bonded together. Each CFC molecule has three chlorine atoms. When a chlorine atom from a CFC molecule comes near a molecule of ozone, the ozone molecule breaks apart. One atom of chlorine can destroy about 100,000 ozone molecules. As a result, more harmful ultraviolet rays reach Earth's surface.

### What is the ozone hole?

The destruction of ozone molecules by CFCs seems to cause a seasonal reduction in ozone over Antarctica called the ozone hole. Every year beginning in late August or early September the amount of ozone in the atmosphere over Antarctica begins to decrease. By October, the ozone concentration is at its lowest point. Then it begins to increase again. By December, the ozone hole disappears. ☑

In the mid-1990s, many governments banned the production and use of CFCs. As a result, there are fewer CFC molecules in the atmosphere today.

**Reading Check**

12. **Identify** What does ozone absorb?

______________________

______________________

**Reading Check**

13. **Explain** Name one way CFCs can enter the atmosphere.

______________________

______________________

**Reading Check**

14. **Recall** Where is the ozone hole located?

______________________

______________________

## After You Read

### Mini Glossary

**atmosphere:** Earth's air, which is made up of a thin layer of gases, solids, and liquids; forms a protective layer around the planet and is divided into five distinct layers

**chlorofluorocarbons (CFCs):** group of chemical compounds used in refrigerators, air conditioners, production of foam packaging, and aerosol sprays that may enter the atmosphere and destroy ozone

**exosphere (EK soh sfihr):** top layer of the atmosphere

**ionosphere(I AH nuh sfihr):** layer of electrically charged particles in the thermosphere that absorbs AM radio waves during the day and reflects them back at night

**mesosphere (MEZ uh sfihr):** third layer of the atmosphere that extends from the top of the stratosphere to about 85 km above Earth

**ozone layer:** layer of the stratosphere with a high concentration of ozone; absorbs most of the Sun's harmful ultraviolet radiation

**pressure:** the force exerted on a surface

**stratosphere (STRA tuh sfihr):** layer of Earth's atmosphere directly above the troposphere

**thermosphere (THUR muh sfihr):** fourth layer of Earth's atmosphere, and its thickest layer; has high temperatures

**troposphere (TRO puh sfihr):** layer of Earth's atmosphere that is closest to the ground

**ultraviolet radiation:** type of energy that comes to Earth from the Sun

**1.** Review the terms and definitions in the Mini Glossary. Write a sentence that explains why the ozone layer is important.

___

___

**2.** List the layers of the atmosphere in order. Begin with the top layer and end with the layer closest to Earth's surface.

1. ___
2. ___
3. ___
4. ___
5. ___

Visit **earth.msscience.com** to access your textbook, interactive games, and projects to help you learn more about Earth's atmosphere.

# Atmosphere

## section 2 Energy Transfer in the Atmosphere

## Before You Read

Imagine you are outside on a warm, sunny day. On the lines below, describe how the Sun feels on your skin.

______________________________________________

______________________________________________

______________________________________________

**What You'll Learn**
- what happens to the Sun's energy on Earth
- about radiation, conduction, and convection
- what the water cycle is

## Read to Learn

### Energy from the Sun

The Sun provides most of Earth's energy. The energy from the Sun drives winds and ocean currents. It allows plants to grow and produce food. Plants, in turn, serve as food for the animals that eat them. Three different things can happen to the energy Earth gets from the Sun. Some of the energy is reflected back into space by clouds, particles, and Earth's surface. Some of it is absorbed by the atmosphere. Some of the energy is absorbed by land and water on Earth's surface. The figure shows how much solar radiation is reflected and absorbed at Earth's surface.

6% reflected by the atmosphere
25% reflected from clouds
4% reflected from Earth's surface
15% absorbed by the atmosphere
50% directly or indirectly absorbed by Earth's surface

**Mark the Text**

**Identify the Main Point** Highlight the main point in each paragraph. Highlight in a different color a detail or example that helps explain the main point.

**Picture This**

1. **Interpret** What happens to most of the Sun's energy?

______________________

______________________

______________________

FOLDABLES™

**B Illustrate and Label** Make a three-tab Foldable to illustrate and describe radiation, conduction, and convection.

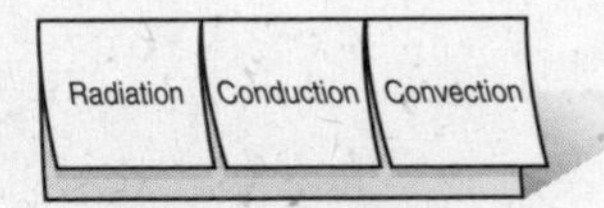


## Heat

Heat is energy that flows from an object with a higher temperature to an object with a lower temperature. Energy from the Sun reaches Earth's surface and heats it. This heat is sent, or transferred, through the atmosphere. Heat is transferred by radiation, conduction, and convection.

### What is radiation?

When the sun is out, you can feel it warming your face. It warms you even though you are not in direct contact with it. How is this possible?

Energy from the Sun reaches Earth in the form of radiant energy, or radiation. **Radiation** is energy that is transferred in the form of rays or waves. Earth radiates, or sends, some of the energy it absorbs from the Sun back toward space. Radiant energy from the Sun warms your face.

### What is conduction?

Walking barefoot on a hot beach will heat up your feet. Heat is transferred to your feet because of conduction. **Conduction** is the transfer of energy that occurs when molecules bump into each other. Molecules are always in motion. But molecules in warmer objects move faster than molecules in cooler objects. When objects are in contact, energy is transferred from warmer objects, like hot sand, to cooler objects, like your feet.

Radiation from the Sun heated the sand. But direct contact with the sand warmed your feet. In the same way, Earth's surface conducts, or transfers, energy directly to the atmosphere. When air moves over warm land or water, molecules in air are heated by direct contact.

### What is convection?

Convection is another way heat is transferred. It occurs after the atmosphere is warmed by radiation or conduction. **Convection** is the transfer of heat by the flow of material. Convection circulates, or moves, heat throughout the atmosphere. How does this happen?

When air is warmed, the molecules in it move apart. The air becomes less dense. Air pressure decreases because fewer molecules are in the same space. In cold air, molecules move closer together. The air becomes more dense and air pressure increases. Cooler, denser air sinks while warmer, less dense air rises. This forms a convection current.

**Think it Over**

2. **Explain** How do different air temperatures form a convection current?

______________________

______________________

______________________

## The Water Cycle

**Hydrosphere** is a term that describes all the water on Earth's surface. As shown in the figure below, there is a constant cycling of water between the hydrosphere and the atmosphere. This constant exchange of water helps to determine Earth's weather patterns and climate types.

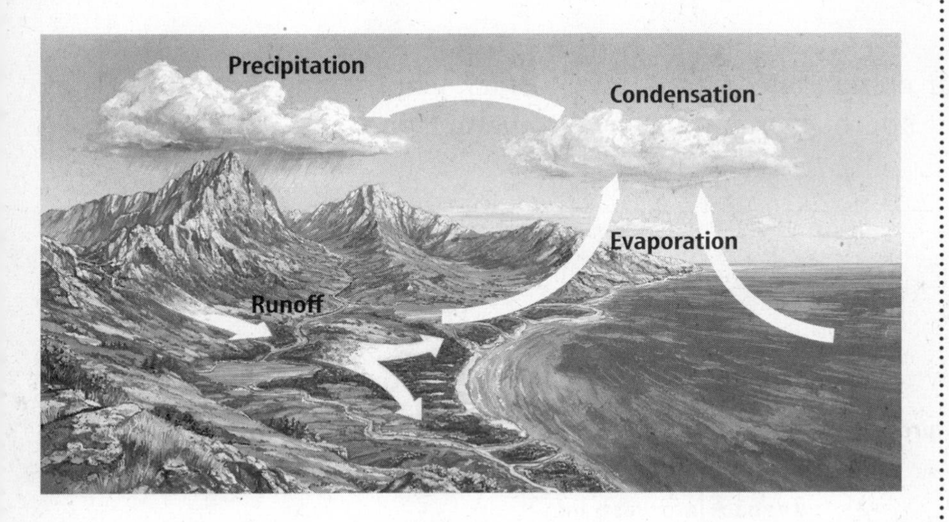


Here's how the water cycle works. Energy from the Sun causes water from lakes, streams, and oceans to change from a liquid to a gas by a process called evaporation. The gas, or water vapor, enters Earth's atmosphere. If the water vapor cools enough, it turns back into water. **Condensation** is the process that occurs when water vapor cools and changes back into a liquid.

Clouds form when condensation occurs high in the atmosphere. Clouds are made up of tiny droplets of water. As these tiny droplets run into each other, they form larger drops. As the water drops grow, they become too large to be held in the air. The drops fall to earth as precipitation, or rain, snow, or sleet. This completes the cycle of returning water to the hydrosphere. ☑

## Earth's Atmosphere is Unique

Why doesn't life exist on Mars or Venus? The atmosphere on Mars is too thin to hold much of the Sun's heat. As a result, it is too cold on Mars for living things to survive. On the other hand, Venus is too hot to support life. Gases in Venus's dense atmosphere trap far too much heat from the Sun. Earth is neither too hot nor too cold. Its atmosphere holds just the right amount of the Sun's energy to support life.

### Picture This

3. **Determine** Circle the process which occurs when water falls as rain, snow, or sleet. What is the process?

______________________

4. **Identify** What process forms clouds?

______________________

## After You Read

### Mini Glossary

**condensation:** process in which water vapor changes to a liquid

**conduction:** transfer of energy that occurs when molecules bump into each other

**convection:** transfer of heat by the flow of material

**hydrosphere:** all the water on Earth's surface

**radiation:** energy transferred by waves or rays

1. Review the terms and their definitions in the Mini Glossary. Choose one term and use it in a sentence telling how it affects energy transfer in the atmosphere.

_______________________________________________

_______________________________________________

2. Fill in the cycle map below to show the water cycle. Use these words to help you.

**Water vapor** **precipitation** **condensation** **clouds** **energy**

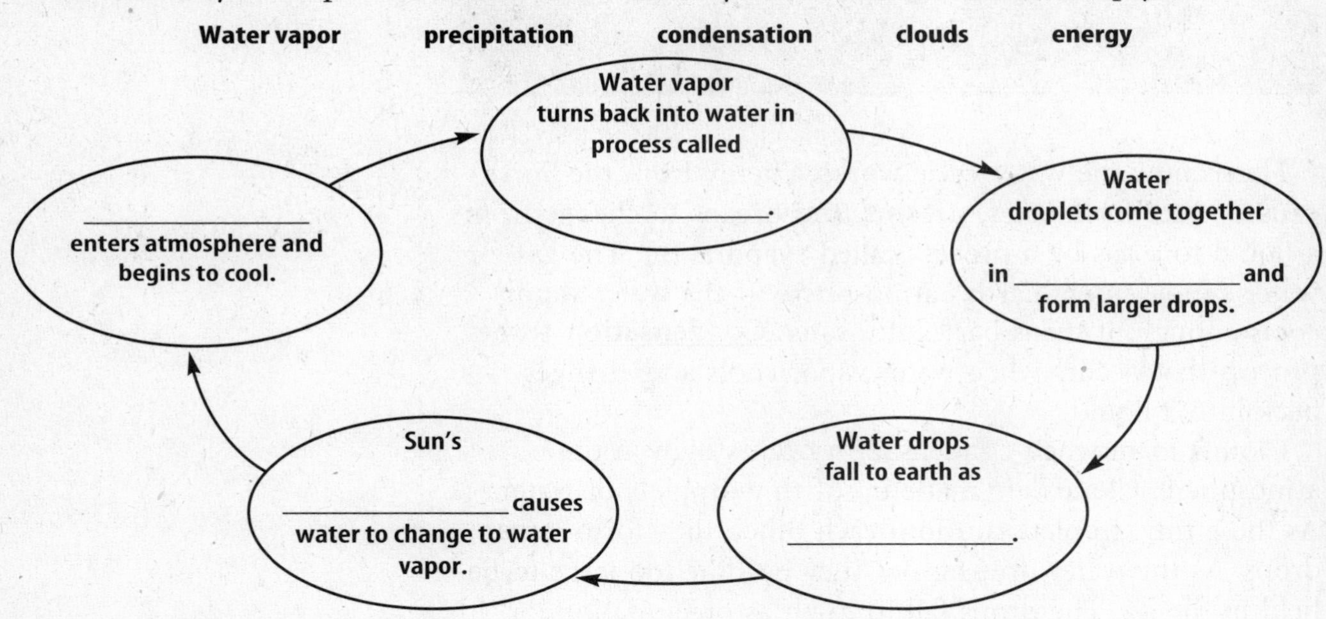


3. Think about what you have learned. How would highlighting the main points and details help you to study for a test?

_______________________________________________

_______________________________________________

**Science Online** Visit **earth.msscience.com** to access your textbook, interactive games, and projects to help you learn more about energy transfer in the atmosphere.

# Atmosphere

## section 3 Air Movement

## Before You Read

When you think of the word *wind* what comes to mind? Brainstorm some words and write them on the lines below.

___________________________________________

___________________________________________

### What You'll Learn

- why different altitudes receive different amounts of solar energy
- about Coriolis effect
- how air is affected by land and water surfaces below it

## Read to Learn

### Forming Wind

Earth is mostly rock or land. Three-fourths of Earth's surface is covered by the oceans. These two areas strongly affect wind systems all over Earth.

Because the Sun heats Earth unevenly, some areas are warmer than others. Remember that warmer air expands and becomes less dense than cold air. As a result, air pressure is lower in areas where air is heated. **Wind** is the movement of air from an area of higher pressure to an area of lower pressure.

### What is heated air?

Different areas of Earth receive different amounts of radiation from the Sun. Why? Because Earth's surface is curved. The equator receives more radiation than areas north or south of it. The Sun's rays hit the equator more directly. Because air at the equator is warm, it is less dense. So it is displaced, or moved, by denser, colder air. Remember that when cooler, denser air sinks while warmer, less dense air rises, a convection current forms. The cold, dense air comes from the poles. They receive less radiation from the Sun, because its rays strike the poles at an angle, spreading out the energy. The resulting dense, high-pressure air sinks and moves along Earth's surface. However, there is more to wind than dense air sinking and less dense air rising.

Study Coach

**State the Main Ideas** As you read this section, stop after each paragraph and put what you have just read into your own words.

FOLDABLES™

**C Classify** Make a three-column Foldable to help you understand the main causes of air movement.

Convection currents
Coriolis effect
Global winds
Polar jet stream
Land breeze
Sea breeze

## What is the Coriolis effect?

What would happen if you threw a ball to a person sitting across from you on a moving merry-go-round? By the time the ball got to the opposite side, the other person would have moved and the ball would appear to have curved.

Like the merry-go-round, the rotation of Earth causes the Coriolis (kohr ee OH lus) effect. The **Coriolis effect** causes moving air and water to appear to turn to their left in the southern hemisphere (south of the equator) and to turn to the right in the northern hemisphere (north of the equator) due to Earth's rotation. This effect is illustrated in the figure below. The flow of air caused by the Coriolis effect and by differences in the amount of solar radiation received on Earth's surface creates wind patterns on Earth's surface. These wind patterns influence the weather. ☑

**Reading Check**

1. **Determine** What causes moving air and water to appear to turn one way in the southern hemisphere and the opposite way in the northern hemisphere?

______________________

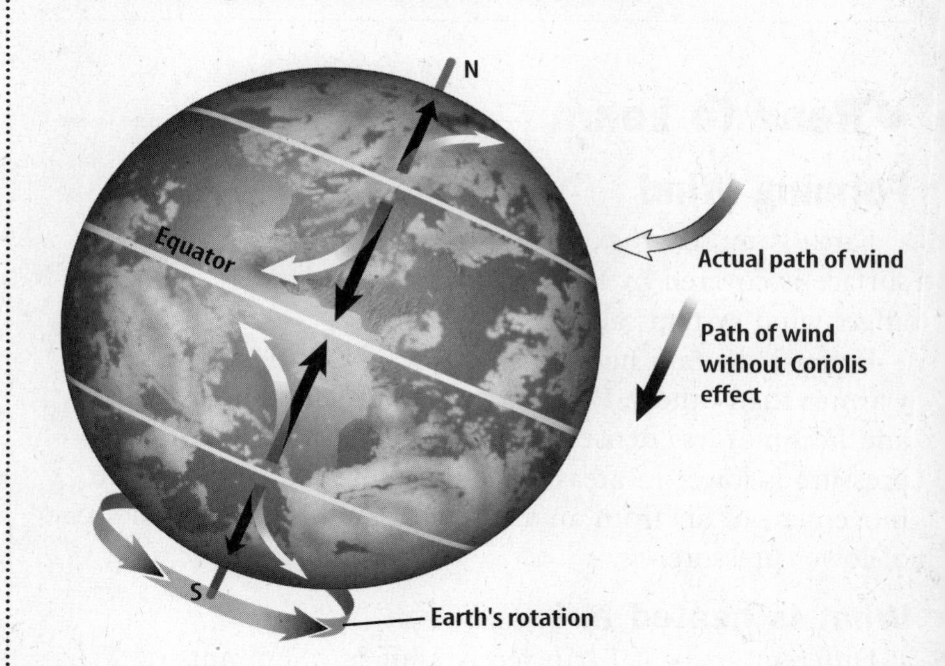


**Picture This**

2. **Explain** Do winds turn to their left or their right north of the equator?

______________________

## Global Winds

How did Christopher Columbus get from Spain to the Americas? The *Nina*, the *Pinta*, and the *Santa Maria* had no source of power other than the wind in their sails.

Early sailors used wind patterns to help them navigate the oceans. Near the equator, there sometimes was little or no wind to fill the sails of their ships. It also rained nearly every afternoon. Why? Because air near the equator has been heated by the Sun. Warm air rises, creating low pressure and little wind. The rising air then cools and causes rain. This windless, rainy zone near the equator is called the doldrums. ☑

**Reading Check**

3. **Identify** What is the name of the windless, rainy zone near the equator?

______________________

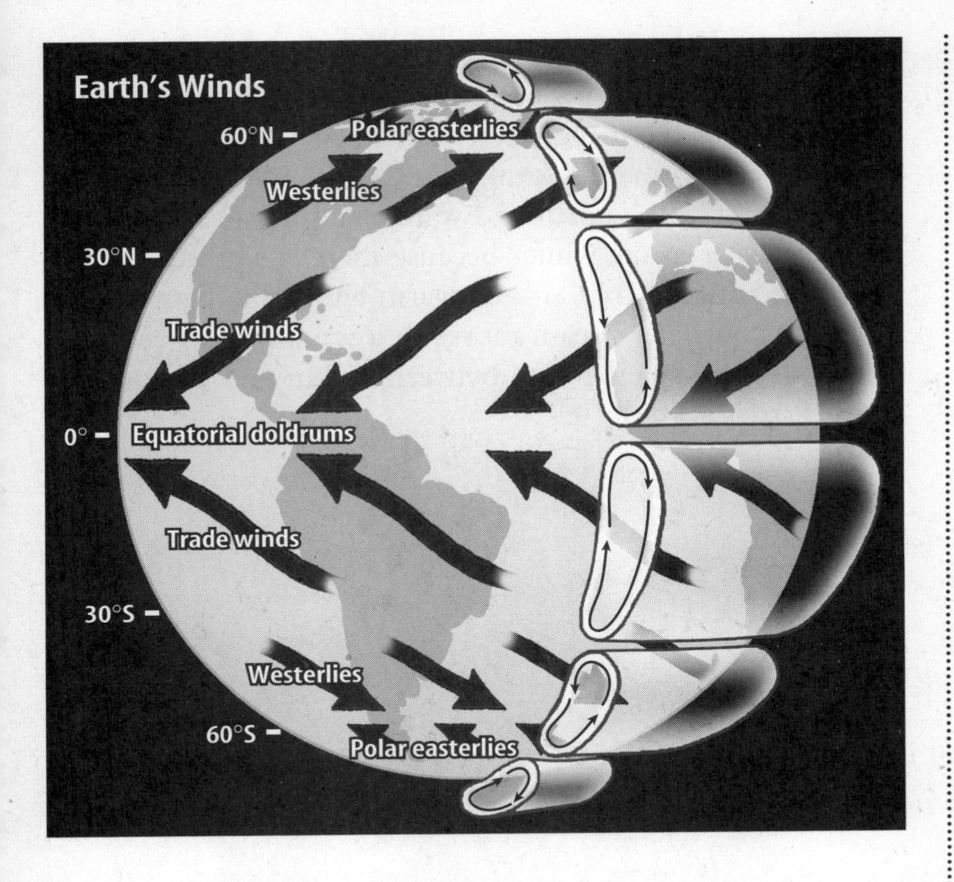


## What winds blow near Earth's surface?

The figure above shows some of the winds that blow near Earth's surface. These prevailing winds move heat and moisture around Earth.

**Trade Winds** Air descending to Earth's surface near 30° north latitude and 30° south latitude creates steady winds. These winds blow in tropical regions. Early sailors liked them because they moved their ships along quickly. Sailors named them trade winds because they relied on these winds to help them sail to many places to trade goods. ☑

**Prevailing westerlies** Between 30° latitude and 60° latitude in the northern and southern hemispheres, winds called the prevailing westerlies blow. These winds blow in the opposite direction from the trade winds. Prevailing westerlies cause much of the movement of weather across North America.

**Polar easterlies** Another surface wind, polar easterlies, are found near the poles. Near the north pole, easterlies blow from northeast to southwest. Near the south pole, polar easterlies blow from the southeast to the northwest.

### Picture This

4. **Identify** Which winds are located on either side of the Equatorial doldrums?

______________________

**Reading Check**

5. **Explain** Why would sailors like trade winds?

______________________

______________________

______________________

## What winds are in the upper troposphere?

**Jet streams** are narrow bands of strong winds that blow near the top of the troposphere. The polar jet stream affecting North America forms along a boundary where colder air lies to the north and warmer air lies to the south. It moves faster in the winter because there is a greater difference between cold air and warm air. As the figure below shows, the polar jet stream moves in a wavy west-to-east direction. It is usually found between 10 km and 15 km above Earth's surface.

**Reading Check**

**6. Define** What are the narrow bands of strong winds that blow near the top of the troposphere?

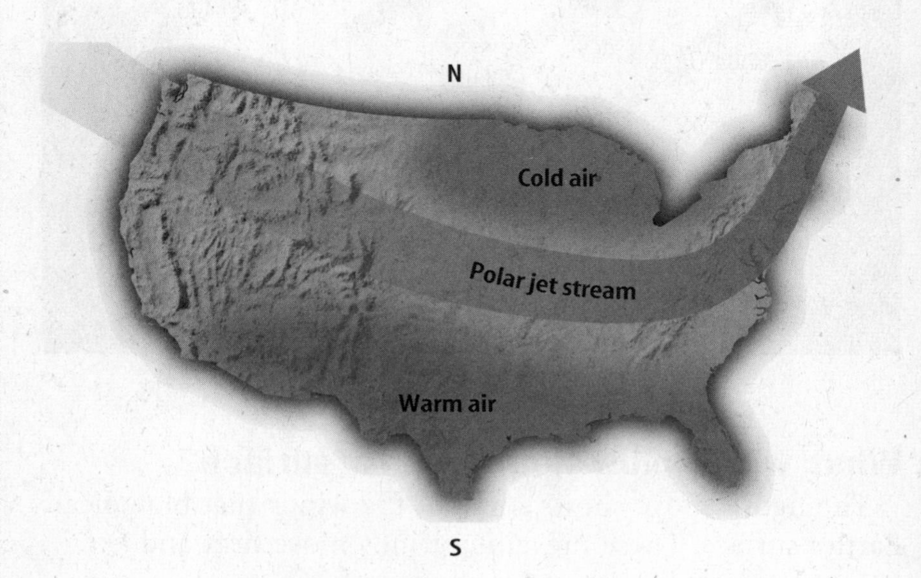


**Picture This**

**7. Determine** Trace with your pencil the direction of the polar jet stream. Is it moving east to west or west to east?

## What are the effects of the jet stream?

The jet stream helps move storms across the country from the west to the east. Jet pilots use information about jet streams to help them fly. When flying to the east, planes save time and fuel. Going west, planes avoid the jet stream by flying at a different altitude. Flying from Boston to Seattle may take 30 minutes longer than flying from Seattle to Boston.

**Think it Over**

**8. Infer** Why would it take longer to fly from east to west than from west to east?

## Local Wind Systems

Major weather patterns for the entire planet are determined by global wind systems. Local weather is affected by smaller wind systems. Those who live near large bodies of water experience two such wind systems. They are sea breezes and land breezes.

**Sea Breeze**

**Land Breeze**

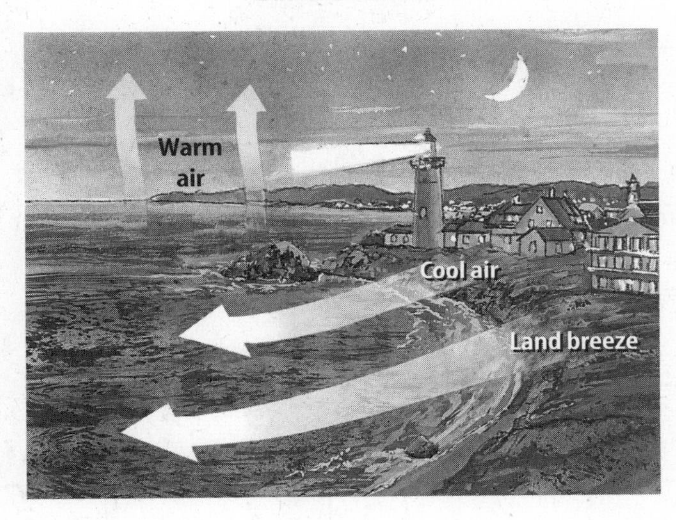


## What causes sea breezes and land breezes?

Convection currents over areas where the land meets the sea can cause wind. During the day, the Sun's heat warms the land more than it warms the water. A **sea breeze** is the movement of air from sea to land during the day. Air over the land is heated by conduction. This heated air is less dense and has lower pressure. Cooler, denser air over the water has higher pressure and flows towards the warmer, less dense air above the land. A convection current results, and wind blows from the sea toward the land. ☑

At night, the land and the air above it cools much faster than ocean water. Cooler, denser air above the land moves over the water, as the warm air over the water rises. The movement of air from land to sea is a **land breeze.**

### Picture This

**9. Interpret** What is happening to the warm air in both figures?

_______________

**Reading Check**

**10. Identify** What causes wind over areas where the land and sea meet?

_______________

## ● After You Read

### Mini Glossary

**Coriolis (kohr ee OH lus) effect:** causes moving air and water to appear to turn left in the southern hemisphere and turn right in the northern hemisphere due to Earth's rotation

**jet streams:** narrow bands of strong winds that blow near the top of the troposphere

**land breeze:** movement of air from land to sea at night, created when cooler, denser air from the land forces warmer air over the sea

**sea breeze:** movement of air from sea to land during the day when cooler air above the water moves over the land forcing the heated, less dense air above the land to rise

**wind:** the movement of air from an area of higher pressure to an area of lower pressure

1. Review the terms and definitions in the Mini Glossary. Then choose one of the definitions and write it in a sentence in your own words.

_______________________________________________

_______________________________________________

2. Fill in the boxes with the correct word, *cooler* or *warmer*, to show what occurs in a sea breeze and a land breeze.

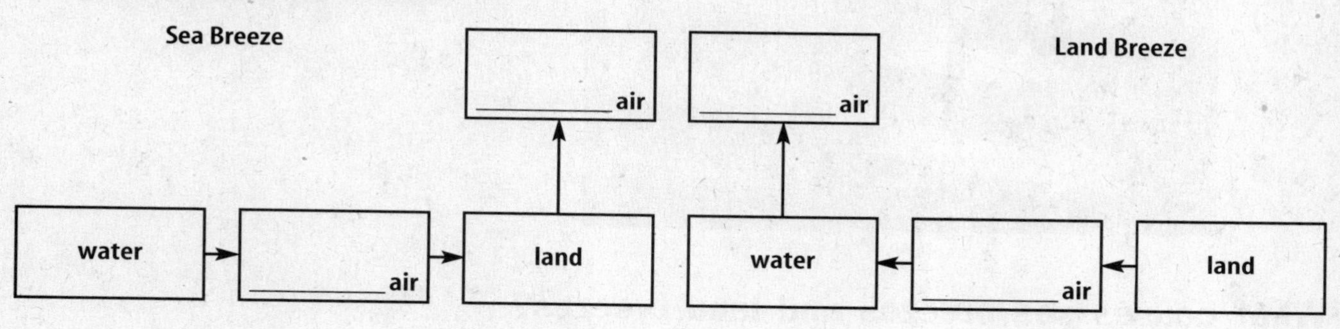


3. Think of Earth's shape. How does the shape of Earth affect the amount of heat different areas receive?

_______________________________________________

_______________________________________________

_______________________________________________

Visit **earth.msscience.com** to access your textbook, interactive games, and projects to help you learn more about air movement.

# Weather

## section 1 What is weather?

## Before You Read

Have you ever flown a kite or watched someone else fly one? On the lines below, describe how the kite moves in the air.

_______________________________________________

_______________________________________________

### What You'll Learn

- how pressure, wind, temperature, and moisture content of air affect weather
- how clouds form and how they are classified
- how rain, hail, sleet and snow develop

## Read to Learn

### Weather Factors

Everybody talks about the weather. It may seem like small talk, but weather is very important to some people. Pilots, truck drivers, farmers, and other professionals study the weather because it can affect their jobs.

Study Coach

**Think-Pair-Share** Work with a partner. As you read this section, discuss what you already know about the topic and what you learn.

### What is weather?

You can look out the window and see that it's raining, or snowing, or windy. But do you really know what weather is? **Weather** is the state of the atmosphere at a specific time and place. Weather describes conditions such as air pressure, wind, temperature, and moisture content in the air.

### How does the Sun affect weather on Earth?

The Sun provides almost all of Earth's energy. Energy from the Sun evaporates water on Earth. Evaporated water enters the atmosphere and forms clouds. Later, the water falls back to Earth as rain or snow.

The Sun also heats Earth. Heat from the Sun is absorbed by Earth's surface, which then heats the air above it. Because of differences in Earth's surface, some places in Earth's atmosphere are warmer and other places are cooler. Air currents and water currents move the heat to different places around Earth. Weather is the result of heat and Earth's air and water.

FOLDABLES™

**A Organize** Use four quarter sheet note cards to record information about the factors that determine weather.

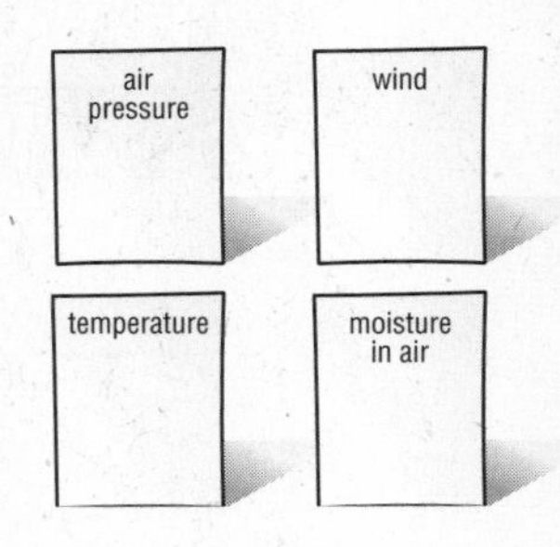

## What affects temperature?

Air is made up of molecules that are always moving randomly, or without any set pattern, even when there is no wind. Temperature is a measure of the average amount of motion of molecules. When the temperature is high, air molecules move rapidly and it feels warm. When the temperature is low, air molecules move more slowly and it feels cold. ☑

**Reading Check**

1. **Determine** When the temperature is high, how do air molecules move?

_______________

## What causes wind?

Have you ever flown a kite? What do you need in order to get the kite off the ground and into the air? Kites fly because air is moving. Air that moves in one direction is called wind. The Sun heats Earth unevenly, but wind helps spread the heat around.

As the Sun warms the air, the air expands and becomes less dense. Warm, expanding air has low atmospheric pressure. Cooler air is denser and sinks, which brings high atmospheric pressure. Wind is the result of air moving from areas of high pressure to areas of low pressure.

The temperature of air can affect air pressure. When air is cooler, molecules are closer together, creating high pressure. When air is heated, it expands and becomes less dense. This creates lower pressure. Beaches are often windy as a result of air moving from areas of high pressure to areas of lower pressure, as shown in the figure below.

**Picture This**

2. **Label** one side of the figure *high pressure* and one side *low pressure.*

_______________ pressure _______________ pressure

## What tools are used to measure wind?

Some instruments measure wind direction and others measure wind speed. A wind vane, sometimes seen on houses or barns, has an arrow that points in the direction from which the wind is blowing. A wind sock, another tool that shows wind direction, has an open end to catch the wind. The wind sock fills and points in the direction toward which the wind is blowing. ☑

An anemometer (a nuh MAH muh tur) is an instrument that measures wind speed. Anemometers have four open cups that catch the wind and cause the anemometer to spin. The faster the wind blows, the faster the anemometer spins.

**Reading Check**

**3. Explain** Name one tool for measuring wind direction and tell how it works.

______________________

______________________

______________________

## What is humidity?

Heat evaporates water into the atmosphere. Where does the water go? Water vapor molecules fit into spaces among the molecules that make up air. The amount of water vapor held in the air is called **humidity.**

Air does not always hold the same amount of water vapor. More water vapor can be present when the air is warm than when it its cool. At warm temperatures, the molecules of water vapor in the air move quickly. As a result, the molecules do not come together easily, as shown on the left in the figure below.

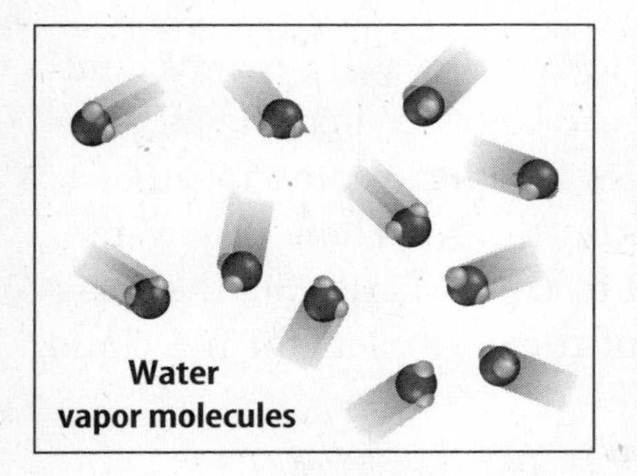


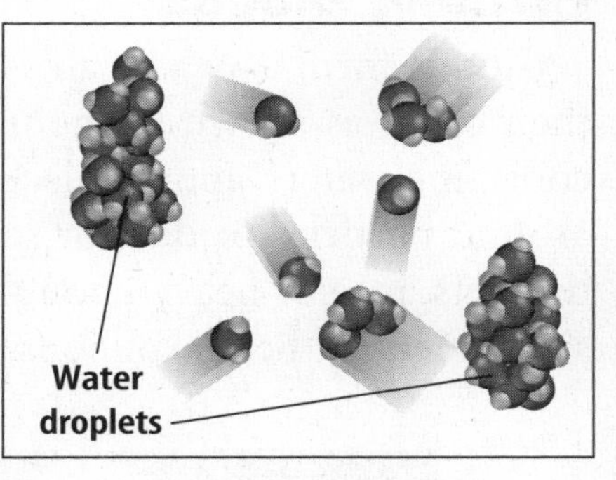


**Picture This**

**4. Determine** Circle the figure that shows droplets of water forming.

At cooler temperatures, the molecules in air move more slowly. This slower movement allows the water vapor molecules to stick together. Droplets of liquid water form, as shown on the right in the figure above. This process of liquid water forming from water vapor is called condensation. If enough water is present in the air for condensation to take place, the air is saturated.

## What is relative humidity?

Weather forecasters report the amount of moisture in the air as relative humidity. **Relative humidity** is a measure of the amount of moisture held in the air compared with the amount of moisture the air can hold at a given temperature. If the weather forecaster says that the relative humidity is 50 percent, this means that the air contains 50 percent of the water needed for the air to be saturated at that temperature.

## Dew Point

When the temperature drops, less water vapor can be present in the air. If temperatures are low enough, water vapor will condense to a liquid or form ice crystals. The temperature at which the air is saturated and condensation forms is the **dew point**. Dew point changes as the amount of water vapor in the air changes. ☑

**Reading Check**

5. **Identify** What is the temperature at which condensation forms called?

_______________

You've probably seen water droplets form on the outside of a can of cold soda. The cold can cooled the air around it to its dew point. The water vapor in the air condensed, forming water droplets on the soda can. Something similar occurs when you see dew. Air near the ground cools to its dew point, and then water vapor condenses and forms dew. If temperatures are near 0° C, frost may form.

## Forming Clouds

Clouds form as warm air is forced upward, expands, and then cools, as shown in the figure below. When the air cools, the water vapor molecules in the air come together around particles of dust or salt in the air. These tiny water droplets are not heavy enough to fall to Earth. So, they stay suspended in the air. Billions of these droplets form a cloud.

**Picture This**

6. **Interpret** Trace the arrows that show moist warm air rising.

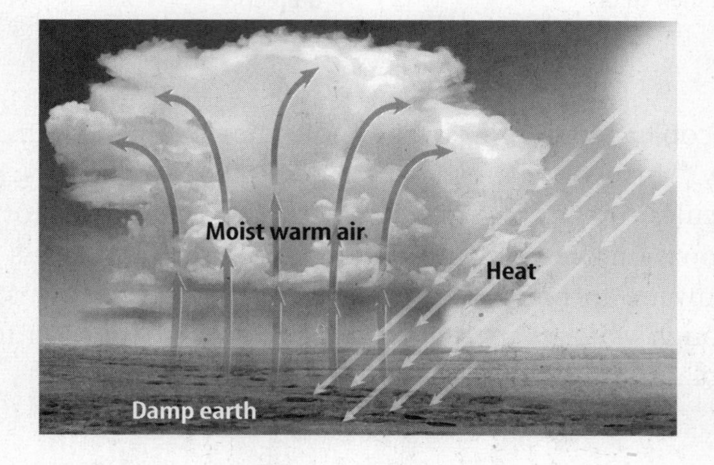

## Classifying Clouds

Clouds are grouped, or classified, by shape and height. Some clouds are tall and rise high into the sky. Some clouds are low and flat. Dense clouds can bring snow or rain. Thin clouds usually appear on sunny days. Three main factors determine the shape and height of clouds—temperature, pressure, and the amount of water vapor in the air.

### What are the different types of clouds?

Stratus clouds are layered in smooth, even sheets across the sky and may be seen on fair, rainy, or snowy days. Usually stratus clouds form low in the sky. **Fog** is a stratus cloud that forms when air is cooled to its dew point near the ground.

Cumulus (KYEW myuh lus) clouds are large, white, puffy clouds that are often flat on the bottom and sometimes tower high into the sky. Cumulus clouds can be seen either in fair weather or in thunderstorms.

Cirrus (SIHR us) clouds are thin, white, feathery clouds. They form high in the atmosphere and are made of ice crystals. Although cirrus clouds are linked with fair weather, they sometimes appear before a storm.

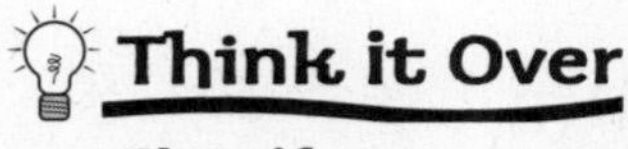


**7. Classify** What are the three main cloud types?

______________________

______________________

______________________

### How is height used to name clouds?

Cloud names are sometimes given prefixes to describe the height of the cloud base. Three common cloud prefixes are *cirro-*, *alto-* and *strato-*. *Cirro-* describes high clouds. *Alto-* is used for clouds that form at middle levels. *Strato-* is used for clouds that form closer to the ground.

Cirrostratus clouds are made of ice crystals and form high in the air. Usually cirrostratus clouds are a sign of fair weather. Sometimes they signal a storm is on the way. Altostratus clouds form at middle levels. If these clouds are not too thick, sunlight can filter through them.

### What types of clouds produce rain and snow?

Dark clouds that contain rain or snow are called nimbus clouds. *Nimbus* is a Latin word meaning "dark rain cloud." The water content of nimbus clouds is so high that only a little sunlight can pass through them.

When a cumulus cloud grows into a thunderstorm, it is called a cumulonimbus (kyew myuh loh NIHM bus) cloud. These high clouds can tower almost 18 km. Nimbostratus clouds are layered clouds that usually bring long, steady rain or snowfall. ☑

**Reading Check**

**8. Determine** When a cumulus cloud becomes a thunderstorm, what is it called?

______________________

**FOLDABLES™**

**B Compare and contrast** Make a four-tab Foldable as shown. As you read, take notes on how the four forms of precipitation are similar and different.

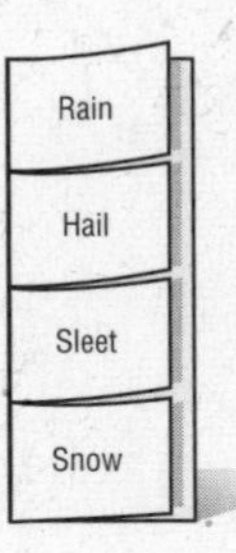


## Precipitation

**Precipitation** is water falling from clouds. Precipitation occurs when cloud droplets combine and grow large enough to fall to Earth. The cloud droplets form around tiny particles like salt and dust in the air.

### Why are some raindrops bigger than others?

You have probably noticed that some raindrops are bigger than others. One reason for this size difference is the strength of updrafts in a cloud. If strong updrafts of wind keep drops in the air longer, they can combine with other drops. As a result, they grow larger.

Another factor which affects raindrop size is the rate of evaporation as the drop falls to Earth. If the air is dry, the raindrop will get smaller as it falls. Sometimes the raindrop will evaporate completely before it even hits the ground.

### How does temperature affect precipitation?

Air temperature determines what kind of precipitation will fall—rain, snow, sleet, or hail. How air temperature affects precipitation is shown in the figures below. When the air temperature is above freezing, water falls as rain. If the air temperature is so cold that water vapor changes to a solid, it snows. Sleet forms if raindrops fall through a layer of freezing air near Earth's surface, forming ice pellets.

During thunderstorms, hail forms in cumulonimbus clouds. Hailstones form when water freezes around tiny centers of ice. Hailstones get larger as they're tossed up and down by rising and falling air. Most hailstones are small, but sometimes they can get larger than softballs. Of all forms of precipitation, hail causes the most damage.

**Picture This**

**9. Identify** In the figures, circle the name of each type of precipitation.

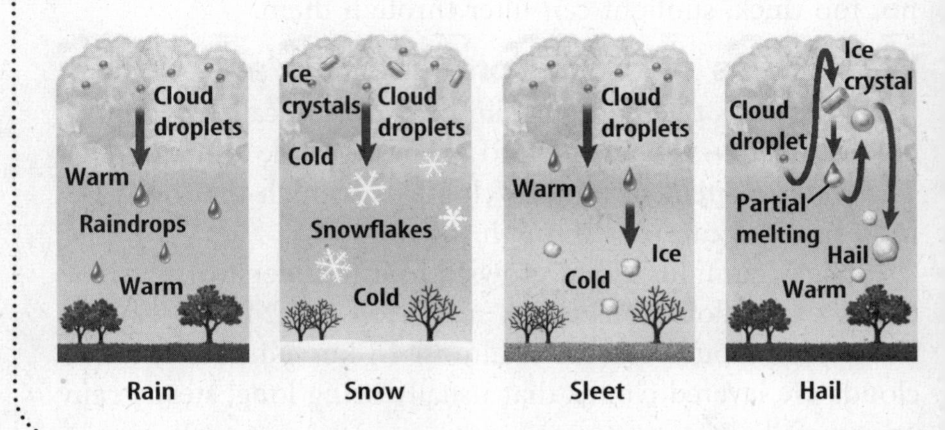

# After You Read

## Mini Glossary

**dew point:** temperature at which air is saturated and condensation forms

**fog:** stratus cloud that forms when air near the ground is cooled to its dew point

**humidity:** amount of water vapor held in the air

**precipitation:** water falling from clouds—including rain, snow, sleet, and hail—whose form is determined by air temperature

**relative humidity:** measure of the amount of moisture held in the air compared with the amount it can hold at a given temperature

**weather:** state of the atmosphere at a specific time and place; determined by air pressure, wind, temperature, and how much moisture is in the air

1. Review the terms and their definitions in the Mini Glossary. Then write one sentence describing today's weather. Use at least two of the terms.

_______________________________________________

_______________________________________________

2. Use these words to fill in the blanks and tell about clouds forming and precipitation: snow, hail, warm moist air, stratus, cumulus, rain, cirrus, sleet, water vapor, clouds

_______________________ **rises, expands, and cools.**

_______________________ **condenses into tiny droplets.**

**Droplets suspend in the air, forming** _______________________.

**Three types of clouds are** _______________, _______________, **and** _______________.

**Four kinds of precipitation come from clouds:** _______________, _______________, _______________, **and** _______________.

3. You were asked to discuss and study this section with a partner. Was this a helpful strategy for learning the information? Why or why not?

_______________________________________________

_______________________________________________

**Science Online** Visit **earth.msscience.com** to access your textbook, interactive games, and projects to help you learn more about weather.

**End of Section**

# Weather

## section 2 Weather Patterns

**What You'll Learn**
- how weather is related to fronts and high and low pressure areas
- about different types of severe weather

## Before You Read

Have you ever gone into a basement or an attic? Describe how the temperature felt compared to the rest of the building.

______________________________

______________________________

______________________________

**Mark the Text**

**Key Terms** Highlight the key terms and their meanings as you read this section.

## Read to Learn

### Weather Changes

Sometimes when you leave school in the afternoon, the weather is different from what it was earlier in the morning. Weather constantly changes.

#### What are air masses?

An **air mass** is a large body of air that has the same temperature and moisture content as the area over which it formed. For example, an air mass that develops over land is drier than one that develops over water. An air mass that develops in the tropics is warmer than one that develops over northern regions. When weather changes from one day to the next, it is because of the movement of air masses.

**FOLDABLES™**

**C Classify** Make a four-tab Foldable as shown. As you read, take notes on the four different fronts.

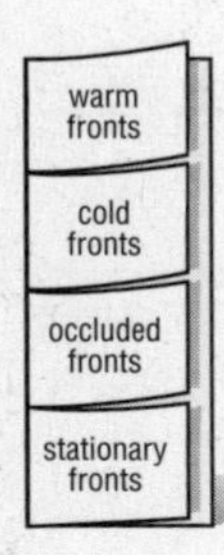


#### How does air pressure affect the weather?

Pressure in the atmosphere varies over Earth's surface. You may have heard a weather forecaster talk about high- and low-pressure systems. Low-pressure systems are masses of rising air. When air rises and cools, clouds form. That's why areas of low pressure usually have cloudy weather. But high-pressure air masses have a sinking motion. As a result, it's hard for air to rise and for clouds to form. So, high pressure usually means nice weather.

## What are cyclones and anticyclones?

Winds blow from areas of high pressure to areas of low pressure. In the northern hemisphere, when wind blows into a low-pressure area, Earth's rotation causes the wind to swirl in a counterclockwise direction. These large, swirling areas of low pressure are called cyclones. Cyclones are associated with stormy weather. ☑

Winds blow away from an area of high pressure. In the northern hemisphere, Earth's rotation causes these winds to swirl in a clockwise direction. High-pressure areas are associated with fair weather and are called anticyclones.

**Reading Check**

1. **Describe** What type of weather are cyclones associated with?

______________________________

## Fronts

A boundary between two air masses that have different temperature, density, or moisture is called a **front**. There are four main types of fronts, including cold, warm, occluded, and stationary.

### What is a cold front?

A cold front occurs when cold air moves toward warm air, as shown on the left in the figure below. The cold air goes under the warm air and lifts it. As the warm air is lifted, it cools and water vapor condenses, forming clouds. If there is a large difference in temperature between the cold air and the warm air, thunderstorms and tornadoes may form.

### What is a warm front?

Warm fronts form when lighter, warmer air moves over heavier, colder air, as shown on the right in the figure below. In a warm front, wet weather may last for days.

**Picture This**

2. **Identify** Color the arrow showing cold air movement in the cold front blue. Color the arrow showing warm air movement in the warm front red.

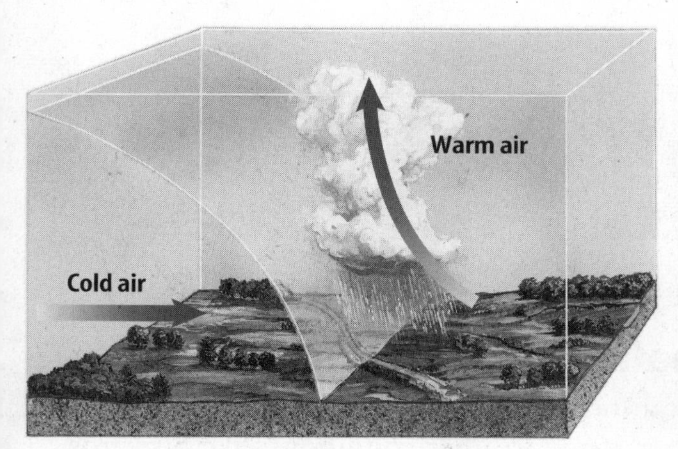


Cold Front

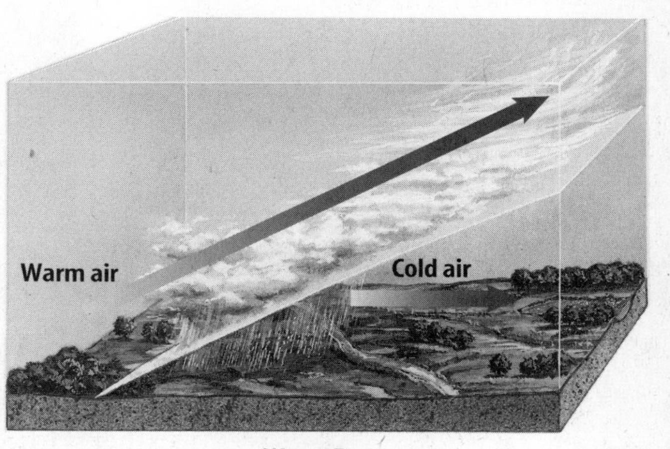


Warm Front

## What is an occluded front?

Most fronts involve two air masses. But occluded fronts involve three air masses—cold air, cool air, and warm air. An occluded front, as shown in the figure below, may form when a cold air mass moves toward cool air with warm air in between. The cold air forces the warm air up. The warm air is then closed off from the surface. The term *occlusion* means "closure."

**Picture This**

3. **Interpret** Color the arrows red that show where the warm air is closed off from the surface in the occluded front.

Occluded Front

## What is a stationary front?

A stationary front occurs when a boundary between air masses stops moving, as shown in the figure below. Stationary fronts can stay in the same place for several days. Often there is light wind and precipitation at the stationary front.

**Picture This**

4. **Identify** Circle the area in the stationary front where neither the cold air nor warm air is moving.

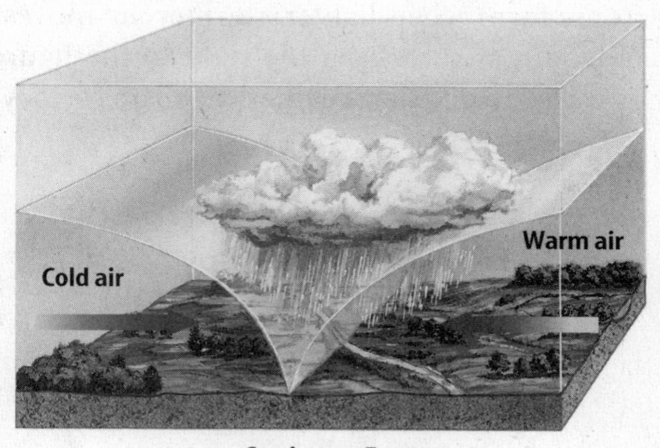


Stationary Front

# Severe Weather

You usually can do your daily activities regardless of the weather. However, some weather conditions, like blizzards, tornadoes, and hurricanes, can force you to change your plans.

## What causes thunderstorms?

During thunderstorms, heavy rain falls, lightning flashes, and thunder rumbles. Hail might fall. What causes these weather conditions?

Thunderstorms occur in warm, moist air masses and along fronts. Warm, moist air is forced up. It cools and condensation begins, forming cumulonimbus clouds. When rising air cools, water vapor condenses into water droplets or ice crystals. Smaller droplets collide and form larger ones. The larger, heavier droplets fall through the cloud toward Earth's surface. The falling droplets collide with more droplets and get bigger. Raindrops cool the air around them. The cool, dense air sinks. Sinking, rain-cooled air and strong updrafts of warmer air cause the strong winds that often come during thunderstorms. Hail may form as ice crystals fall. ☑

## What damage do thunderstorms cause?

Sometimes thunderstorms stall in one area, causing heavy rains. When streams can no longer hold all the water running into them, flash floods occur. Because they occur with little warning, flash floods are dangerous.

Thunderstorms often bring strong winds that can cause damage. If a thunderstorm has winds over 89 km/h, it is called a severe thunderstorm. Hail from thunderstorms can dent cars, break windows, and flatten crops.

## What causes lightning?

Inside a storm cloud, warm air is lifted rapidly as cooler air sinks. This movement of air can cause different parts of a cloud to have opposite charges. When an electrical current runs between areas with opposite charges, lightning flashes. Lightning can occur between two clouds, inside one cloud, or between a cloud and the ground. ☑

## What causes thunder?

Thunder comes from the rapid heating of air around a bolt of lightning. Lightning can reach temperatures of about 30,000° C. That's five times hotter than the surface of the Sun. This heat causes air around the lightning to expand rapidly. Then the air cools quickly and shrinks. Because of the sudden expanding and shrinking, molecules in the air move more rapidly. The rapid movement of molecules creates sound waves. Thunder is the sound waves you hear.

**✔ Reading Check**

5. **Explain** How do water droplets falling out of a thundercloud get bigger as they fall toward Earth's surface?

______________________

______________________

______________________

**✔ Reading Check**

6. **Determine** What causes different parts of a cloud to have opposite charges?

______________________

______________________

## What are tornadoes?

Some severe thunderstorms produce tornadoes. A **tornado** is a violently rotating column of air that touches the ground. Severe thunderstorms produce wind at different heights which blow at different speeds and in different directions. This difference in wind speed and direction is called wind shear. Wind shear creates a rotating column parallel to the ground. Updrafts in a thunderstorm can tilt the rotating column upward, creating a funnel cloud. If the funnel cloud touches the ground, it is called a tornado. ☑

**Reading Check**

**7. Identify** What is a violently rotating column of air that touches the ground called?

_______________

The figure below shows a diagram of a tornado. Notice the different levels of winds and the rotating updraft. The strong updraft usually forms at the base of a type of cumulonimbus cloud called a wall cloud.

**Picture This**

**8. Identify** Find the updraft and trace over it with your pencil.

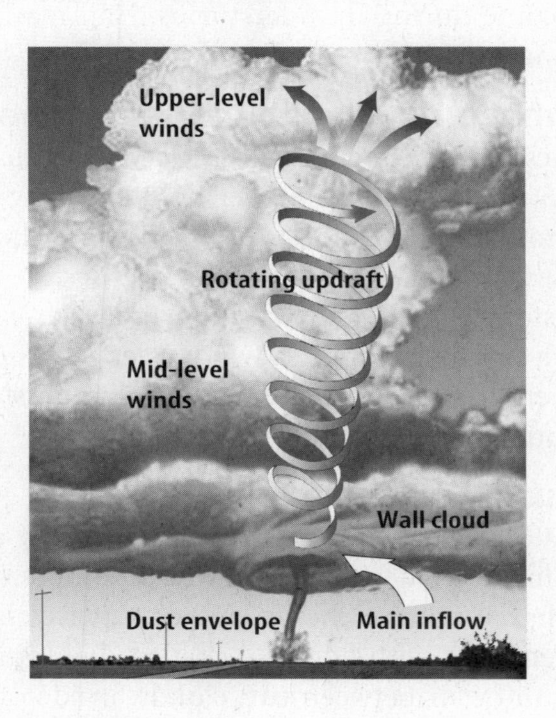


## How much damage can a tornado do?

Winds from tornadoes can rip apart buildings and tear trees from the ground. If the winds of a tornado blow through a house, they can lift off the roof and blow out the walls. It can look as though the building exploded. In the center of a tornado is a powerful updraft. The updraft can lift animals, cars, and even houses into the air. Tornados do not last long, but they are very destructive. In May of 1999, thunderstorms produced more than 70 tornadoes in Kansas, Oklahoma, and Texas. These tornadoes caused 40 deaths, 100 injuries, and more than $1.2 billion in damage.

## How are tornadoes ranked?

As you have read, winds from tornadoes can cause severe damage. Theodore Fujita, a tornado expert, created a scale to describe and rank tornadoes. The scale, named the Fujita Scale after him, is shown below. The Fujita Scale ranks tornadoes based on how much damage they cause. Tornadoes range from F0 which cause only light damage to F5 which cause incredible damage. Luckily, only about one percent of all tornadoes are in the category of F4 and F5.

**The Fujita Scale**

| Rank | Wind speed (km/h) | Damage |
|---|---|---|
| F0 | <116 | Light: broken branches and chimneys |
| F1 | 116–180 | Moderate: roofs damaged, mobile homes upturned |
| F2 | 181–253 | Considerable: roofs torn off homes, large trees uprooted |
| F3 | 254–332 | Severe: trains overturned, roofs and walls torn off |
| F4 | 333–419 | Devastating: houses completely destroyed, cars picked up and carried elsewhere |
| F5 | 420–512 | Incredible: total demolition |

**Picture This**

**9. Determine** Circle the category that describes severe damage.

## What is a hurricane?

The most powerful storm is a hurricane. A **hurricane** is a large, low-pressure system that forms over the warm Atlantic Ocean and has winds of at least 119 km/h. It is like a machine that turns heat energy from the ocean into wind. Similar storms are called typhoons in the Pacific Ocean and cyclones in the Indian Ocean. ☑

**Reading Check**

**10. Identify** What are two storms similar to hurricanes?

____________________

____________________

Hurricanes are similar to low-pressure systems over land—only stronger. In the Atlantic and Pacific Oceans, low-pressure systems sometimes develop near the equator. In the northern hemisphere, winds around this low pressure rotate counterclockwise. As the storms move across the ocean, they gain strength from the heat and moisture of warm ocean water.

## What happens when a hurricane reaches land?

Hurricanes can strike land with great force. The high winds sometimes produce tornadoes. Heavy rains and high waves cause large amounts of damage. Sometimes floods follow the heavy rains and cause additional damage. Hurricanes can destroy crops, tear down buildings, and kill humans and animals.

## What happens to the hurricane on land?

As long as the hurricane remains over water, it gets energy from the warm moist air rising from the ocean. In the figure below, small rising arrows show the movement of warm air from the water below. Cool air goes down through the eye, or center, of the hurricane. The storm needs this energy from the ocean water. When a hurricane reaches land, it loses its energy supply and the storm loses its power.

**Picture This**

**11. Identify** Highlight all the arrows moving counterclockwise.



## What is a blizzard?

Severe storms also can occur in the winter. If you live in the northern United States, you may have experienced the howling wind and blowing snow of a blizzard. A **blizzard** is a winter storm with conditions that include very cold temperatures, high winds, and blowing snow that makes it difficult to see. A blizzard usually lasts at least three hours.

## How can you stay safe during severe storms?

When severe weather approaches, the National Weather Service issues a watch or a warning. A watch tells you that even though the weather isn't dangerous yet, it may become dangerous soon. During a watch, stay tuned to a radio or television station that is reporting the weather. ☑

When a warning is given, the weather is already severe. During a severe thunderstorm or tornado warning, go to a basement or to a room in the middle of the house away from windows. When a hurricane or flood watch is given, be prepared to leave home. During a blizzard, stay indoors.

**12. Explain** What does a weather watch tell you?

______________________

______________________

______________________

# After You Read

## Mini Glossary

**air mass:** large body of air that has the same characteristics of temperature and moisture content as the area where it formed

**blizzard:** severe winter storm with temperatures below −12° C, winds of at least 50 km/h, and blowing snow that causes poor visibility that lasts at least three hours

**front:** boundary between two air masses with different temperature, density, or moisture

**hurricane:** large, severe storm that forms over tropical oceans and has winds of at least 119 km/h

**tornado:** violently rotating column of air in contact with the ground

1. Review the terms and their definitions in the Mini Glossary. Then write a sentence explaining how hurricanes get and keep their strength.

___

___

___

2. Write the name of the correct weather front above each description.

**warm front, stationary front, occluded front, cold front**

| ___ | ___ |
|---|---|
| **Cold air goes under warm air. Warm air is lifted.** | **3 air masses: cold, cool, warm Warm air closed off from Earth.** |
| ___ | ___ |
| **Neither warm nor cold air is moving.** | **Lighter, warmer air moves over cold air.** |

3. Did highlighting key terms and their meanings help you learn the information about weather patterns? Would you use this study strategy again?

___

___

___

Visit **earth.msscience.com** to access your textbook, interactive games, and projects to help you learn more about weather patterns.

# Weather

## section 3 Weather Forecasts

**What You'll Learn**
- how data are collected for weather maps and forecasts
- what symbols are used on a weather map

### Before You Read

How good are you at predicting the weather? On the lines below, list things you consider when you're deciding what the day's weather might be like.

_______________________________________________

_______________________________________________

_______________________________________________

Study Coach

**Sticky Notes** As you read this section, mark the pages you find interesting or where you have a question. Share these pages with another student or with the teacher.

### Read to Learn

#### Weather Observations

By looking at the thermometer or at clouds in the sky, you can tell things about the weather. Certain things about weather you know just from where you live. For example, if you live in Florida, you know that it will probably be warm and sunny.

#### What does a meteorologist do?

A **meteorologist** (mee tee uh RAH luh jist) studies the weather. A meteorologist gathers information about temperature, air pressure, wind, humidity and precipitation. By using tools like computers, Doppler radar, satellites, and weather balloons, a meteorologist makes weather maps and forecasts the weather.

FOLDABLES™

**C Organize** Make a Foldable like the one shown below to help you learn about weather forecasts.

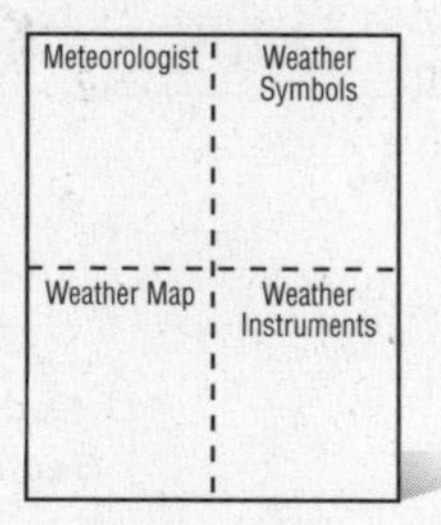


#### Forecasting Weather

Meteorologists gather information and make predictions about weather in the future. Because storms can be dangerous, it is important to know if a storm is coming. The National Weather Service uses two sources to predict the weather. They collect information, or data, from the upper atmosphere. They also collect data from Earth's surface.

**Station Models** Meteorologists gather data from Earth's surface. Then this data is recorded on a map. A **station model** shows weather conditions at a specific location using symbols on a map. Information coming from station models and from instruments in Earth's atmosphere is put into computers and helps forecast weather. ☑

## How do maps show temperature and pressure?

Weather maps have lines that connect locations with the same temperature or the same pressure. An **isotherm** (I suh thurm) is a line that connects places with the same temperature. Iso means "same." Therm means "temperature." You may have seen isotherms on weather maps on TV.

Weather maps, like the one below, also have isobars. An **isobar** is a line that connects two places with the same atmospheric pressure. Isobars show how fast wind is blowing in an area. When isobars are drawn close together, there is a big difference in air pressure. This means a strong wind is blowing. When isobars are drawn farther apart, there is little difference in pressure. Winds in this area are gentler. Isobars also show locations of high- and low-pressure areas.

On the weather map below, the pressure areas are drawn as circles with the word *High* or *Low* in the middle of the circle. Fronts are drawn as lines and symbols. This information helps meteorologists forecast the weather.

**Reading Check**

1. **Describe** What does a station model show?

______________________

______________________

______________________

**Picture This**

2. **Locate** Find the low pressure area by Portland and trace over the circle.

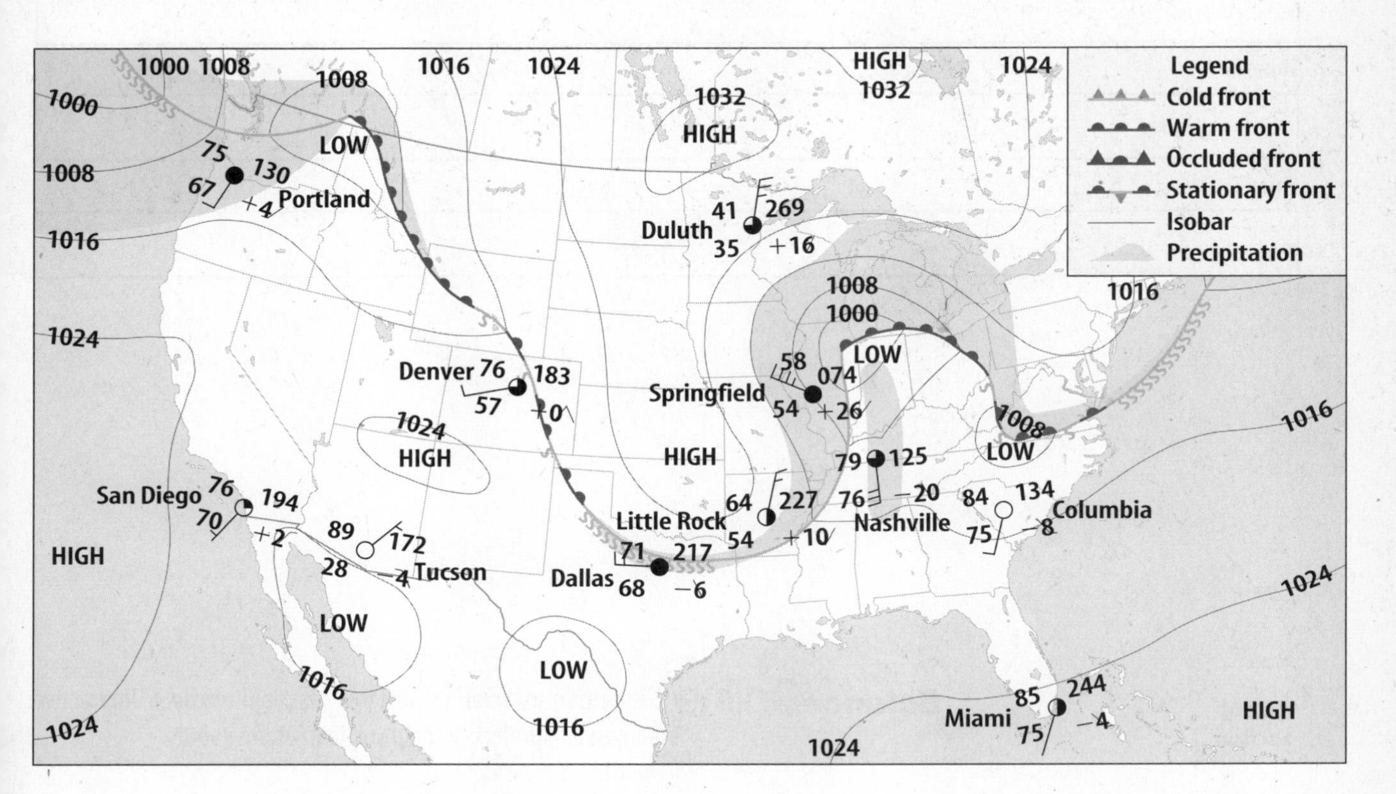

## After You Read

### Mini Glossary

**isobar:** line drawn on a weather map that connects two places with the same atmospheric pressure

**isotherm:** line drawn on a weather map that connects locations with the same temperature

**meteorologist:** person who studies the weather and uses information from Doppler radar, weather satellites, computers, and other instruments to make weather maps and provide forecasts

**station model:** indicates weather conditions at a specific location by using symbols on a map

1. Review the terms and their definitions in the Mini Glossary. Then write a sentence explaining the difference between an isobar and an isotherm.

______________________________________________________________________

______________________________________________________________________

2. Arrange the following events in order to show how a meteorologist studies weather and uses information.

**A meteorologist:**
**forecasts weather**
**gathers data on weather conditions**
**makes weather maps**

**First**

|   |
|---|
|   |

**Second**

|   |
|---|
|   |

**Third**

|   |
|---|
|   |

Visit **earth.msscience.com** to access your textbook, interactive games, and projects to help you learn more about weather forecasts.

# Climate

## section 1 What is climate?

### Before You Read

What do you think of when you hear the word *climate*? On the lines below, describe the climate where you live.

_______________________________________________

_______________________________________________

_______________________________________________

**What You'll Learn**

- what climate is and what determines a region's climate
- how latitude, oceans, and other factors affect climate

### Read to Learn

#### Climate

Imagine you are wandering through a rain forest. You see beautiful pink and purple flowers under towering trees. Unusual birds fly through the air, and animals leap through the tree branches. All of these organisms grow well in hot temperatures with plenty of rainfall. Rain forests have a hot, wet climate. **Climate** is the pattern of weather that occurs in an area over many years. An area's climate determines which plants and animals can survive and how people live.

Study Coach

**Two-Column Notes** As you read, organize your notes in two columns. In the left-hand column, write the main idea of each paragraph. Next to it, in the right-hand column, write details about the main idea.

#### Latitude and Climate

Latitude is a measure of how far north or south of the equator a place is. A place's latitude affects its climate.

#### How does latitude affect climate?

Areas nearest the equator, the tropics, have warmer temperatures than areas farther away from the equator. The **tropics** are the climate zones that get the most radiation from the Sun, or solar radiation. The tropics are located between latitude 23.5° N and latitude 23.5° S. This area receives the most direct solar energy. The tropics have temperatures that are always hot, except in the mountains.

FOLDABLES™

**A Classify** Make a half-book Foldable as shown. Record information about the different zones as you read this section.

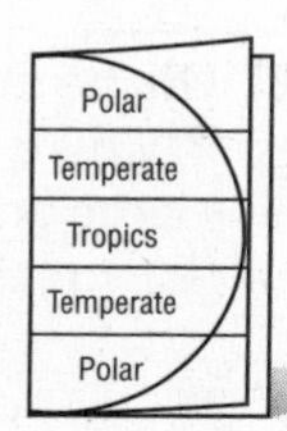

**FOLDABLES™**

**B Cause and Effect** Divide one sheet of paper into three parts and label as shown. As you read, write down the effect that oceans, mountains, and large cities have on weather.

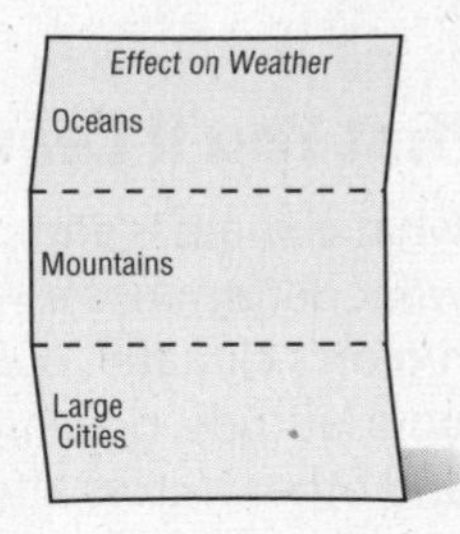


## Where are the polar and temperate zones?

There are two polar zones. The north polar zone stretches from latitude 66.5° N to the north pole. The south polar zone stretches from latitude 66.5° S to the south pole. A **polar zone** receives solar radiation at a low angle and is never warm.

There are also two temperate zones. A **temperate zone** is located between the tropics and the polar zones and has a climate with moderate temperatures. Most of the United States is in a temperate zone.

# Other Factors

Besides latitude, other factors influence a region's climate. Some natural features that affect climate are large bodies of water, ocean currents, and mountains. Large cities also can change weather patterns and affect local climate.

## How do large bodies of water affect climate?

Water takes longer to heat up than land. Water also cools down more slowly than land. Large bodies of water affect the climate of nearby areas by absorbing and giving off heat. In general, areas near a large body of water are warmer in the winter and cooler in the summer than areas that are not near a large water body. ☑

**Reading Check**

1. **Explain** How does a large body of water affect the climate of nearby land?

______________________

______________________

______________________

Find San Francisco on the map below. Now look inland and find Wichita, Kansas. Although these two cities are at the same latitude, they do not have the same temperatures. San Francisco is warmer in winter and cooler in summer because it is near the ocean.

**Picture This**

2. **Label** On the map, mark the region where you live. Does any large body of water affect your climate?

______________________

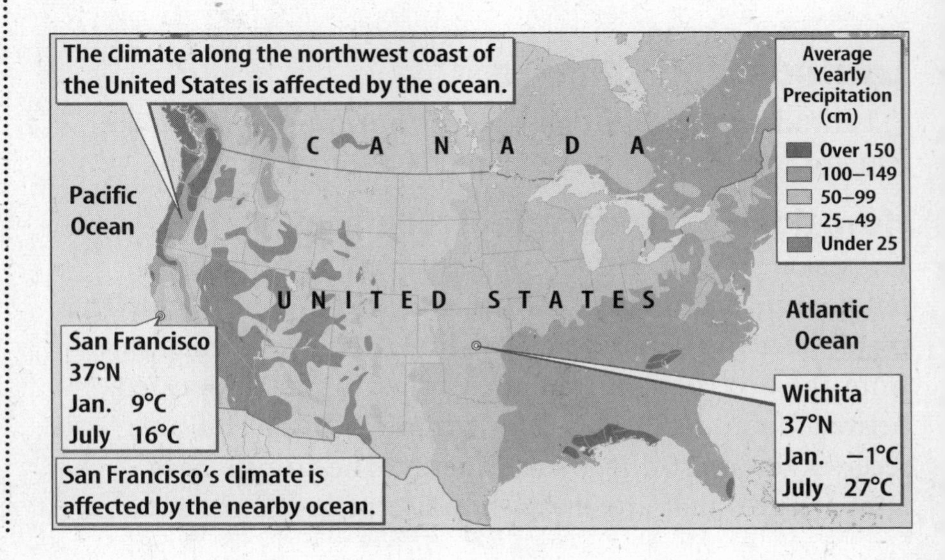

## How do ocean currents affect climate?

Ocean currents affect climates near the coast. Warm ocean currents begin near the equator. From the equator, they flow toward the north and south poles, warming the land areas they pass. As they near the poles, the currents cool off. The cool currents flow back toward the equator, cooling the air and climates of nearby land.

Winds blowing from the sea are often moister than those blowing from land. The moist sea air passes over the land, bringing rain. The sea air gives coastal areas a wetter climate than inland areas far from the sea. Look again at the map on the previous page. The northwest coast of the United States, including Washington, Oregon, and northern California, has a wet climate. This area receives a lot of moisture from the Pacific Ocean.

## How do mountains affect climate?

At the same latitude, the climate is colder in the mountains than at sea level. Radiation from the Sun heats Earth's surface. Heat from Earth then warms the atmosphere. Thin mountain air has fewer molecules to absorb heat than air near sea level. As a result, mountain air tends to be cooler than air at sea level.

Mountain ranges have a windward side which faces the wind and a leeward side which faces away from the wind. On the windward side of a mountain range, air rises, cools, and then drops its moisture on the land. On the leeward side, air flows down, heats up, and dries the land. Deserts are common on the leeward side of mountains because of this warm, dry air. ☑

## How do large cities affect climate?

Streets, parking lots, and buildings in large cities absorb the Sun's rays and heat up. The heat is transferred to the air where air pollution traps it. This trapped heat creates what is known as the heat-island effect. Temperatures in large cities can be 5°C higher than in the nearby countryside.

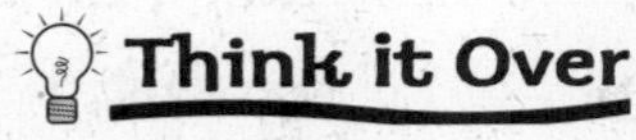


**3. Explain** Why does mountain air absorb less heat than air at sea level?

______________________

______________________

**Reading Check**

**4. Identify** Is the leeward side of a mountain the wet side or the dry side?

______________________

# After You Read

## Mini Glossary

**climate:** pattern of weather that occurs in an area over many years

**polar zone:** climate zone that gets the least solar radiation and is never warm

**temperate zone:** climate zone with moderate temperatures that is located between the tropics and the polar zones

**tropics:** climate zones that get the most solar radiation and are always hot, except at high elevations

1. Review the terms and their definitions in the Mini Glossary. Then write a sentence that explains the difference between the tropics and the polar climate zones.

______________________________________________

______________________________________________

2. Write a sentence in each box to explain how mountains affect climate.

**The cool air releases moisture as rain or snow on the windward side of the mountain.**
**The dry air passes over the mountain to the leeward side.**
**The dry air flows down the leeward side of the mountain and heats up.**
**Moist air flows toward a mountain and is forced upward, where it cools.**

**How Mountains Affect Climate**

| **First** |
|---|
| |

| **Second** |
|---|
| |

| **Third** |
|---|
| |

| **Fourth** |
|---|
| |

3. You organized your notes in two columns, one for main ideas and one for details. How did this help you understand climate?

______________________________________________

______________________________________________

End of Section

Science Online Visit **earth.msscience.com** to access your textbook, interactive games, and projects to help you learn more about climate.

# Climate

## section 2 Climate Types

### Before You Read

Imagine that you and your family move to Death Valley, one of the hottest deserts on Earth. Describe what you might do to adapt to this climate.

---

---

---

**What You'll Learn**
- how climates are classified
- how organisms adapt to a climate

### Read to Learn

**Mark the Text**

**Highlight** Highlight the key terms and their meanings as you read this section.

## Classifying Climates

How would you classify the climate in your region? Is it warm most of the time, or is it usually wet and cold? Humans classify many things around them. For example, music can be classified as rap, rock, pop, jazz, country, and others.

### How are Earth's climates classified?

You have read about Earth's climates and what affects them. Climatologists—people who study climates—usually use a system developed by Wladimir Köppen in 1918 to classify Earth's climates.

Köppen observed that the types of plants found in a certain area depended on that area's climate. Köppen studied the average temperature and average precipitation in different regions around the world. He then studied the types of plants that grew in the different areas. Based on the information he gathered, Köppen classified the world's climates into six main groups—tropical, mild, dry, continental, polar, and high elevation. Each group is divided further into types. For example, dry climates are broken down into semiarid (partly dry) and arid (very dry). The map on the next page shows regions of each climate group on Earth.

**FOLDABLES™**

**C Main Ideas** Make a two-tab Foldable as shown to help you understand how organisms are adapted to different climates.

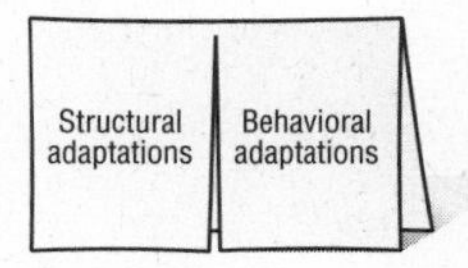

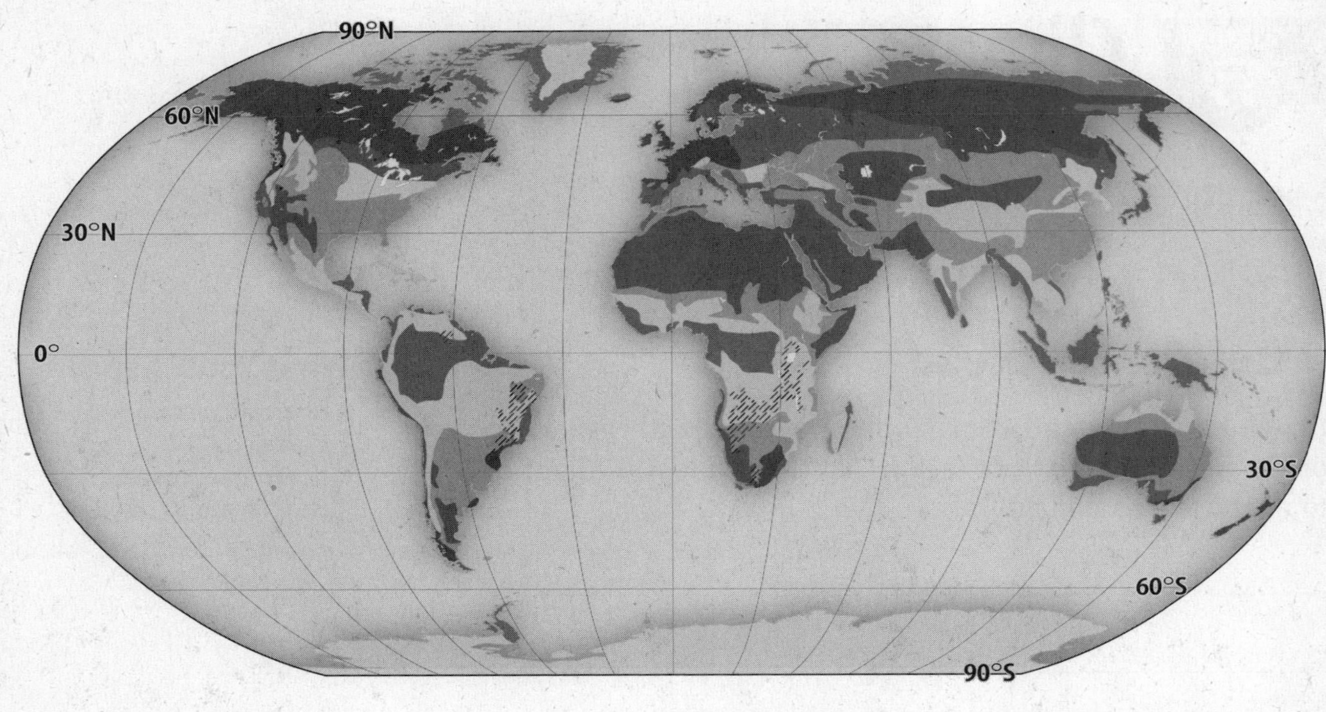


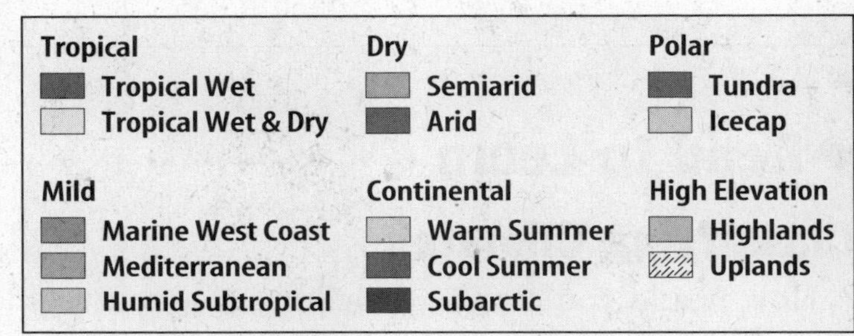


## Picture This

1. **Locate and Label** Find your region on the map of North America. Label your region with the name of its climate classification.

## Adaptations

Köppen observed that an area's climate determines the types of plants that grow there. For example, there are no cacti in rain forests or fir trees in deserts. Living things grow best in certain climates.

Organisms are adapted to their environment. An **adaptation** is any structure or behavior that helps an organism survive in its environment. Organisms that are adapted to a particular climate may not be able to survive in other climates.

## What are structural adaptations?

Structural adaptations are physical characteristics that are inherited. They develop over a long period of time. Fur on mammals is a structural adaptation. Fur keeps animals warm in cold climates. A cactus has a thick, fleshy stem that prevents water in the cactus from evaporating. This structural adaptation helps the cactus survive in hot, dry climates.

## What are behavioral adaptations?

Some organisms have behavioral adaptations that help them survive. Rodents and other mammals, such as bears, hibernate in winter. **Hibernation** is a period in which an animal's activity is greatly reduced, its body temperature drops, and body processes slow down. Factors such as cooler temperatures, shorter days, and lack of food are thought to trigger hibernation.

Other animals have adapted differently. In cold weather, bees gather together in a ball to stay warm. On hot, sunny days, desert snakes rest under rocks. At night, when it's cooler, the snakes come out to search for food.

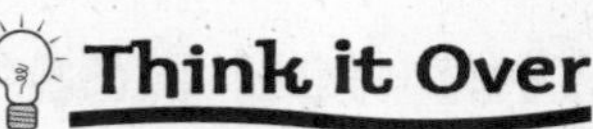


**2. Apply** Think of an animal. What is one structural or behavioral adaptation that helps this animal survive?

______________________

______________________

______________________

## What is estivation?

During periods of extreme heat or dryness, some animals enter a state called estivation (es tuh VAY shun). Body processes slow down during estivation. Lungfish live in small lakes or ponds. During the long, hot, dry months of summer, the lake or pond water may evaporate. Then the lungfish burrows into mud at the bottom of the lake. When the dry season is over and the lake again fills with water, the lungfish emerges from the mud. The figure below shows a lungfish during estivation.

### Think it Over

**3. Explain** What happens to an animal's body processes during estivation?

______________________

______________________

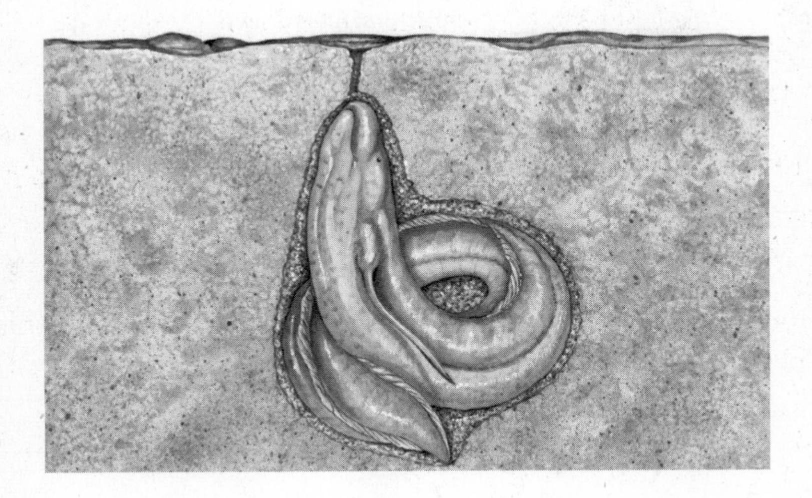

### Picture This

**4. Identify** Trace the outline of the lungfish's body as it estivates in the mud.

## What adaptations do humans have?

Like lungfish and hibernating rodents, humans have adaptations that help them adjust to their environment. In hot weather, you sweat. Sweat glands release water onto your skin. When the water evaporates, it takes some heat with it and cools you off. In cold weather, you may shiver. Shivering moves your muscles and this action helps warm you.

# After You Read

## Mini Glossary

**adaptation:** any structural or behavioral change that helps an organism survive in its particular environment

**hibernation:** behavioral adaptation for winter survival in which an animal's activity is greatly reduced, its body temperature drops, and its body processes slow down

**1.** Review the terms and their definitions in the Mini Glossary. Then write a sentence explaining what kind of adaptation hibernation is. Name one animal that hibernates.

______________________________________________

______________________________________________

______________________________________________

**2.** Animals have both structural and behavioral adaptations that help them survive in their environment. Fill in the boxes below with examples of how animals adapt.

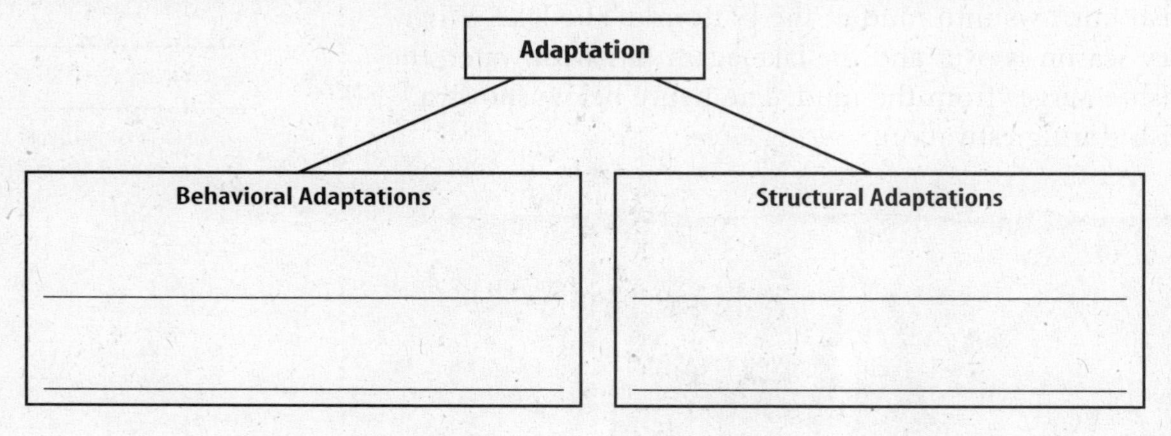


**3.** You highlighted the key terms and their meanings as you read this section. How did this help you understand climate types and adaptations?

______________________________________________

______________________________________________

______________________________________________

**ScienceOnline** Visit **earth.msscience.com** to access your textbook, interactive games, and projects to help you learn more about types of climate and adaptations.

# Climate

## section 3 Climatic Changes

## Before You Read

If you lived in an extremely hot desert climate, could you be active outdoors during the summer? Explain your answer.

_______________________________________________

_______________________________________________

### What You'll Learn

- what causes the seasons
- what are the possible causes of climatic change

## Read to Learn

### Earth's Seasons

Weather changes as the seasons change. In temperate zones, people can play softball in the summer and go sledding in the winter. **Seasons** are short periods of climatic change caused by changes in the amount of solar radiation an area receives.

Why does the amount of sunlight an area receives change? Earth is tilted on its axis. As Earth revolves around the Sun, different parts of the planet are tilted toward the Sun. Areas tilted toward the Sun get more solar radiation than areas tilted away from the Sun. This gives Earth its seasons.

### Are seasons the same all over the world?

Some areas of the world experience extreme changes at different seasons, but other parts of the world do not. The tropics are near the equator. As a result, they get a fairly constant amount of solar radiation all year long. The tropics do not experience seasonal changes in temperature.

Temperate areas in middle latitudes get different amounts of solar radiation during different parts of the year. When a temperate area is tilted toward the Sun, it is summer. When it is tilted away from the Sun, it experiences winter. Temperate zones experience seasonal differences in temperature.

Study Coach

**Authentic Questions** As you read this section, write down questions or comments you have about the text. Discuss them with your class after you have finished this section.

FOLDABLES™

**D Explain** Make a four-tab Foldable as shown to help you learn about seasons and climate changes.

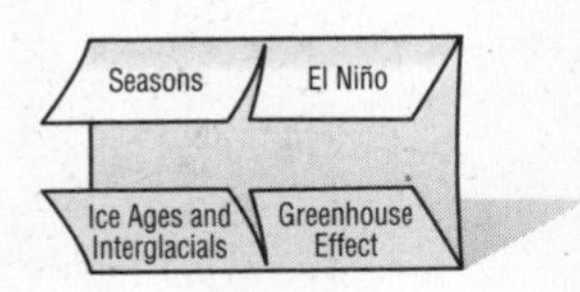

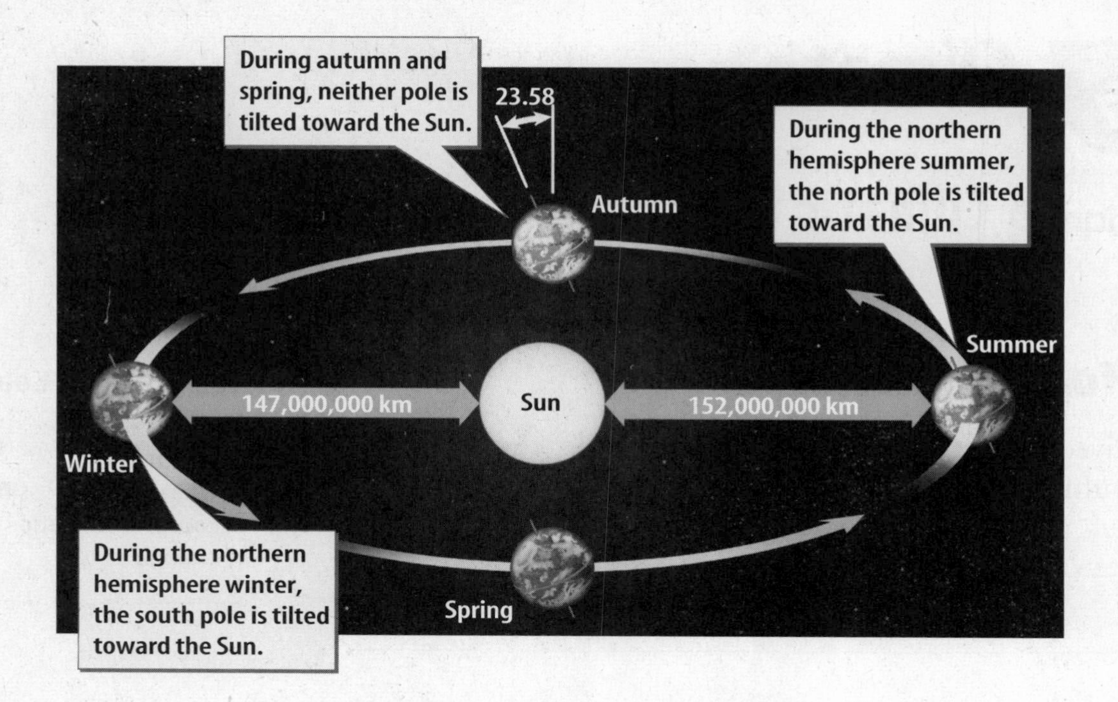


**Picture This**

1. **Locate** Use a marker to highlight the tilt of Earth's axis during each season as Earth revolves around the Sun.

**Reading Check**

2. **Determine** When it is summer at the north pole, what part of Earth is tilted toward the Sun?

## Why do high latitudes have extreme seasonal changes?

High latitudes near the north and south poles experience huge changes in temperature and in the number of daylight hours. During summer in the northern hemisphere, the north pole is tilted toward the Sun all the time. The Sun does not set for about six months and temperatures are fairly warm. When it is winter in the northern hemisphere, the north pole is in complete darkness for about six months and temperatures are extremely cold. Find the north pole in the figure above. Notice that its tilt toward the Sun changes completely from summer to winter. ☑

The seasons at the south pole are the opposite of those at the north pole. When the north pole gets months of sunlight, the south pole is in frozen darkness. Both poles are at high latitudes and experience seasonal extremes.

## El Niño and La Niña

**El Niño** (el NEEN yoh) is a climate event that involves the tropical Pacific Ocean and the atmosphere. El Niño, and its opposite, La Niña, affect weather. Both of these climate events can disrupt normal temperature and precipitation patterns around the world.

## How does El Niño affect weather patterns?

During normal years, strong trade winds blow east to west along the equator. These normal winds push warm water west across the tropical Pacific Ocean and away from South America. Cold water is forced up from the ocean depths along the coast of South America. ☑

During an El Niño, the east-to-west blowing winds weaken. Sometimes the winds even blow in the opposite direction, and then warm water flows east toward South America. Cold water is no longer forced up along the South American coast. The water temperature of the Pacific Ocean off the coast of South America becomes unusually warm, rising between 1°C and 7°C above normal.

El Niño can affect weather patterns, such as the jet stream that carries weather from west to east across North America. Wind and rainfall patterns around the world are affected. This can cause droughts in Australia and Africa and floods in South America.

## How does La Niña affect weather patterns?

La Niña is the opposite of El Niño. During La Niña, the winds that blow from east to west across the Pacific Ocean are stronger than usual. As a result, warm water gathers in the western Pacific. The water in the eastern Pacific Ocean near South America becomes cooler than usual. La Niña also affects weather patterns in North America. It can cause drought in the southern United States and heavy rain storms in the northwestern United States.

## Climatic Change

If you explored the south pole, you might find a 3-million-year-old fossil of a tropical plant or animal. Today, the south pole is far too cold for tropical organisms to survive. This fossil tells you that over millions of years, Earth's climate has changed. Scientists now know that at one time, the planet's climate was much warmer than it is today. At other times, the worldwide climate was much colder than today's climate.

**Reading Check**

3. **Describe** In what direction do trade winds blow normally?

______________________

______________________

**Think it Over**

4. **Explain** How can fossils show that Earth's climate has changed?

______________________

______________________

______________________

**Applying Math**

**5. Calculate** If our present interglacial period lasts 15,000 years, about how many years will it be until the next ice age? Use the box to show your work.

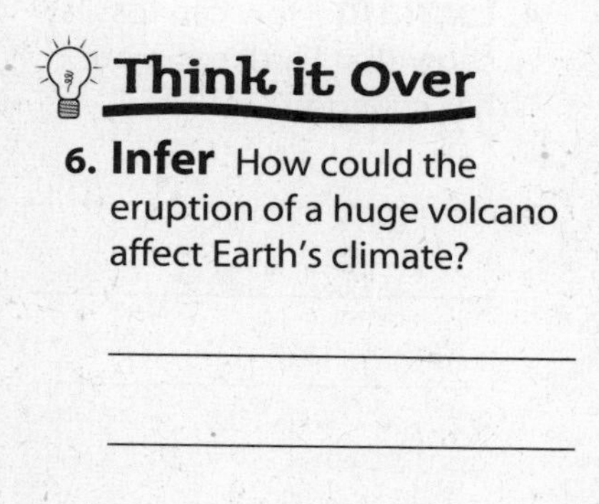

**Think it Over**

**6. Infer** How could the eruption of a huge volcano affect Earth's climate?

## What is an ice age?

Sediments in many parts of the world show that several times in the past 2 million years, large sheets of ice, or glaciers, covered much of Earth's surface. These times are called ice ages. During ice ages Earth's climate was much colder than it is today. Ice ages seem to last between 60,000 and 100,000 years.

Ice ages are followed by warm periods, when much of Earth has a tropical climate. These warm periods, called interglacial periods, last between 10,000 and 15,000 years. Today, we are in an interglacial period that began about 11,500 years ago.

## What other evidence exists for climatic change?

Other evidence shows that climate can change even more quickly. Ice cores—samples of ice drilled out of glaciers—reveal information about climate change the way tree rings reveal information about yearly rainfall. Scientists have drilled ice cores through glaciers in Greenland. These ice cores show that during the last ice age, some cold periods lasted only 1,000 to 2,000 years. They were followed by warmer periods lasting about the same length of time.

## What causes climatic change?

Earth's climate changes for many reasons. Terrible events, such as large volcanic eruptions or meteorite collisions, can affect Earth's climate over short periods of time, such as a year or several years. These events change climate by adding solid particles and liquid drops to the upper atmosphere.

The amount of energy given off by the Sun also affects Earth's climate. Over a short or a long period of time, the amount of solar energy given off may change. Because the Sun's energy warms Earth, any changes in the amount of solar energy Earth receives may change its climate.

Changes in Earth's movements in space affect climate over many thousands of years. The movement of Earth's tectonic plates can also result in climate change over millions of years. All of these factors can work separately or together to change Earth's climate.

## Where do atmospheric solids and liquids come from?

Earth's atmosphere normally contains small solid and liquid particles. Some of these particles enter the atmosphere naturally from volcanic eruptions, soot from fires, and wind eroding the soil. Other particles enter the atmosphere as pollution from car exhaust and smokestacks.

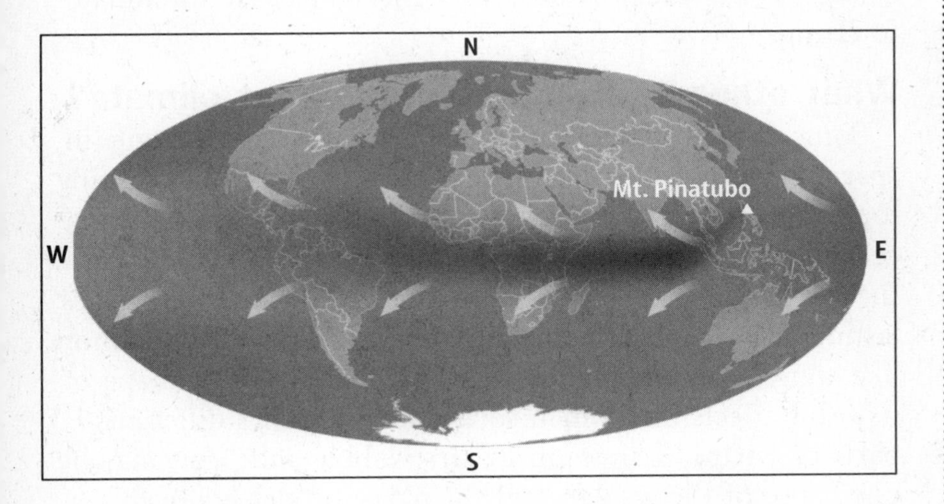


**Picture This**

7. **Determine** In what direction did the ash from Mt. Pinatubo travel as it circled the globe?

## How do atmospheric solids and liquids affect climate?

Particles in the atmosphere affect how much sunlight hits Earth. Volcanic eruptions put so many ash particles in the atmosphere, some sunlight is blocked and Earth's climate cools. The figure above shows how particles from the huge eruption of Mt. Pinatubo in the Philippines were carried around the world in the atmosphere. Meteors crashing into Earth from space raise enormous clouds of dust and particles that block the Sun and cool the planet.

In cities, particles that enter the air as pollution can change the local climate. Particles from pollution may increase cloud cover near cities. Studies show that pollution might also reduce rainfall in areas near large cities.

## Can sunspots affect Earth's climate?

If the amount of radiation from the Sun changes, Earth's climate can change. Sunspots may affect the amount of energy given off by the Sun. Sunspots are dark spots on the Sun's surface. Sunspots may be related to the cold period that occurred in Europe between 1645 and 1715. During this time, few sunspots appeared on the Sun. WARNING: *Never look directly at the Sun. It can damage your eyes.*

**Think it Over**

8. **Predict** What might Europe's climate have been like between 1645 and 1715 if many sunspots had appeared on the Sun?

Applying Math

**9. Calculate** What is the difference between Earth's greatest angle of tilt and smallest angle of tilt?

## How does Earth's tilt affect climate change?

Some climate changes may result from changes in the tilt of Earth on its axis. Today, Earth's axis is tilted at an angle of 23.5°. At times, the angle of Earth's tilt has been about 21.5°. At other times, it has been about 24.5°. When Earth's angle of tilt is greatest, there probably are more extreme seasonal changes between summer and winter. Earth's tilt changes about every 41,000 years and may cause its climate to change.

## What other Earth movements affect climate?

Some climate changes may involve Earth's movements in space. Earth wobbles a bit on its axis, like a slowly spinning top. These wobbles affect how much solar radiation hits certain areas on Earth, and this affects Earth's climate. Also, the shape of Earth's orbit around the Sun changes in cycles lasting about 100,000 years. Sometimes Earth's orbit is more like a circle, and sometimes its orbit is flatter. The shape of its orbit affects how much solar energy reaches different parts of Earth. Changes in Earth's wobble and orbit may be one cause of the growth and shrinking of glaciers over millions of years.

## What effect do Earth's plate movements have on climate?

The movement of Earth's plates may explain some long-term climate changes. As continents and oceans move on their tectonic plates, the transfer of heat around the globe is altered. The changes in heat transfer affect global wind and precipitation patterns. Over time, the change in wind and precipitation patterns can change climate. For example, the Himalaya Mountains were formed when two tectonic plates crashed into each other about 40 million years ago. The formation of these huge mountains changed Earth's climate.

Many theories exist about why Earth's climate has changed over time. Probably all of these things play some role in changing climates.

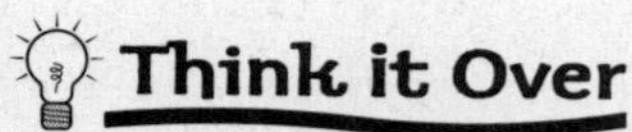


**10. Draw Conclusions** Is Earth's climate the result of one factor or many factors?

# Climatic Changes Today

Since 1992, officials from many countries have met to discuss the greenhouse effect and global climate change. Some people are concerned that the greenhouse effect is causing Earth's atmosphere and oceans to get warmer.

## What is the greenhouse effect?

The **greenhouse effect** is the natural heating process that occurs when certain gases in Earth's atmosphere trap heat. Solar energy heats Earth's surface. Some of this heat is reflected into space. Certain gases in Earth's atmosphere, called greenhouse gases, trap some of the heat and keep it near Earth's surface, as shown in the figure.

Greenhouse Effect

Water vapor, carbon dioxide, and methane are some of the most important greenhouse gases. Without these greenhouse gases in the atmosphere, life on Earth would be impossible because the planet would be too cold. However, if there are too many greenhouse gases in the atmosphere, the climate may get too warm.

**Picture This**

11. **Identify** and circle the arrows that show the heat trapped near Earth's surface by greenhouse gases in the atmosphere.

## Global Warming

Over the last 100 years, Earth's surface temperature has increased by about 0.6°C. **Global warming** is the increase in the average global temperature of Earth. During this same time, carbon dioxide levels in the atmosphere have increased 20 percent. Most scientists hypothesize that the rising carbon dioxide levels are causing global warming. Other scientists think global warming might be caused by changes in the amount of energy radiated from the Sun. ☑

**Reading Check**

12. **Determine** What is the increase in the average global temperature of Earth called?

_______________

### What are the effects of global warming?

If global warming continues, many glaciers could melt. When glaciers melt, extra water enters the oceans, causing sea levels to rise. If sea levels rise, some low-lying coastal areas will be covered with water. Some ice caps and glaciers are already melting, and in some places sea levels are rising. Studies show that these events are related to global warming.

Organisms are adapted to their environment. How can organisms cope with rapidly changing environments? In some tropical waters around the world, warming sea temperatures are killing coral reefs. The corals cannot survive in the warmer water. Many organisms that depend on coral reefs will not be able to survive if the coral dies.

## Human Activities

Human activities affect the air in Earth's atmosphere. Burning fossil fuels and clearing trees and plants increase the amount of carbon dioxide in the atmosphere. Recall that carbon dioxide is a greenhouse gas. More carbon dioxide in the atmosphere may contribute to global warming. Each year, the amount of carbon dioxide in the atmosphere continues to increase.

### How does burning fossil fuels increase carbon dioxide levels?

Natural gas, coal, and oil are examples of fossil fuels. When these fuels are burned, carbon dioxide is produced and enters Earth's atmosphere. When there is more carbon dioxide in the atmosphere, more heat is trapped near Earth's surface. The more these fuels are burned, the more carbon dioxide is added to the atmosphere. Studies show that levels of carbon dioxide have increased by about 25 percent over the last 150 years. ☑

**Reading Check**

13. **Identify** What greenhouse gas forms when fossil fuels are burned?

______________________________

### How does cutting down forests affect carbon dioxide levels?

**Deforestation** is destroying or cutting down large numbers of trees. Large areas of forest have been cleared in every country on Earth. Trees are cut down for roads, mining, paper, and farming. Every year for the last 20 years, about one percent of all tropical forests have been cut down.

Deforestation increases levels of carbon dioxide in the atmosphere. As trees grow, they take carbon dioxide out of the air. When trees are cut down, the carbon dioxide they could have removed from the air remains in the atmosphere. After they are cut down, some trees are burned for fuel. Burning trees produces carbon dioxide, so even more of this greenhouse gas enters the atmosphere.

**Think it Over**

14. **Apply** What can humans do to solve the problem of deforestation?

______________________________

______________________________

The Carbon Cycle

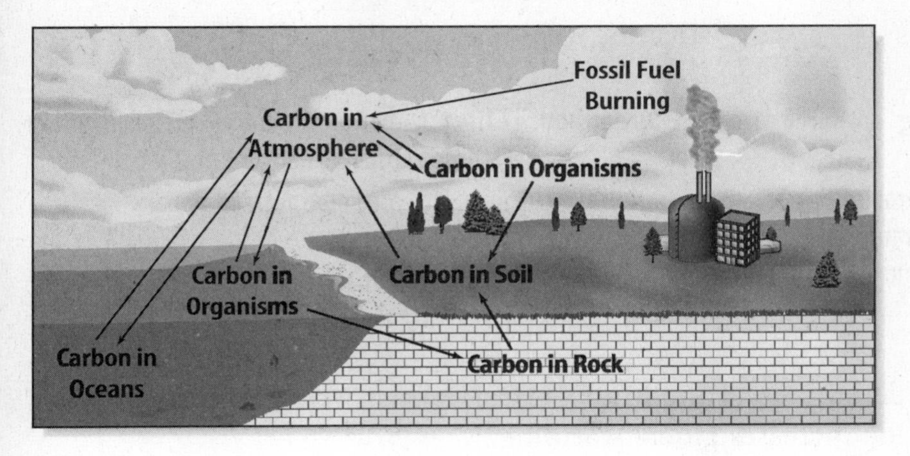


## The Carbon Cycle

Carbon, mainly in the form of carbon dioxide, is constantly recycled among the atmosphere, oceans, and organisms. The recycling of carbon in nature is shown in the figure above. Organisms, such as plants, use photosynthesis to make their own food. They take in carbon dioxide from the atmosphere and produce and store carbon-based foods. The plants, and the carbon-based food they contain, are eaten by animals. Some carbon dioxide is released as food is broken down to release energy. When organisms die and decay, some of the carbon in their bodies is stored in the soil and forms humus. Some carbon dioxide is released into the air. The constant movement of carbon in nature is called the carbon cycle. ☑

### What happens to carbon in the ocean?

Some carbon dioxide from the atmosphere dissolves in sea water. Algae and other photosynthetic organisms in the ocean use the carbon dioxide to make their own food. Ocean animals give off carbon dioxide, just as you do when you breathe out. At this time, Earth's oceans absorb more carbon dioxide than they give off.

### How does climate affect the carbon cycle?

When Earth's climate changes, the amount of carbon dioxide that recycles among the atmosphere, oceans, and land also may change. Some people hypothesize that if Earth's climate gets warmer, the oceans and land will absorb more carbon dioxide. Scientists continue to collect data to study changes in the carbon cycle.

**Picture This**

15. **Identify** Circle the area where burning fossil fuels adds carbon dioxide to the atmosphere.

**Reading Check**

16. **Identify** What is the main form of carbon that moves through the carbon cycle?

_______________________

# After You Read

## Mini Glossary

**deforestation:** destruction or cutting down of large numbers of trees

**El Niño:** climatic event that begins in the tropical Pacific Ocean and can disrupt normal temperature and precipitation patterns around the world

**greenhouse effect:** natural process that occurs when certain gases in Earth's atmosphere trap heat near the surface

**global warming:** increase in the average global temperature of Earth

**season:** short period of climatic change caused by the tilt of Earth's axis as Earth revolves around the Sun

**1.** Review the terms and their definitions in the Mini Glossary. Then write a sentence explaining how carbon dioxide increases global warming.

______________________________________________

______________________________________________

**2.** Explain how each detail supports the main idea.

| Main Idea: Several factors cause Earth's climate to change. | |
|---|---|
| **Detail 1:** Earth's movement in space | |
| **Detail 2:** El Niño | |
| **Detail 3:** Burning fossil fuels | |

**3.** As you read this section, you wrote down some questions and comments you had for later discussion. How did this help you understand the information in this section?

______________________________________________

______________________________________________

Visit **earth.msscience.com** to access your textbook, interactive games, and projects to help you learn more about climatic changes.

chapter 18

# Ocean Motion

## section 1 Ocean Water

### Before You Read

What comes to mind when you think about the ocean? What resources do you know about that come from the ocean?

### What You'll Learn

- the origin of the water in Earth's oceans
- how substances such as dissolved salts get into seawater
- the composition of seawater

### Read to Learn

## Importance of Oceans

It is easy to enjoy the ocean on a sunny day. You can lie on the beach and listen as the waves roll onto the shore. But the ocean is also useful. It provides us with many of the resources we need.

Mark the Text

**Identify the Main Point** Highlight the main point in each paragraph. Then use a different color to highlight a detail or example that helps explain each main point.

### What resources come from oceans?

Oceans are important sources of food, energy, and minerals. Humans get foods such as shrimp, fish, crabs, and clams from oceans. Kelp, a seaweed found in the ocean, is used in making ice cream, salad dressing, and medicines. Sources of energy can also come from the ocean. Oil and natural gas are often found beneath the ocean floor. Mineral resources can be found in the oceans too. Gold and copper are often mined in shallow waters. Almost one-third of all the table salt in the world comes from oceans. It is removed from the seawater through evaporation. ☑

**Transportation** Oceans also allow for the efficient transportation of goods. Millions of tons of oil, coal, and grains are shipped over the oceans each year.

Reading Check

1. **Identify** Name some foods that humans get from oceans.

## Origin of Oceans

Earth had many volcanoes during its first billion years. When the volcanoes erupted, they sent out lava, ash, and large amounts of water vapor, carbon dioxide, and other gases. This is shown in the figure on the left below. Scientists hypothesize that about 4 billion years ago, this water vapor began to be stored in Earth's early atmosphere.

Over millions of years, the water vapor cooled enough to condense into storm clouds. Rain fell from these clouds and filled low areas on Earth called **basins**. You can see this in the figure on the right below. Earth's oceans formed in the basins. Today, oceans cover over 70 percent of Earth's surface.

**Picture This**

**2. Describe** How did land forms affect where oceans were formed?

______________________

______________________

______________________

**How Water Vapor Formed Earth's Oceans**

## Composition of Oceans

Ocean water contains dissolved gases such as oxygen, carbon dioxide, and nitrogen. These gases enter the ocean from the atmosphere, the air that surrounds Earth. Oxygen also enters the ocean from organisms that photosynthesize. Organisms that use oxygen to breathe produce carbon dioxide which is released when they exhale. This is another way carbon dioxide enters the ocean.

**Salts** If you've ever tasted ocean water, you know that it is salty. Ocean water contains many dissolved salts. Chloride, sodium, sulfate, magnesium, calcium, and potassium are some of the ions in seawater. An ion is a charged atom or group of atoms. Some of these ions come from rocks that are dissolved slowly by rivers and groundwater. These include calcium, magnesium, and sodium. Rivers carry these chemicals to the oceans. Erupting volcanoes add other ions, such as bromide and chloride.

**FOLDABLES™**

**A Classify** Make the following Foldable to help you organize information about the composition of seawater.

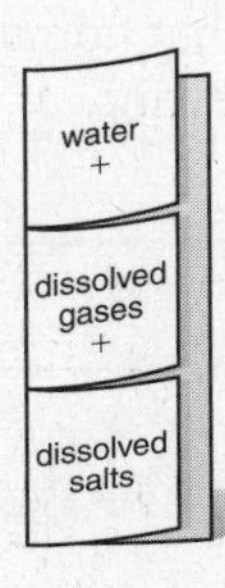

## Is seawater always salty?

Seawater is about 96.5 percent pure water. Many ions are found dissolved in seawater. When seawater evaporates, these ions combine to form minerals called salts. The chart below shows that chloride and sodium are the most common ions in seawater. During evaporation, these two ions combine. They form a salt called halite. Halite is the salt used every day to flavor foods. The dissolved halite and other salts give ocean water its salty taste.

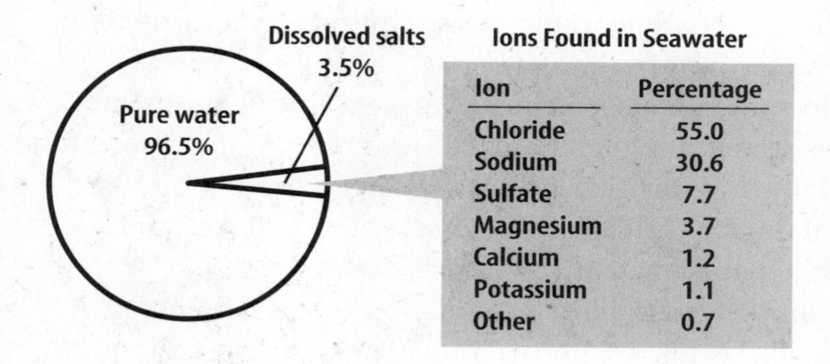


Ions Found in Seawater

| Ion | Percentage |
|---|---|
| Chloride | 55.0 |
| Sodium | 30.6 |
| Sulfate | 7.7 |
| Magnesium | 3.7 |
| Calcium | 1.2 |
| Potassium | 1.1 |
| Other | 0.7 |

**Applying Math**

**3. Calculate** What percentage of all the ocean's ions are either chloride or sodium? Use the information shown on the chart to help you.

_______________

## What is salinity?

<u>Salinity</u> (say LIH nuh tee) is a measure of the amount of salts dissolved in seawater. Salinity is usually measured in grams of dissolved salt in one kilogram of seawater. One kilogram of ocean water contains about 35 g of dissolved salts, or 3.5 percent. The amount of salts in seawater has stayed nearly constant for hundreds of millions of years. This tells you that the oceans are not getting saltier.

**Think it Over**

**4. Summarize** How many grams of salts will be left if you evaporate 1 kg of ocean water?

**a.** 1,000 g
**b.** 35 g
**c.** 85.6 g
**d.** 3.5 g

## How does the ocean stay in balance?

New elements and ions are constantly being added to water. Rivers, volcanoes, and the atmosphere all add material to the oceans. But, the ocean is said to be in balance. Elements are being removed from the ocean at the same rate that they are being added. Dissolved salts are removed when they become solid salts. Then they become part of the ocean sediment.

Some marine organisms, such as oysters, use dissolved calcium to make shells. Because many organisms use calcium, it is removed more quickly from seawater. Chloride and sodium are removed from seawater more slowly.

## How can salt be removed from ocean water?

If you have ever swum in the ocean, you may have noticed a white, flaky substance on your skin when it dried. That substance is salt. When seawater evaporates, salt is left. As the demand for fresh drinking water increases around the world, scientists are working on creating an efficient way to remove salt from seawater. Desalination (dee sa luh NAY shun) is the process of removing salt from seawater. Some desalination plants use evaporation to remove the salt.

## How is ocean water desalinated?

Scientists are working on several methods for turning saltwater into freshwater.

In some plants, seawater is passed through a membrane. The membrane filters out salt ions. Freshwater also can be obtained by melting frozen seawater. Frozen seawater contains less salt than liquid seawater. The ice can be washed and melted to produce freshwater.

Another method of desalination uses solar energy. A building is filled with seawater. The building has a glass roof. Solar energy heats the water until it evaporates. Then, the freshwater is collected from the glass roof. Look at the figure below. It shows a solar desalination plant.

**Picture This**

**5. Interpret Scientific Illustrations** Use a marker to trace the path taken by seawater in a desalination plant. Then, use another color to trace the path taken by freshwater.

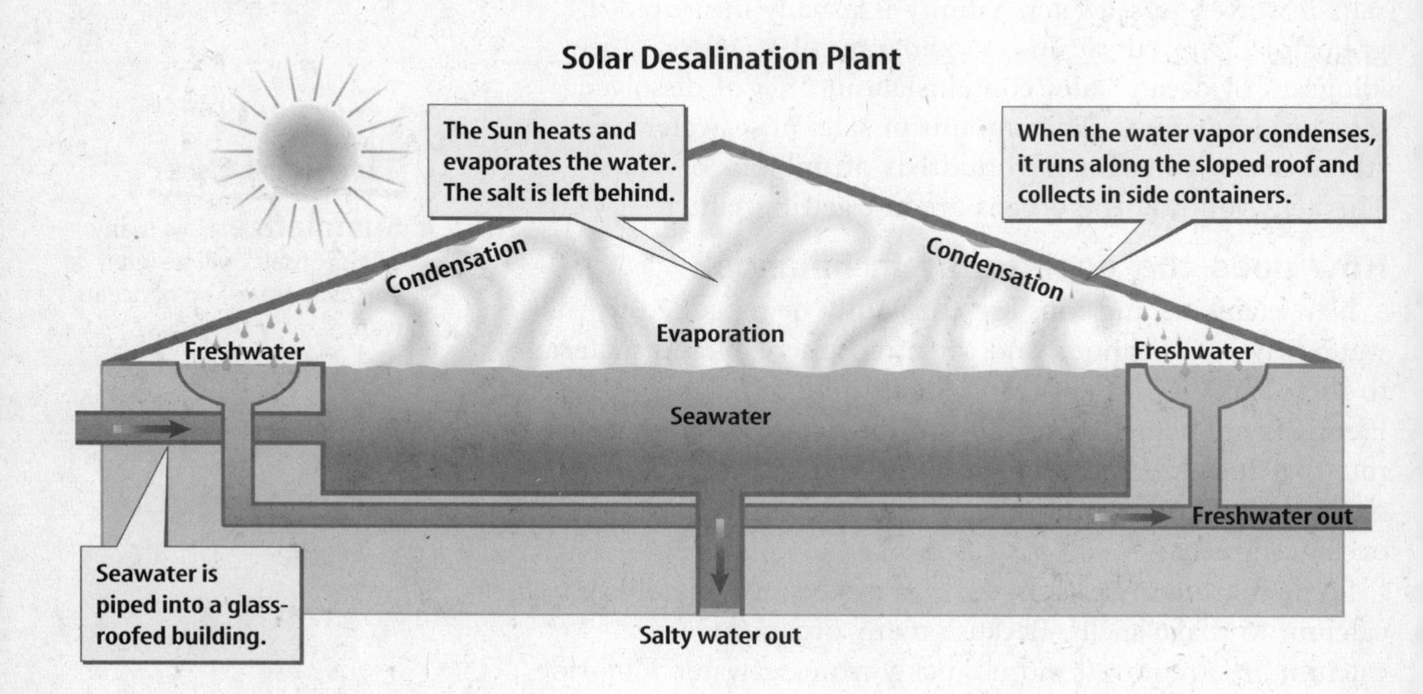

## After You Read

### Mini Glossary

**basin:** low area on Earth in which an ocean formed

**salinity:** a measure of the amount of salts dissolved in seawater

1. Review the terms and their definitions in the Mini Glossary. Choose one word and write a sentence to explain the meaning in your own words.

_______________________________________________

_______________________________________________

_______________________________________________

2. Use the chart to describe how a solar desalination plant works.

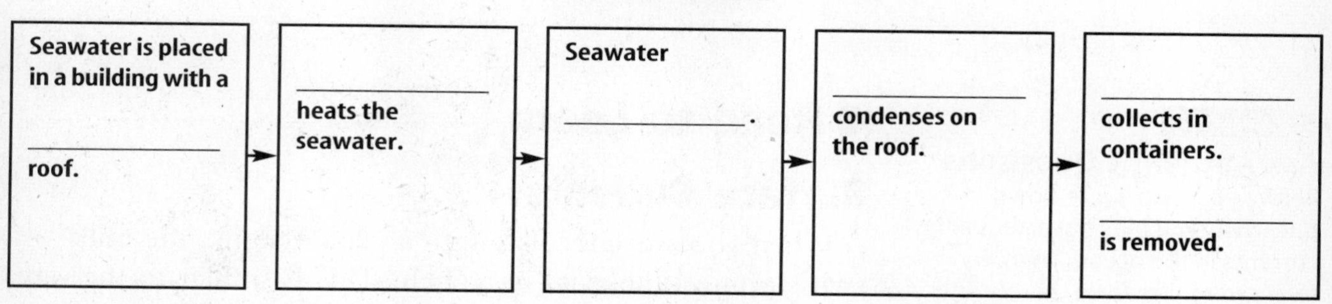


3. Choose one of the question headings in the Read to Learn section. Write the question in the box. Then answer it in your own words.

**Write your question here.**

_______________________________________________

_______________________________________________

_______________________________________________

Science Online Visit **earth.msscience.com** to access your textbook, interactive games, and projects to help you learn more about ocean water.

**End of Section**

# Ocean Motion

## section 2 Ocean Currents

### What You'll Learn

- the Coriolis effect
- what influences surface currents
- the temperature of coastal waters
- about density currents

## Before You Read

Imagine that you are stirring chocolate into a glass of milk with a spoon. How does the milk move? What happens when you stir faster?

_______________________________________________

_______________________________________________

_______________________________________________

**Study Coach**

### Ask Authentic Questions

Before you read, write down questions you may have about currents in the ocean. Then, try to answer them from the material in this section.

**FOLDABLES™**

### B Organize Information

Make the following Foldable from a half sheet of notebook paper to summarize information about currents in the northern and southern hemispheres.

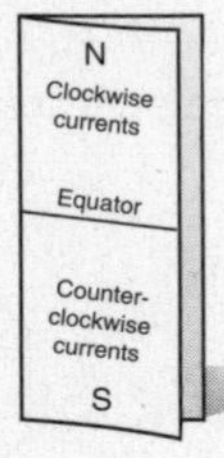


## Read to Learn

### Surface Currents

When you stir chocolate into a glass of milk, the milk swirls around the glass in a circle. This is similar to the way an ocean current moves. Ocean currents are a mass movement, or flow, of ocean water. Think of an ocean current as a river moving within the ocean.

A **surface current** is a current that moves water horizontally, or parallel to Earth's surface. Surface currents are powered by wind blowing over the water. The wind forces the water in the ocean to move in huge, circular patterns. In fact, the currents on the ocean's surface are related to the circulation of the winds on Earth. However, these currents don't affect the deep sections of the ocean. They move only the upper few hundred meters of seawater.

Some seeds and plants are carried between continents by surface currents. Sailors have relied on surface currents and winds to make sailing easier. You can see some surface currents in the figure on the next page. The arrows show the circular direction that the currents follow. Some of the currents are caused by warm winds and some are caused by cool winds.

## Major Surface Currents of Earth's Oceans

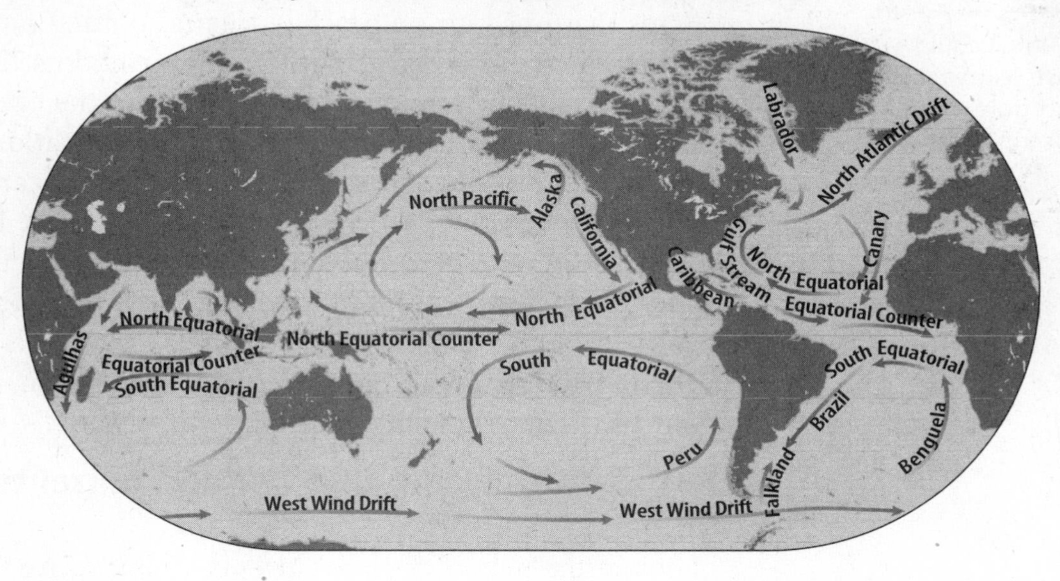


## How do surface currents form?

Surface ocean currents and surface winds are affected by the Coriolis (kor ee OH lus) effect. The **Coriolis effect** is the shifting of winds and surface currents from their expected paths because of Earth's rotation.

Earth rotates toward the east. Because of this, winds in the northern hemisphere turn to their right and winds in the southern hemisphere turn to their left. These surface winds can cause water to pile up in certain parts of the ocean. When gravity pulls water off the pile, the Coriolis effect turns the water. This causes surface water in oceans to spiral, or circle, around the piles of water.

Look again at the map of major surface currents. The circular patterns that you see are caused by the Coriolis effect. The currents north of the equator circle to their right. Currents south of the equator circle to their left.

## What is the Gulf Stream?

Much of what is known about surface currents comes from records that were kept by sailors in the nineteenth century. Sailors always have used surface currents to make traveling easier. Sailors heading west use surface currents that flow west. Sailors heading east use currents such as the Gulf Stream. The Gulf Stream is a 100-km-wide surface current in the Atlantic Ocean. When America was still a colony of England, sailors noticed that trips to England were faster than trips from England. Going eastward with the Gulf Stream made the journey quicker. ☑

### Picture This

1. **Identify** Name one current that affects the oceans around North America's coasts.

______________________

______________________

### Reading Check

2. **Summarize** How can surface currents be helpful to ships?

______________________

______________________

## Think it Over

3. **Infer** What could scientists learn about currents from a drift bottle's trip?

______________________

______________________

### How are surface currents tracked?

Items that wash up on beaches, such as bottles, can provide information about ocean currents. One method used to track surface currents is to release drift bottles into the ocean. Drift bottles are released from a variety of coastal locations.

Inside each bottle, a message and a numbered card state where and when the bottle was released. When the bottle washes ashore, the person who finds it may notice the card inside. The person will fill out the card with the information about when and where it washed ashore. The card is returned to the research team and provides valuable information about the surface currents that carried the bottle.

### How do warm and cold surface currents affect the climate?

Look at the map of surface currents again. Notice that some currents start near the north and south poles, and other currents start near the equator. Currents on the west coasts of continents begin near the poles where the water is colder. The California Current is an example of such a current. It starts near the north pole and is a cold surface current.

Currents on the east coast of continents start near the equator where the water is warmer. The Gulf Stream starts in waters near the equator and is a warm surface current.

As a warm surface current flows away from the equator, heat is released to the atmosphere. The atmosphere is warmed. The transfer of heat helps determine climate.

## Upwelling

Recall that surface currents carry water horizontally—parallel to Earth's surface. Water also travels vertically, from the bottom to the top of the ocean. **Upwelling** is a vertical circulation in the ocean that brings deep, cold water to the ocean surface. ✔

**Reading Check**

4. **Explain** What happens during upwelling?

______________________

______________________

______________________

Along some coasts of continents, wind blowing parallel to the coast carries water away from the land because of the Coriolis effect. Cold water from deep in the ocean rises up to replace it. The cold water is full of nutrients from organisms that died, sank to the bottom, and decayed. Fish are attracted to these nutrient-rich areas. Areas of upwelling are important fishing grounds. The figure on the next page illustrates upwelling off the coast of Peru.

## Upwelling

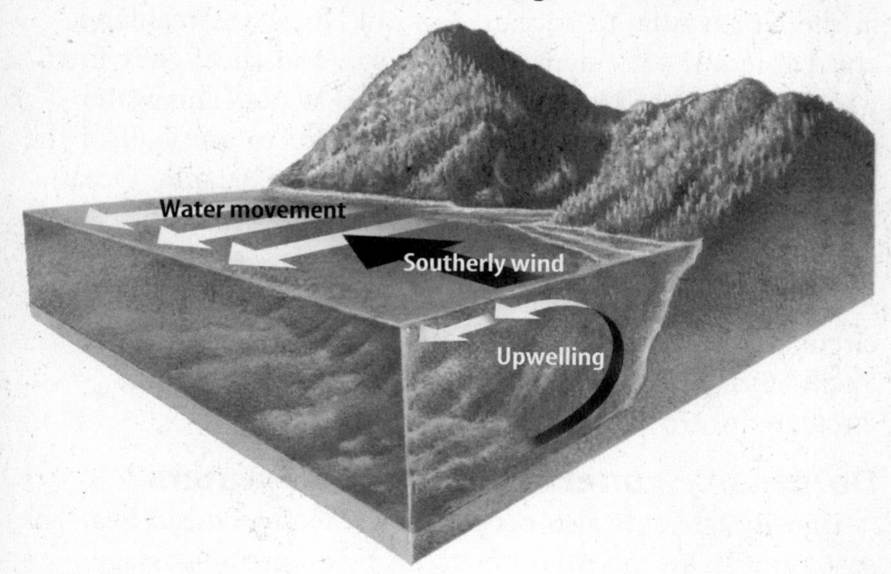


## Picture This

**5. Infer** Why does upwelling around Peru make Peru a rich fishing ground?

______________________

______________________

______________________

## Density Currents

Deep in the ocean, there is no wind to move the water. Instead, differences in density cause water to circulate or move. Cold water is more dense than warm water. Salty water is more dense than less salty water.

A **density current** forms when a mass of seawater becomes more dense than the surrounding water. Gravity causes this dense water to sink beneath less dense seawater. The deep, dense water spreads to the rest of the ocean. Changes in temperature and salinity work together to create density currents. A density current moves water very slowly.

### Where are density currents found?

One important density current begins in Antarctica. In winter, the seawater there is more dense than at any other time. When seawater freezes, the salt is left behind in the unfrozen water. This extra salt increases the seawater's density and causes it to sink. Slowly, the water begins to spread along the ocean bottom toward the equator forming a density current. In the Pacific Ocean, it could take up to 1,000 years for the water in this density current to reach the equator.

## Think it Over

**6. Sequence of Events** Number the events to show the order in which a density current forms in Antarctica.

_____ seawater freezes

_____ unfrozen seawater sinks

_____ dense seawater spreads along ocean floor

**North Atlantic Deep Water** Another density current starts in the North Atlantic Ocean. Around Norway, Greenland, and Labrador, cold, dense waters form and sink. They form what is known as North Atlantic Deep Water. This water covers the floor of the northern one-third to one-half of the Atlantic Ocean. In the southern part of the Atlantic Ocean, this current meets the density current from Antarctica. The Antarctic density current is colder and denser. The North Atlantic Deep Water floats just above it. Density currents circulate more quickly in the Atlantic Ocean than in the Pacific Ocean. In the Atlantic, a density current could circulate in 275 years. ☑

**Reading Check**

7. **Think Critically** Which is more dense, the Antarctic current or the North Atlantic Deep Water?

______________________

______________________

## Do density currents affect other waters?

Density currents also occur in the Mediterranean Sea. The sea connects to the Atlantic Ocean by a narrow passage called the Strait of Gibraltar. Warm temperatures and dry air in the Mediterranean region cause the seawater to evaporate. The salts remain behind. This increases the salinity and density of the sea. The dense, salty water travels through the Straits of Gibraltar into the Atlantic Ocean. Because it is much denser than water at the surface of the ocean, it sinks. However, it is not as dense as the very cold, salty water of the North Atlantic Deep Water. So, the water from the Mediterranean floats above it. It forms a middle layer known as the Mediterranean Intermediate Water. You can see the different water layers in the figure below.

**Picture This**

8. **Interpret Scientific Illustrations** Which layer of water shown in the figure is most dense?

______________________

______________________

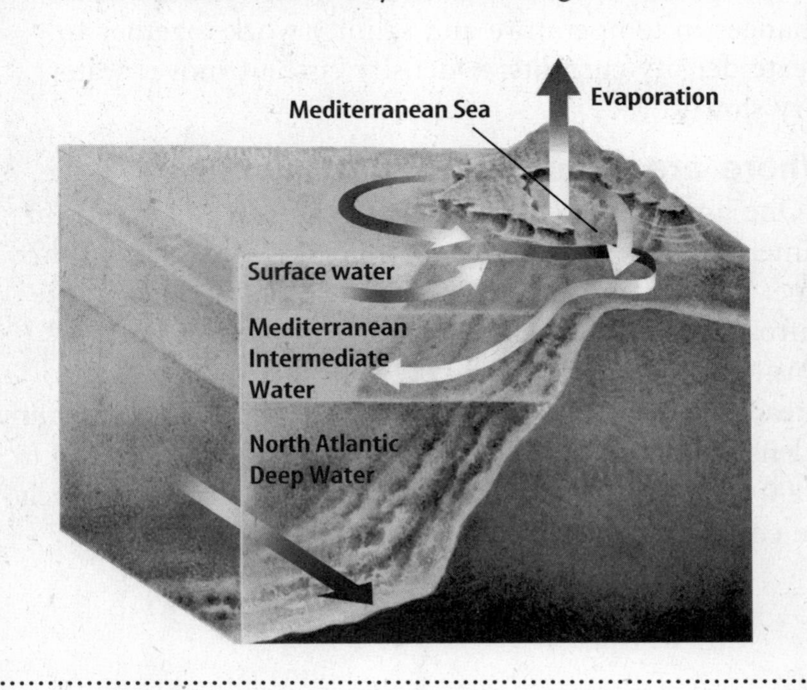

# After You Read

## Mini Glossary

**Coriolis effect:** the shifting of winds and surface currents from their expected paths that is caused by Earth's rotation

**density current:** a current that forms in the ocean because a mass of seawater becomes more dense than the surrounding water and sinks

**surface current:** a current in the ocean that moves water horizontally, or parallel to Earth's surface

**upwelling:** a vertical circulation in the ocean that brings deep, cold water to the ocean surface

1. Review the terms and their definitions in the Mini Glossary. Write a sentence that explains where density currents and surface currents are found.

______________________________________________

______________________________________________

______________________________________________

2. Complete the spider map about the Coriolis effect. List some of the results on ocean currents of the Coriolis effect.

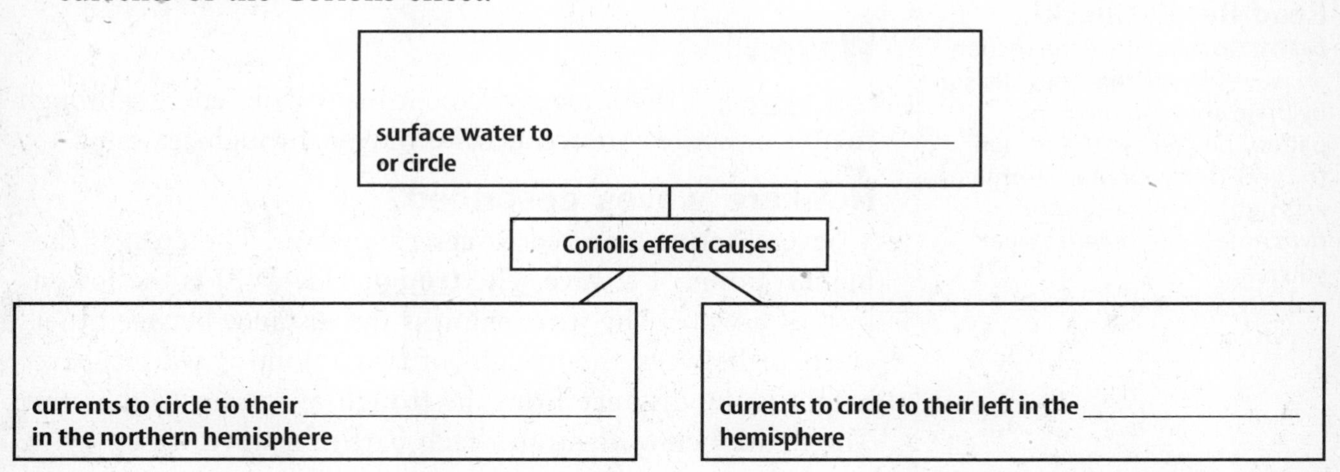


3. Before you read this section, you wrote down questions you had about ocean currents. Were you able to answer any of those questions? What information would you still like to learn about ocean currents?

______________________________________________

______________________________________________

Visit earth.msscience.com to access your textbook, interactive games, and projects to help you learn more about ocean currents.

# chapter 18 Ocean Motion

## section 3 Ocean Waves and Tides

### What You'll Learn

- wave formation
- how water particles move within waves
- the movement of a wave
- how ocean tides form

## Before You Read

Where have you seen a water wave? How would you describe the wave?

_______________

_______________

_______________

Study Coach

**Read-Recall-Check-Summarize** *Read* the section on waves. *Recall* the main ideas by brainstorming them on paper. Then reread the section to *check* that your brainstorming was right. Finally, use your own notes to *summarize* what you read.

## Read to Learn

### Waves

A **wave** is a rhythmic movement that carries energy through matter or space. An ocean wave moves through seawater.

### How are waves described?

Several terms are used to describe waves. The **crest** is the highest point of a wave. The **trough** (TRAWF) is the lowest part of a wave. The wavelength is the distance between the crests or between the troughs of two adjoining waves. Wave height is the distance from the trough of a wave to the crest. The figure below illustrates each part of a wave.

**Picture This**

1. **Identify** Highlight the crests in the figure. Then use another color to highlight the troughs.

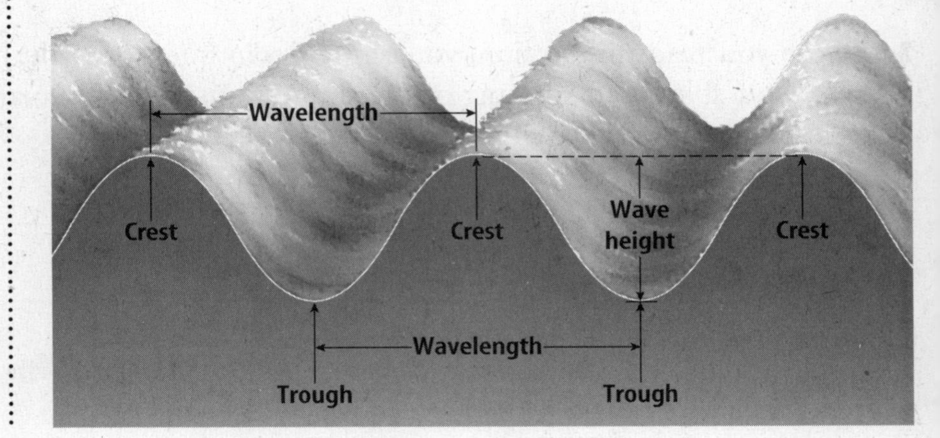

## How is a wave's energy measured?

Waves carry energy. This energy can be measured. Half of a wave's height is called the amplitude (AM pluh tewd). To measure the energy carried by a wave, the amplitude is squared. A wave with twice the amplitude of another wave carries four times ($2 \times 2 = 4$) the energy. Small waves have small amplitudes. Amplitude increases as waves grow larger. Large waves can damage ships and coastal property.

## How do waves move?

A bobber on a fishing line moves up and down in the water as a wave passes. It does not move outward with the wave. It returns to near its original position after the wave has passed.

Like the bobber, each molecule of water in a wave returns to its original position after a wave passes. The molecule may be pushed forward by the next wave, but it will return again to its original position. The water in waves does not move forward unless the wave is crashing onto shore. The water molecules in a wave move around in circles, coming back to about the same place. Only the energy moves forward. ☑

Below a depth equal to about half the wavelength, water movement stops. Below that depth, water is not affected by waves.

**Breakers** As a wave reaches the shore, it changes shape. In shallow water, friction with the ocean bottom slows water at the bottom of the wave. The top of the wave keeps moving because it has not been slowed by friction. Eventually, the top of the wave outruns the bottom and it collapses. Water tumbles over on itself, and the wave breaks onto the shore. A **breaker** is a collapsing wave. After a wave breaks onto shore, gravity pulls the water back to sea.

## How do water waves form?

On windy days, waves form on lakes and oceans. When wind blows across a body of water, wind energy is transferred to the water. When wind speed is great enough, water piles up forming a wave. As the wind continues to blow, the wave grows in height.

Wave height depends on the speed of the wind, the distance over which the wind blows, and the length of time the wind blows. When the wind stops blowing, waves stop forming. But waves that have already formed will continue to move for long distances. Waves reaching one shore could have formed halfway around the world.

**FOLDABLES™**

**C Classify** Cut a sheet of paper into eight note cards. Use the cards to record important information about waves and tides.

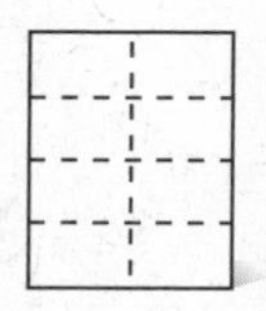

**Reading Check**

**2. Summarize** What moves forward in a wave?

**a.** molecules
**b.** energy
**c.** water
**d.** bobbers

**Think it Over**

**3. Recognize Cause and Effect** Would a wave be higher or lower on a very windy day? Why?

______________________

______________________

______________________

## Tides

Throughout the day, the level of the sea rises and falls. This rise and fall in sea level is called a **tide**. A tide is caused by a giant wave produced by the gravitational pull of the Sun and the Moon. Although this wave is only 1 m or 2 m high, its wavelength is thousands of kilometers long. As the crest of this wave approaches the shore, sea level seems to rise. This rise in sea level is called high tide. When the trough of this huge wave nears the shore, sea level appears to drop. This drop in sea level is referred to as low tide.

### What is the tidal range?

As Earth rotates, Earth's surface passes through the crests and troughs of this giant wave. Many coastal areas, such as the Atlantic and Pacific coasts of the United States, have two high tides and two low tides each day. But because ocean basins vary in size and shape, some coastal locations, such as many along the Gulf of Mexico, have only one high and one low tide each day. A **tidal range** is the difference between the level of the ocean at high tide and the level of the ocean at low tide.

**Applying Math**

4. **Calculate** A sea has a high tide that measures 15 m high. At low tide, it measures 12.5 m high. What is its tidal range?

______________________

### Why do tidal ranges vary in different locations?

Most shorelines have tidal ranges between 1 m and 2 m. However, tidal ranges can be as small as 30 cm or as large as 13.5 m.

The shape of the seacoast and the shape of the ocean floor both affect the ranges of tides. A wide seacoast allows water to spread out farther. At high tide, the water level might only rise a few centimeters. In a narrow gulf or bay, the water cannot spread out. The water will rise many meters at high tide. A narrow gulf or bay will have a greater tidal range than a smooth, wide area of shoreline.

**Think it Over**

5. **Describe** When a tidal bore enters a river, why does it cause the flow of water in the river to reverse?

______________________

______________________

______________________

### What are tidal bores?

Sometimes, a rising tide enters a river from the sea. If the river is narrow and shallow and the sea is wide, a wave called a tidal bore forms. A tidal bore can have a breaking crest or it can be a smooth wave.

Tidal bores usually are found in places with large tidal ranges. When a tidal bore enters a river, its force causes water in the river to reverse its flow. Waves in a tidal bore might reach 5 m in height and speeds of 65 km/h.

## How does the Moon affect tides?

The Moon and the Sun exert a gravitational pull on Earth. The Sun is much bigger than Earth, but the Moon is much closer. The Moon has a stronger pull on Earth than the Sun. Earth and the water in Earth's oceans respond to this pull. The water bulges outward as the Moon's gravity pulls it. This results in a high tide. The process is shown in the figure below.

The Moon's gravity pulls at Earth. This creates two bulges of water. One bulge is on the side of Earth closest to the Moon. The other bulge is on the opposite side. The high tide on the side of Earth near the Moon happens because the water is being pulled away from Earth towards the Moon. The high tide on the side of Earth opposite the Moon happens because the gravitational pull on that part of Earth is greater than the pull on the water on that side of Earth. The areas of Earth's oceans that are not toward or away from the Moon are the low tides. As Earth rotates, the bulges follow the Moon. This results in high and low tides happening around the world at different times.

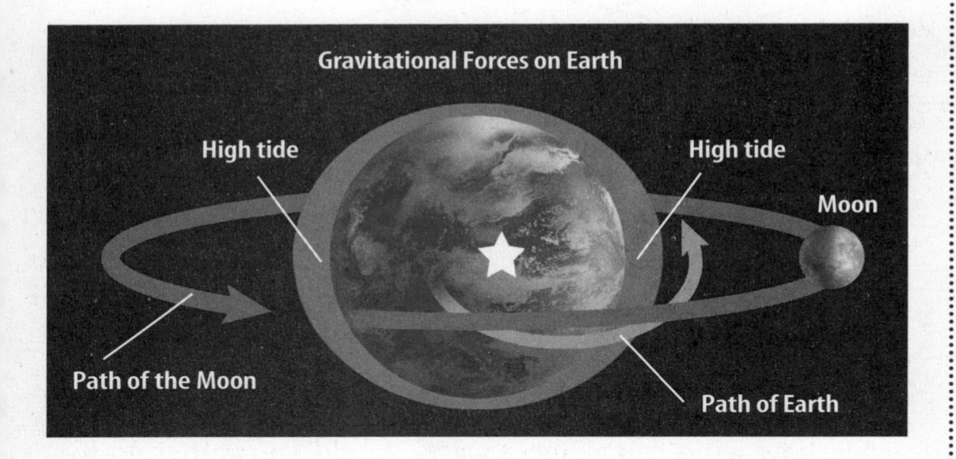


**Picture This**

6. **Interpret Scientific Illustrations** Describe the sea level of the ocean in the area of the star in the figure.

______________________

______________________

______________________

## What effect does the Sun have on tides?

The Sun's pull can add to or subtract from the gravitational pull of the Moon. Occasionally during Earth's revolution, Earth, the Sun, and the Moon are lined up together. Then, the Moon and the Sun both pull at Earth. The combined gravitational pull results in spring tides on Earth. During spring tides, high tides are higher and low tides are lower than usual. When the Sun, Earth, and the Moon form a right angle, high tides on Earth are lower than usual. Low tides are higher than usual. These are called neap tides.

# After You Read

## Mini Glossary

**breaker:** a collapsing wave
**crest:** the highest point of a wave
**tidal range:** the difference between the level of the ocean at high tide and the level of the ocean at low tide
**tide:** rise and fall in sea level
**trough:** the lowest part of a wave
**wave:** a rhythmic movement that carries energy through matter or space

1. Review the terms and their definitions in the Mini Glossary. Write a sentence or two describing a wave's crest and its trough.

_______________________________________________

_______________________________________________

_______________________________________________

2. Complete the Venn diagram to compare and contrast the effects that the Moon and the Sun have on Earth. Use the words *high, low, Moon, strong,* and *weak.*

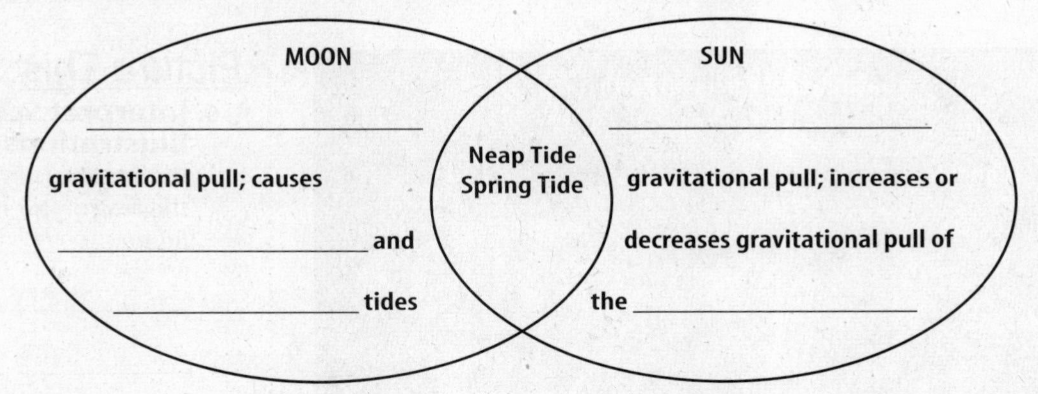


3. Earlier in this section, you learned how wind can shape a wave. Would you rather go to the beach on a windy day, a calm day, or during a storm? Predict what the waves will be like each day.

_______________________________________________

_______________________________________________

_______________________________________________

Visit **earth.msscience.com** to access your textbook, interactive games, and projects to help you learn more about waves.

# Oceanography

## section 1 The Seafloor

## Before You Read

Imagine you are in a deep-sea submersible skimming along the ocean floor. What do you think you might see? How deep do you think the water might be?

_______________

_______________

_______________

**What You'll Learn**

- the difference between a continental shelf and a continental slope
- differences between a mid-ocean ridge, an abyssal plain, and an ocean trench
- mineral resources found in the ocean

## Read to Learn

### The Ocean Basins

While on your imaginary trip along the ocean floor, the lights of your vessel may have shown a mountain range. You may have seen a huge opening in the seafloor that is so deep you can't see the bottom.

#### What is the continental shelf?

Ocean basins are low areas of Earth that are filled with water. They have many different features. Beginning at the ocean shoreline is the continental shelf. The **continental shelf** is the gradually sloping end of a continent that extends under the ocean.

On some coasts, the continental shelf extends a long distance. This is true for North America's Atlantic and Gulf coasts. The continental shelves on these coasts stretch 100 km to 350 km into the sea.

On the Pacific Coast, the coastal range mountains are close to the shore. The shelf there is only 10 km to 30 km wide. The ocean covering the continental shelf can be as deep as 350 m.

Study Coach

**Create-a-Quiz** As you read this section, create a quiz question for each topic. When you have finished reading, see if you can answer your own questions correctly.

FOLDABLES™

**A Summarize** Make the following four-tab Foldable to help you summarize information about ocean features.

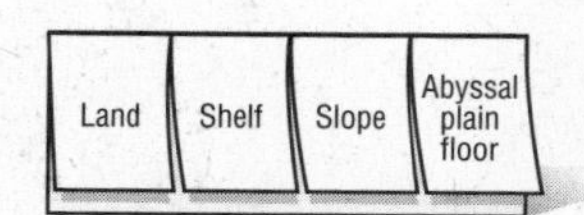

## What other features are found in the ocean?

The figure below shows that beyond the shelf, the ocean floor drops more steeply, forming the continental slope. The **continental slope** extends from the outer edge of the continental shelf down to the ocean floor. Beyond the continental slope lie the trenches, valleys, plains, mountains, and ridges of the ocean basin.

**Picture This**

1. **Identify** the features on the ocean floor that appear as small hills or mountains.

Sea surface
Abyssal plain
Continental shelf
Volcanic island
200 m
Continental slope
Seamount
4,000 m
6,000 m
Mid-ocean ridge
Oceanic trench
10,000 m

## How do abyssal plains form?

In the deep ocean, sediment that comes mostly from land, settles constantly on the ocean floor. This sediment fills in valleys in the ocean floor and creates flat seafloor areas called **abyssal** (uh BIH sul) **plains.** The figure above shows an abyssal plain. Abyssal plains are from 4,000 m to 6,000 m below the surface of the ocean.

In the Atlantic Ocean, some abyssal plains are large and extremely flat. The Canary Abyssal Plain has an area of approximately 900,000 km$^2$. Sometimes abyssal plains have small hills and seamounts. Seamounts are underwater, inactive volcano peaks. Seamounts are most commonly found in the Pacific Ocean.

2. **Identify** What is a seamount?

## Ridges and Trenches

A **mid-ocean ridge** is a part of an ocean basin area where new ocean floor is formed. Mid-ocean ridges are found at the bottom of all oceans and form a continuous underwater ridge about 70,000 km long. Crustal plates, large section of Earth's crust and upper mantle, are constantly moving. As these plates move, the ocean floor changes. When plates separate, hot magma from inside Earth erupts through small cracks, forming new crust. This process is called seafloor spreading. The hot lava is cooled by ocean water and hardens to solid rock.

## What are subduction zones?

While seafloor is being formed along mid-ocean ridges, it is being destroyed in other parts of the ocean. Areas where old ocean floor slides beneath another plate and moves into Earth's mantle are called subduction zones.

On the ocean floor, subduction zones are found within deep ocean trenches, as shown in the figure on the previous page. A **trench** is a long, narrow, steep-sided depression where one plate sinks beneath another. Most trenches are found in the Pacific Basin. Ocean trenches are usually longer and deeper than any valley on land. The Mariana Trench reaches about 11 km below the ocean surface and is the deepest place in the Pacific Ocean. The Mariana Trench is so deep that Mount Everest, the tallest mountain on Earth, could easily fit into it. The figure below shows what it would look like if Mt. Everest were placed in the Mariana Trench.

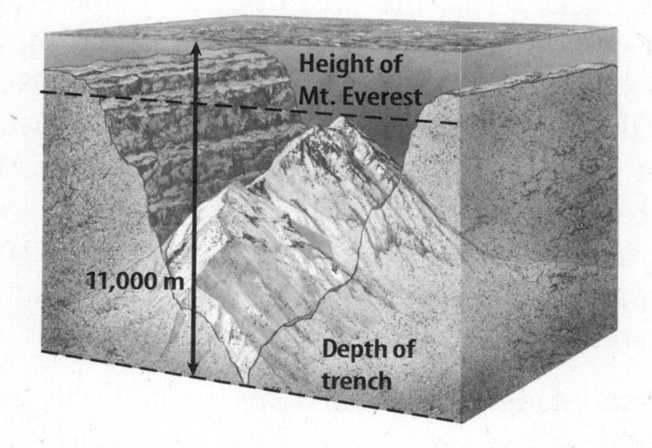


**Picture This**

**3. Interpret Scientific Illustrations**
Approximately how deep is the Mariana Trench?

_______________

## Mineral Resources from the Seafloor

Resources can be found in many places in the ocean. Some deposits on the continental shelf are easy to remove. Other resources can be found only in the deep abyssal regions of the ocean floor. People are still trying to figure out how to get these valuable resources to the surface.

### What resources are found on the continental shelf?

Many organisms live along continental shelves. When organisms die, they sink to the bottom and decay. Over millions of years, accumulations of organic matter form deposits of useful resources. Oil and natural gas are found under the seafloor. Wells are drilled into the seafloor to remove these resources.

**FOLDABLES™**

**B Compare and Contrast**
Make the following two-tab Foldable to show how mid-ocean ridges and ocean trenches are similar and different.

| Mid-ocean Ridge | Ocean Trenches |
|---|---|

**FOLDABLES™**

**C Identify** Make a three-tab Foldable to explain the resources that are found in the ocean.

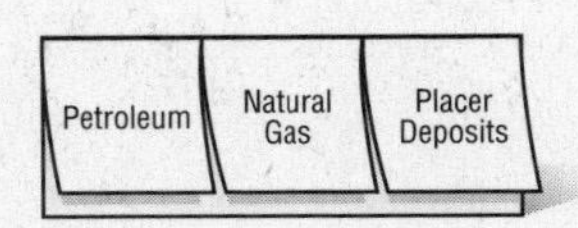


## What other resources can be found?

Other deposits on the continental shelf include phosphorite, which is used to make fertilizer, and limestone, which is used to make cement. Sand and gravel, both important in construction, also can be removed from the continental shelf.

Rivers that flow into oceans carry important minerals to the continental shelf. Sometimes ocean waves and currents cause denser mineral grains to come together in one place. These deposits, called placer (PLAHS ur) deposits, can occur in coastal regions where rivers entering the ocean suddenly lose energy, slow down, and drop their sediment. Metals such as gold and titanium, and gems such as diamonds, are mined from placer deposits.

## What are deep-water deposits?

Hot water streams out into surrounding seawater through holes and cracks along mid-ocean ridges. As the extremely hot water cools, minerals deposits sometimes form. Elements such as sulfur, iron, copper, zinc, and silver can be found in these areas. Today, no one mines these areas because it is too expensive. However, in the future, these deposits could become important.

Other mineral deposits can precipitate, or form solids, from seawater. Minerals dissolved in seawater come out of solution and form solids on the ocean floor.

**Manganese** Manganese nodules are small, darkly colored lumps found across large areas of the ocean basins. Manganese nodules form around a core, such as old sharks' teeth or whale bone. These nodules grow very slowly—perhaps as little as 1 mm to 10 mm every million years. Manganese nodules are rich in manganese, copper, iron, nickel, and cobalt. These minerals are used to make steel, paint, and batteries. Most of these modules lay thousands of meters deep in the ocean and are not currently being mined. Suction devices similar to huge vacuum cleaners have been tested to collect them.

**Think it Over**

4. **Infer** Why might a producer of steel want to develop an inexpensive way to mine manganese nodules?

______________________

______________________

______________________

# After You Read

## Mini Glossary

**abyssal plain:** flat seafloor area from 4,000 m to 6,000 m below the ocean surface, formed by the deposits of sediments

**continental shelf:** gradually sloping end of a continent that extends beneath the ocean

**continental slope:** ocean basin feature that dips steeply down from the continental shelf

**mid-ocean ridge:** area where new ocean floor is formed when lava erupts through cracks in Earth's crust

**trench:** long narrow, steep-sided depression in the seafloor formed when one crustal plate sinks beneath another

1. Review the terms and their definitions in the Mini-Glossary. Use two of the terms in a sentence to describe where one ocean basin feature is located.

_______________________________________________

_______________________________________________

2. Label the ocean features in the diagram below.

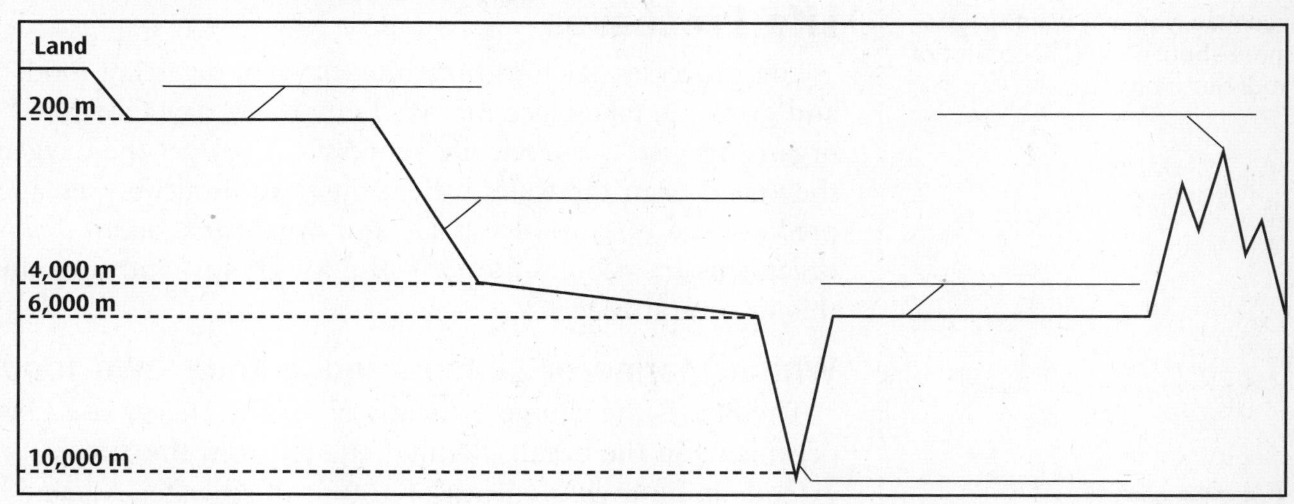


3. As you read this section, you wrote quiz questions for each topic. How did writing a quiz help you understand the information in this section?

_______________________________________________

_______________________________________________

_______________________________________________

Visit **earth.msscience.com** to access your textbook, interactive games, and projects to help you learn more about the seafloor.

## section 2 Life in the Ocean

### What You'll Learn

- how photosynthesis and chemosynthesis occur in oceans
- the key characteristics of plankton, nekton, and benthos

## Before You Read

What kinds of creatures live in the ocean?

______________________________________________

______________________________________________

Study Coach

**Authentic Questions** As you read, you may want to know more about a topic. Keep a list of your questions.

## Read to Learn

### Life Processes

Life processes, such as breathing oxygen, digesting food, and growing, take place in your body every day. Ocean organisms also carry out life processes. They get the oxygen they need from the water, where they eat, hunt prey, escape predators, and reproduce. Like land organisms, ocean organisms must obtain food to use for energy, and they do this in several ways.

#### Which marine organisms make their own food?

The Sun is the source of nearly all of the energy used by organisms in the ocean. Radiant energy from the Sun extends to an average depth of 100 m in seawater. Marine organisms such as plants and algae use the Sun's energy to build their tissues and produce their own food. This process of making food is called **photosynthesis.**

During photosynthesis, carbon dioxide and water are changed to sugar and oxygen in the presence of sunlight. Organisms that undergo photosynthesis are called producers. Producers find the nutrients they need in the surrounding ocean water. Marine producers include sea grasses, algae, seaweeds, and microscopic algae, which are responsible for about 90 percent of all marine production. Organisms that feed on producers are called consumers. Consumers in the ocean include shrimp, fish, dolphins, whales, and sharks.

FOLDABLES™

**D Compare and Contrast** Make the following Venn diagram Foldable to compare photosynthesis and chemosynthesis.

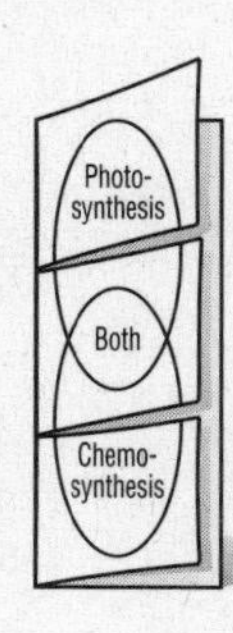

## How is the Sun's energy used in a food chain?

Energy from the Sun is transferred through food chains. Marine organisms use only a small part of the Sun's energy, then pass it from producers to consumers.

Two food chains are shown in the figure below. In the simple food chain, microscopic algae are eaten by small organisms. A whale shark eats only small organisms. The more complex food chain starts with microscopic algae being eaten by small organisms. The organisms are eaten by herring. Cod eat the herring, seals eat the cod, and eventually great white sharks eat the seals. At each stage in the food chain, energy is passed from one organism to another.

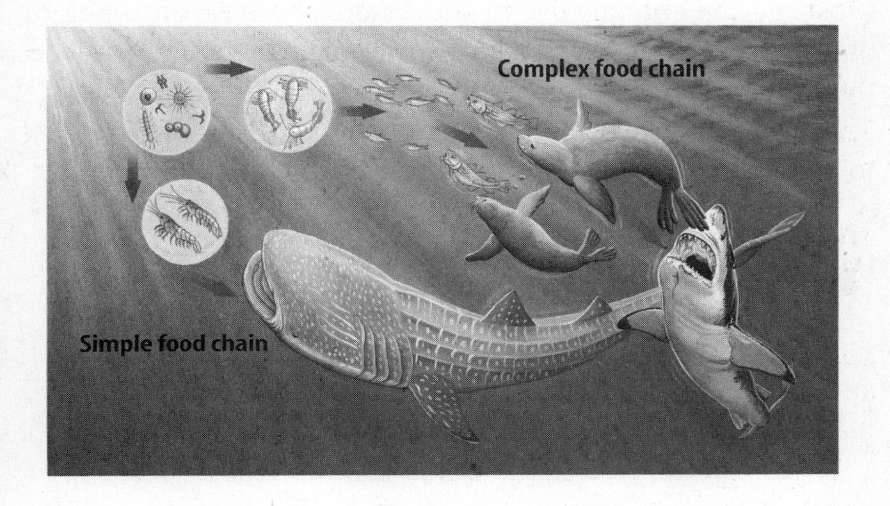


## What is a food web?

An ecosystem is a community of organisms and their environment. Many complex feeding relationships exist in an ecosystem. Most organisms depend on more than one species for food. For example, herring eat other organisms and cod eat more than herring. Ecosystem food chains look like overlapping threads in a spider web. These complicated food chains are called food webs.

## Does all marine life depend on the Sun?

Some food webs do not depend on the Sun and photosynthesis. In these food webs, bacteria perform chemosynthesis (kee moh SIHN thuh sus). **Chemosynthesis** is a food-making process that uses sulfur or nitrogen compounds as an energy source instead of sunlight. Chemosynthetic bacteria form the base of a food chain and support marine organisms such as giant tube worms, clams, and crabs. ☑

### Picture This

1. **Identify** In the figure, which organism is at the base of both food chains?
   a. Copepods
   b. Great White Shark
   c. Microscopic algae

### Reading Check

2. **Summarize** What two compounds are used by chemosynthetic organisms to make food?

_______________

_______________

**Think it Over**

**3. Draw Conclusions** Why can't sponges search for mates to reproduce?

## What other life processes occur in the ocean?

Reproduction is a vital life process. Non-moving organisms such as corals and sponges release their reproductive cells into the water, where they unite to form more organisms of the same type. Other organisms, such as salmon, travel long distances across the ocean to reproduce in a specific location. It is important for successful reproduction that eggs and newly hatched larvae develop in a safe place.

# Ocean Life

Many types of plants and animals live in the ocean. Some live in the open ocean, and others live on the ocean floor. Most marine animals live in the waters above or on the floor of the continental shelf. In this shallow water, sunlight reaches the bottom, so there are many photosynthetic organisms. These waters also contain many nutrients that producers use to carry out life processes. As a result, the greatest amount of ocean food is found in the waters of the continental shelf.

## What are plankton?

Marine organisms that drift with the current are called **plankton.** Plankton range in size from microscopic algae and animals to organisms as large as a jellyfish. Some plankton are shown in the figures below.

Phytoplankton are single-celled producers that perform photosynthesis. Phytoplankton float in the upper layers of the ocean, where there is lots of sunlight for photosynthesis. Diatoms are a form of phytoplankton. Diatoms and other phytoplankton are food for zooplankton, tiny animals that drift with ocean currents.

Zooplankton include newly hatched fish and crabs and other very small organisms. Zooplankton feed on phytoplankton and are usually the second step in ocean food chains.

**Picture This**

**4. Identify** Is the zooplankton in the figure a producer or a consumer?

**Zooplankton**

**Phytoplankton**

## What are nekton?

**Nekton** are ocean animals that actively swim, rather than drift with ocean currents. Nekton can be found from polar regions to the tropics and from shallow water to the deepest part of the ocean. Nekton include all swimming forms of fish and other animals, from tiny herring to huge whales. Sharks, manatees, and deep-ocean fish are all examples of nekton. ☑

As nekton move throughout the oceans, they must be able to control their buoyancy, or how easily they float or sink. For example, when you hold your breath underwater, the air held in your lungs helps you float. When you let the air out, you sink. Many fish have a special organ filled with gas that helps them control their buoyancy.

By changing their buoyancy, organisms can change their depth in the ocean. Being able to move between different depths allows animals to search more areas for food.

**Light-Generating Organs** Some deep-dwelling nekton have light-generating organs that they use in several ways. Some deep-sea fish dangle a shining lure from beneath their jaw. Prey attracted by the lure are swallowed quickly. Some deep-sea organisms use light to briefly blind predators so they can escape. Other use light to attract mates.

## What are bottom dwellers?

**Benthos** (BEN thahs), or bottom dwellers, are plants and animals living on or in the seafloor Benthic animals include crabs, snails, sea urchins, and bottom-dwelling fish such as flounder. These animals move or swim across the bottom searching for food. Sponges and sea anemones are benthic animals that live permanently attached to the seafloor, where they filter food particles from seawater. Certain types of worms live burrowed in the sediment of the ocean floor.

## Where can benthos be found?

Bottom-dwelling animals can be found living from the shallow water of the continental shelf to the deepest areas of the ocean. Benthic plants and algae, however, are limited to the shallow areas of the ocean where sunlight is available for photosynthesis. Kelp is a benthic algae that grows from the seafloor to a height of up to 30 m. ☑

**Reading Check**

5. **Classify** List three examples of nekton.

______________________

______________________

______________________

**FOLDABLES™**

**E Classify** Make a three-tab Foldable to write details about ocean life.

Plankton
Nekton
Benthos

**Reading Check**

6. **Explain** Why do benthic plants and algae live in shallow areas of the ocean?

______________________

______________________

FOLDABLES™

**F Organize Information**
Make a four-tab Foldable to help you organize information about the four types of ocean margin habitats.

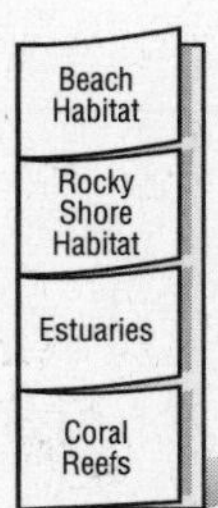


## Ocean Margin Habitats

The area of the environment where a plant or animal normally lives is called a habitat. There is a variety of habitats along the near-shore areas of the continental shelf, called ocean margins. Beaches, rocky shores, estuaries, and coral reefs are examples of the different habitats found along ocean margins.

**Beaches** Many microscopic organisms and worms live their whole life between moist grains of sand on a beach. Burrowing animals, such as small clams, make holes in the sand. When water covers the holes, these animals rise to the surface to filter food from the water. Larger animals like snails and fish live in areas that are always covered with water.

**Rocky Shore Areas** Some regions have rocky shorelines where algae, mussels, and barnacles attach to underwater rocks along the shore. Sea stars and hermit crabs crawl along rock surfaces in search of food.

Tide pools form when water remains on shore, trapped by rocks during low tide. Tide pools are an important, protected habitat for many marine animals, such as octopuses and fish. They contain large amounts of food.

**Estuaries** An **estuary** is an area where a river flows into an ocean. Because estuaries receive fresh water from rivers, they are not as salty as the ocean. Rivers also bring nutrients to estuaries, so they have many phytoplankton. Estuaries teem with life, from oysters and shrimp to fish and manatees. Estuaries are important habitats for many marine organisms because they have lots of food and fewer predators than the ocean. Many newly hatched organisms remain in estuaries until they become adults.

**Think it Over**

7. **Explain** Why do many animals spend their early lives in estuaries?

______________________

______________________

**Coral Reefs** Corals thrive in clear, warm, water that gets a lot of sunlight. Each coral animal builds a hard covering around its body using calcium it removes from seawater. Each covering is attached to others to form a large colony called a reef. A **reef** is a hard, solid structure built by corals. As a coral reef forms, benthos, such as sea stars and sponges, and nekton, such as fish, begin living on it.

In all ocean margin habitats, nutrients, food, and energy are cycled among organisms in complex food webs. Plankton, nekton, and benthos depend on each other for survival.

# After You Read

## Mini Glossary

**benthos:** marine plants and animals that live on or in the ocean floor

**chemosynthesis:** food-making process using sulfur or nitrogen compounds, rather than light energy from the Sun; used by bacteria living near hydrothermal vents

**estuary:** area where a river enters the ocean that contains a mixture of freshwater and ocean water and provides an important habitat for many marine organisms

**nekton:** marine organisms that actively swim in the ocean.

**photosynthesis:** food-making process using light energy from the Sun, carbon dioxide, and water

**plankton:** marine organisms that drift in ocean currents

**reef:** rigid, wave-resistant, ocean margin habitat built by corals from skeletal materials and calcium

1. Review the terms and their definitions in the Mini-Glossary. Write a sentence using one of the following terms: *benthos*, *nekton*, or *plankton*. In your sentence, use another term to describe where these organisms live or how they make their food.

_______________________________________________

_______________________________________________

_______________________________________________

2. The diagram below shows the flow of energy in a simple ocean food web. Fill in each blank with the name of one of the organisms listed here:

   cod    herring    great white sharks

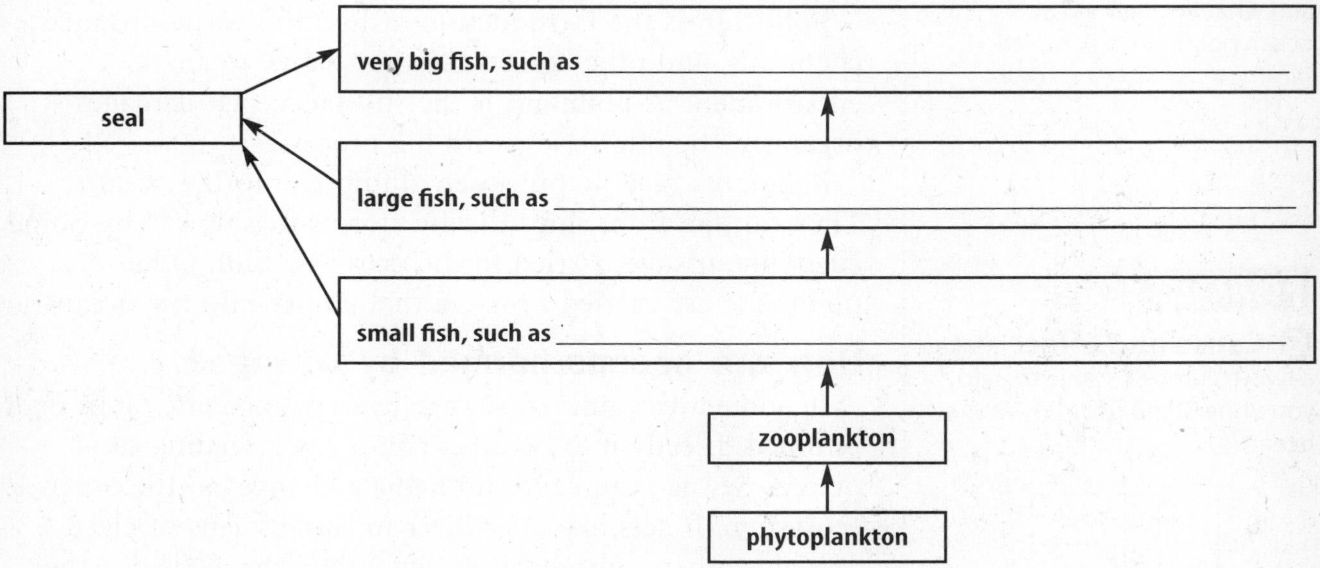

seal

very big fish, such as ____________________

large fish, such as ____________________

small fish, such as ____________________

zooplankton

phytoplankton

**ScienceOnline** Visit **earth.msscience.com** to access your textbook, interactive games, and projects to help you learn more about life in the ocean.

**End of Section**

chapter 19

# Oceanography

## section 3 Ocean Pollution

### What You'll Learn

- five types of ocean pollution
- how ocean pollution affects the entire world
- how ocean pollution can be controlled

## Before You Read

How do oceans become polluted? What can you do to help reduce the amount of pollution in our oceans?

______________________________________________

______________________________________________

______________________________________________

Study Coach

**Know—Want to Know—Learned** Divide a piece of paper into three columns. Use the columns to write what you know about ocean pollution, what you want to learn about it, and what you learned about ocean pollution in this section.

FOLDABLES™

**G Cause and Effect** Make a two-column Foldable to help you understand pollution in the ocean.

| Cause | Effect |
|---|---|
| Sewage | |
| Chemical | |
| Oil | |
| Solid waste | |

## Read to Learn

### Sources of Pollution

How would you feel if someone littered your room with trash? Organisms in the ocean experience these things when people pollute seawater.

**Pollution** is the introduction of harmful waste products, chemicals, and other substances not native to an environment. A pollutant is the substance that damages organisms by interfering with life processes.

Pollutants may be purposely dumped into the ocean. They can fall from ships during storms or shipwrecks. Some air pollutants are carried to the ocean by rain. Other pollutants are carried in rivers that empty into the ocean.

### How are oceans harmed by sewage?

In some areas, human sewage leaks from septic tanks or is pumped directly into oceans or into rivers leading to oceans. Sewage can cause immediate changes in the ocean ecosystem. It acts like a fertilizer to some forms of algae. The algae then reproduce rapidly. When the algae die, they are broken down by huge numbers of bacteria, which use up much of the oxygen in the water. Other organisms, like fish, cannot get enough oxygen and die.

## What happens after long-term exposure?

Entire ocean ecosystems have been changed as a result of long-term, repeated dumping of sewage and fertilizer runoff. In some areas of the world, sewage is dumped directly onto coral reefs. Rapidly growing algae smothers the coral organisms, and the coral reef dies. Other organisms that depend on the reef for food and shelter can be affected.

**Sources of Ocean Pollution**

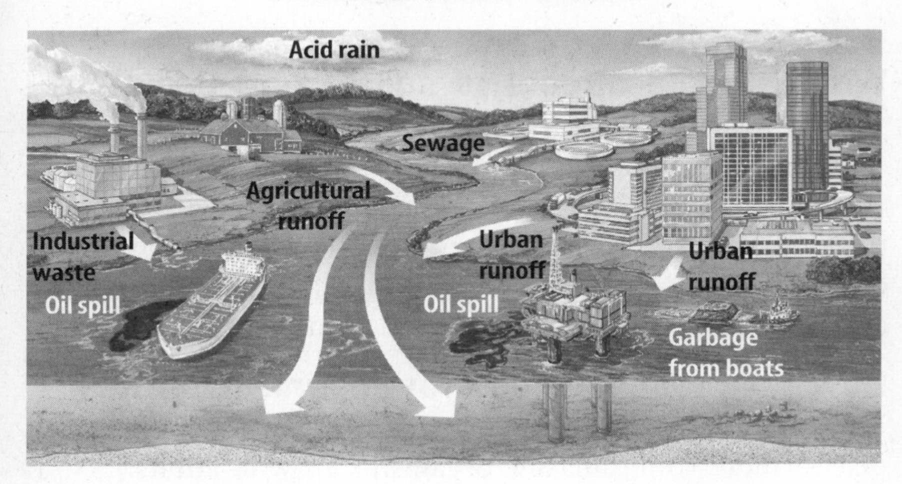


## How does chemical pollution affect marine life?

The figure above shows how pollutants from land can enter the ocean. When it rains, weed killers and insect killers are carried to streams. Eventually, these chemicals reach the ocean and may kill organisms far from where they were applied.

Factories may release chemicals directly into streams that eventually empty into the ocean. Other chemicals are released into the air and settle in the ocean. Industrial chemicals can include metals like mercury and lead. In a process called biological amplification (am plah fah KAY shun), harmful chemicals build up in the tissues of organisms that are at the top of the food chain. Top consumers like dolphins and seabirds accumulate greater amounts of toxins as they feed on smaller organisms. At high levels, some chemicals can damage an organism's immune and reproductive system.

## How does oil pollution enter the ocean?

As much as 44 percent of oil that reaches the ocean comes from land. Oil that washes from cars and streets, or that is poured down drains or into soil, flows into streams and then into the oceans. Other sources of oil pollution are shown in the circle graph on the next page.

### Think it Over

1. **Draw Conclusions** If a coral reef dies, what could happen to the organisms that live on the coral reef?

_______________

_______________

### Picture This

2. **Use Scientific Illustrations** Trace the path each pollutant takes from its source into the water.

**Picture This**

**3. Interpret Data** What is the source of most of the oil polluting the oceans?

______________________

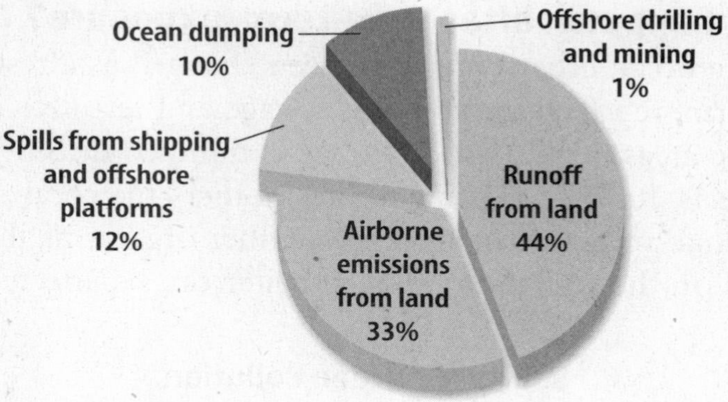

Ocean dumping 10%
Offshore drilling and mining 1%
Spills from shipping and offshore platforms 12%
Runoff from land 44%
Airborne emissions from land 33%

### What are the effects of solid-waste pollution?

Solid wastes and trash do more than litter beaches. They can harm marine life. Solid wastes, such as fishing lines, can entangle animals. Sea turtles often eat plastic bags because they look like their normal prey, floating jellyfish. Illegally dumped medical waste, such as needles and plastic tubing, also are a threat to humans and other animals.

### How can sediment destroy marine habitats?

Coral reefs and saltwater marshes are safe, protected places where young marine organisms grow to adults. However, human activities such as agriculture, deforestation, and construction tear up the soil. Rain washes soil into streams and eventually into the ocean, causing huge amounts of silt to accumulate in coastal areas. When large amounts of silt cover coral reefs and fill marshes, these habitats are destroyed and many organisms will not survive.

**Think it Over**

**4. Infer** What might happen to the habitat of a fish living in a saltwater marsh that is covered by silt?

______________________

______________________

## Effects of Pollution

Today, there is not a single area of the ocean that is not polluted in some way. As pollution from land continues to reach the ocean, scientists are recording dramatic changes in this environment.

Since the late 1980s, estuaries and the rivers that feed into them from Delaware to North Carolina have had huge populations of toxic algae, which kill billions of fish. These algae, a type of plankton, also cause rashes, nausea, and memory loss in humans. The growth of these algae is probably caused by fertilizer runoff and other waste materials. In Florida, toxic red tides kill fish and manatees. Some people blame the red tides on sewage releases and fertilizer runoff. The maps at the top of the next page show an increase in the number of harmful algal blooms since the early 1970s.

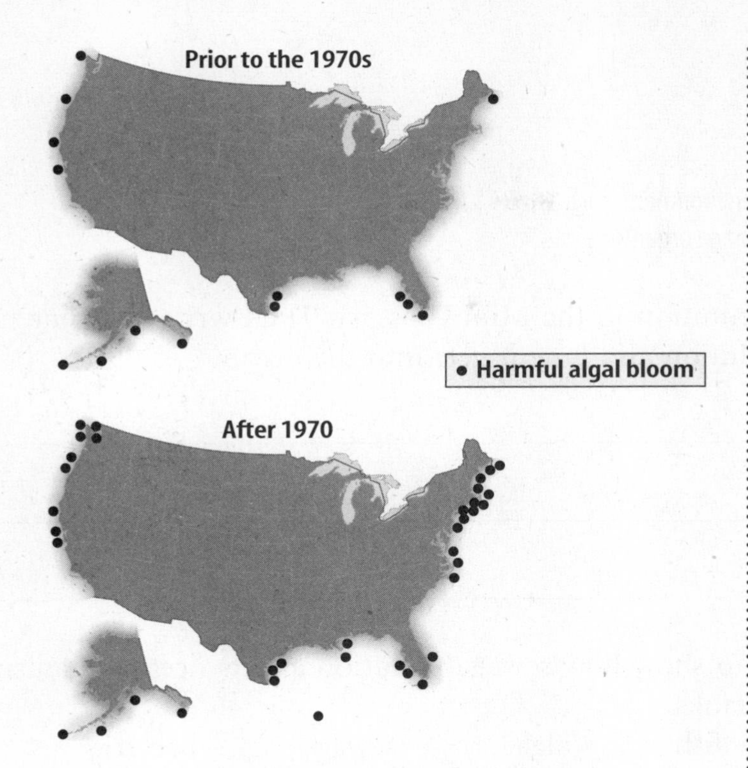


## Picture This

5. **Analyze** What region had the greatest increase in harmful algae blooms?
   **a.** Northwest Coast
   **b.** Northeast Coast
   **c.** Southeast Coast
   **d.** Alaska

## Controlling Pollution

Some people believe that oceans take care of themselves because they are large. However, others view ocean pollution as a serious problem. Many international organizations have met to discuss plans for reducing ocean pollution. Treaties make it unlawful for countries to dump some kinds of harmful wastes from vessels, aircraft, and oil platforms. One treaty requires that some ships and operators of offshore oil wells have the proper equipment to prevent and contain oil spills.

A large amount of pollution enters the ocean from land. Even though reducing land pollution has been discussed, no international agreement exists to prevent and control land-based activities that affect the oceans. ☑

**Reading Check**

6. **Identify** International organizations have established treaties to prevent and control land-based activities affecting the oceans. True or false?

______________________

### What can you do to reduce ocean pollution?

Current international and U.S. laws aren't effective enough to reduce ocean pollution. Further cooperation is needed. You can help by disposing of wastes properly and volunteering for beach or community cleanups. You can recycle materials such as paper, plastic, and glass and never dump chemicals like oil or paint onto soil, down drains, or into water. One of the best things you can do is learn about marine pollution and how people affect the oceans.

# After You Read

## Mini Glossary

**pollution:** introduction of wastes to an environment, such as sewage and chemicals, that can damage organisms

1. Review the term and its definition in the Mini Glossary. Then write one sentence describing one type of pollution and how it gets into the ocean.

2. Complete the chart below to show how sewage pollution affects ocean organisms. Use these words to fill in the blanks:

oxygen    fish    algae    sewage    bacteria

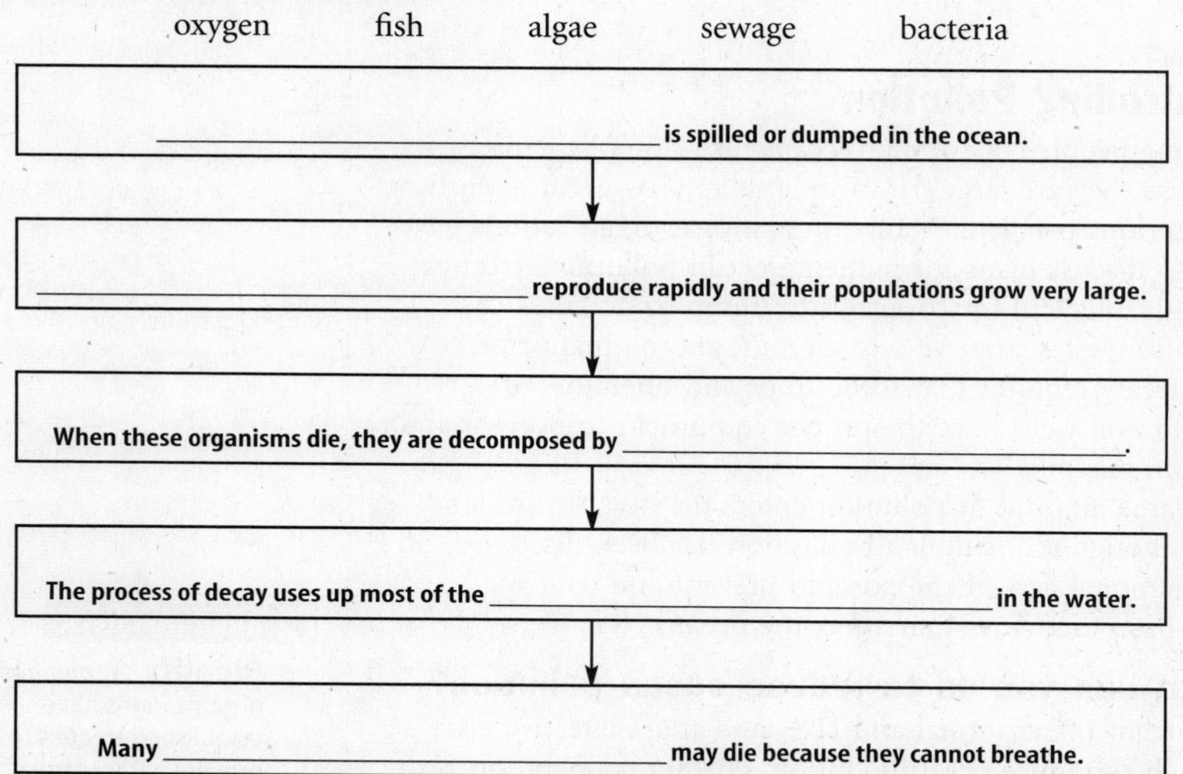

Science Online Visit **earth.msscience.com** to access your textbook, interactive games, and projects to help you learn more about ocean pollution.

# Our Impact on Land

## section 1 Population Impact on the Environment

## Before You Read

How do you think the total number of people on Earth in 1900 compares to the human population in 2000? How will the population change in the next 100 years?

---

---

---

**What You'll Learn**

- how fast the human population is increasing
- why the human population is increasing rapidly
- ways you can affect the environment

## Read to Learn

### Population and Carrying Capacity

Humans share Earth with many other living things. There are large and small animals. Plants also share Earth. Bears, ants, fish, and trees are examples of populations. A **population** is the total number of individuals of one species occupying the same area. That area can be large or small. A human population can be of one community, such as Los Angeles, or the entire planet.

Study Coach

**Create a Quiz** As you read the text, write a quiz question for each topic. When you have finished reading, try to answer your questions correctly.

#### Is Earth's population increasing?

In 2000, there were 6.1 billion humans on Earth. Each day, approximately 200,000 more humans are born. Earth is experiencing a population explosion. The word *explosion* is used because the rate at which the population is growing is rapidly increasing.

The growth rate of the human population is shown in the graph on the next page. Notice the slow population growth through the 1800s. After the mid-1800s, the population increased much faster. By 2050, the population is predicted to be about 9 billion—one and a half times what it is now.

FOLDABLES™

**A Think Critically** Make the following three-tab Foldable to explain the three factors that affect the world's population

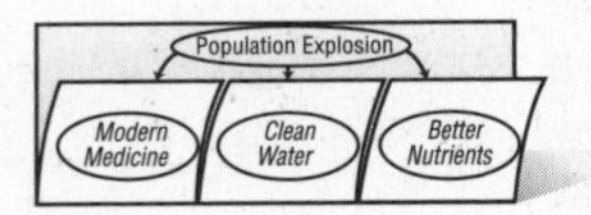

## Applying Math

1. **Interpret Data** What was the approximate human population in 1950?
   - **a.** 2 billion
   - **b.** 2.2 billion
   - **c.** 2.8 billion
   - **d.** 3.1 billion

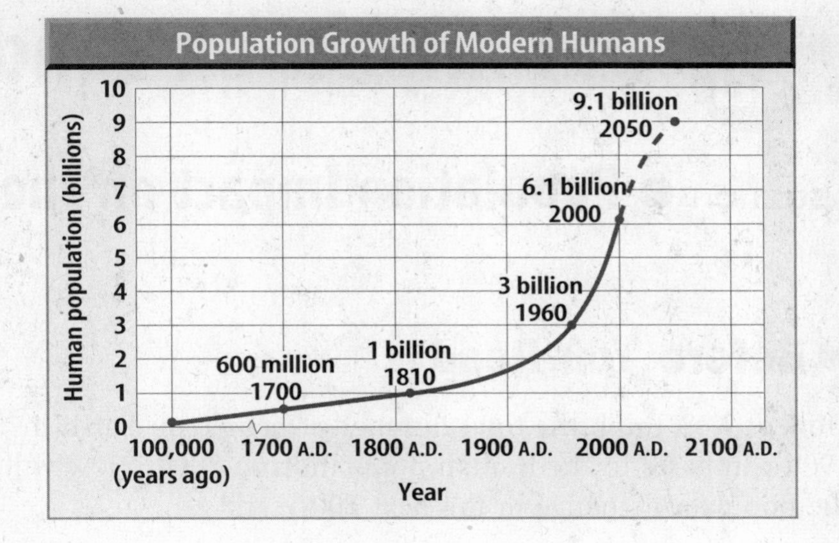


## How much has the human population grown?

Many years ago, few people lived on Earth. Look at the graph of population growth. It took thousands of years for the population to reach 1 billion humans. After the mid-1800s, the population grew much faster. The human population has increased because modern medicine, clean water, and better nutrition have decreased the death rate. Also, more babies are being born since humans now have a better chance of surviving longer.

By 2050, the population is predicted to be about 9 billion, or one and a half times what it is now. What effect will this population increase have on the environment? Will Earth have enough natural resources to support such a large population? What will become of the things that have helped the population grow, such as health care and clean water?

## What are population limits?

Each person uses space and resources. Population size depends on the amount of resources that are available on Earth and how the population uses those resources. A population without resources will suffer and the population size could decrease.

Humans once thought that Earth had an endless supply of natural resources, such as fossil fuels, clean water, rich soils, and metals. It's now known that Earth's resources are limited. The planet has a carrying capacity. **Carrying capacity** is the largest number of individuals of a particular species that the environment can support. Unless Earth's resources are treated with care, they could disappear and Earth could reach its carrying capacity. ☑

2. **Define** What is carrying capacity?

______________________

______________________

## People and the Environment

How will you affect the environment over your lifetime? By the time you're 75 years old, you will have produced enough garbage to equal the mass of 11 African elephants—53,000 kg. You will have consumed 18 million liters of water. That's enough water to fill 68,000 bathtubs. If you live in the United States, you will have used several times as much energy as a person living elsewhere in the world.

### How do daily activities affect the environment?

Every day, the environment is affected by human activity. The energy for electricity often comes from fossil fuels. Fossil fuels must be mined and then burned to create energy. Each of these steps changes the environment.

Water that is used must be treated to make it as clean as possible before being returned to the environment. You eat food, which needs soil to grow. Many vegetables are grown using chemical substances, such as pesticides to kill insects and herbicides to kill weeds. These chemicals can get into water supplies. If the concentration of chemicals is too great, it can threaten the health of living things. ☑

### What are pollutants?

Many of the products humans use are made of plastic and paper. Plastic is made from oil. The process of refining oil can produce **pollutants,** which are substances that contaminate the environment. Several of the steps used in making paper can damage the environment. Trees are cut down. Trucks use oil to transport trees to the paper mill. Water and air pollutants are given off in the papermaking process.

### How can the environment be protected?

After many products are made and used, they are thrown away as garbage. Unnecessary packaging is another waste disposal problem. As the population increases, more resources will be used and more waste will be created. Humans must think carefully before using natural resources and products that come from natural resources. If you conserve resources, you can lessen the impact on the environment.

**FOLDABLES™**

**B Understand Cause and Effect** Make the following three-column Foldable to help you understand how daily activities affect land and the environment.

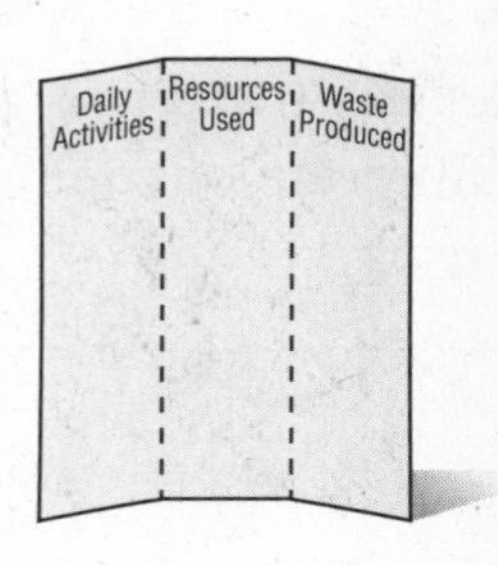


**Reading Check**

3. **Identify** Why are pesticides and herbicides used when growing vegetables?

______________________

______________________

______________________

## After You Read

### Mini Glossary

**carrying capacity:** the largest number of individuals of a given species that the environment can support.

**pollutant:** substance that contaminates the environment

**population:** total number of individuals of one species occupying the same area

1. Review the terms and their definitions in the Mini Glossary above. Then write a sentence using one of the terms to show how humans can or should take care of the environment.

_______________________________________________

_______________________________________________

2. Complete the concept map to show you understand Earth's carrying capacity.

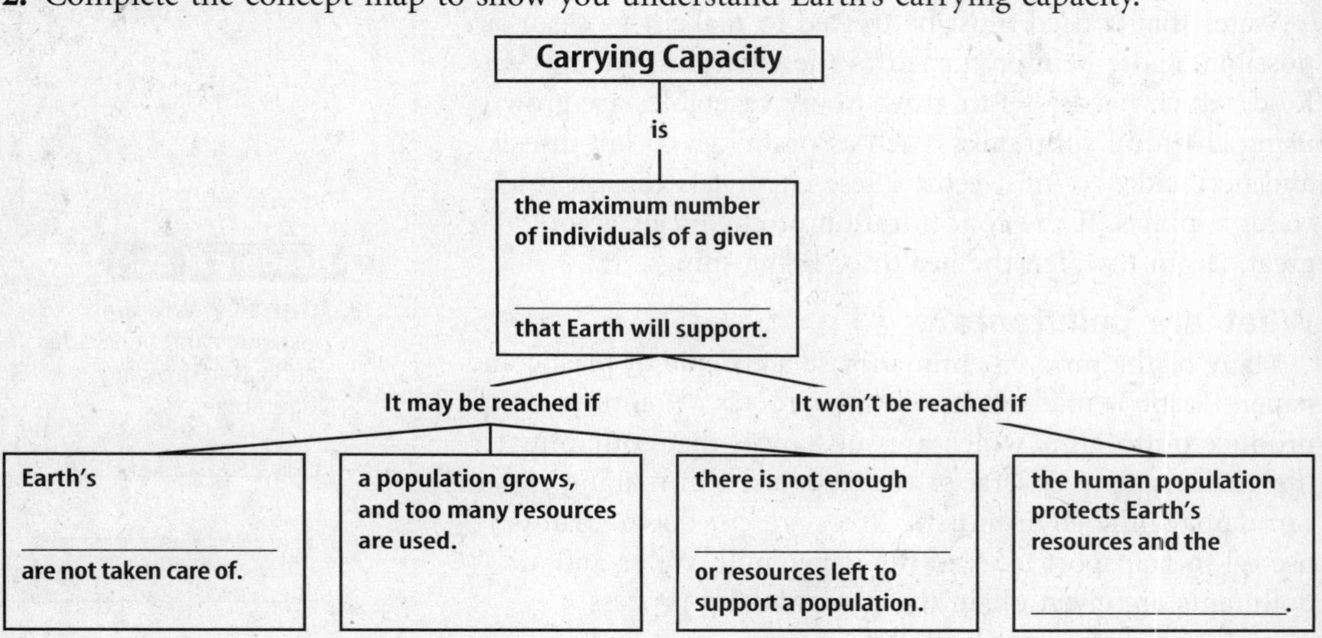


3. Review the questions you wrote as you read this section. What resources could you use to find answers to your questions? Did the questions you wrote help you understand the information in this section?

_______________________________________________

_______________________________________________

_______________________________________________

**ScienceOnline** Visit **earth.msscience.com** to access your textbook, interactive games, and projects to help you learn more about population impact on the environment.

# Our Impact on Land

## section 2 Using Land

### Before You Read

Think of one way that humans use land. Describe how this particular use of land might harm the environment.

_______________________________________________

_______________________________________________

_______________________________________________

**What You'll Learn**
- ways that land is used
- how land use creates environmental problems
- things you can do to help protect the environment

### Read to Learn

## Land Usage

Land is not often thought of as a natural resource. Yet it is as important to humans as clean air, clean water, and oil. Land is used for agriculture, logging, garbage disposal, and building. These activities can affect Earth's land resources.

### How does agriculture affect the land?

About 16 million km$^2$ of Earth's total land surface is used for farming. To feed a growing population, more food needs to be produced. Farmers use chemical fertilizers to increase the amount of food grown on each km$^2$ of land. Herbicides and pesticides are also used to reduce weeds, insects, and other pests that can damage crops.

Organic farming techniques include the use of natural fertilizers, crop rotation, and biological pest controls. These methods help crops grow without using chemicals. However, organic farming cannot currently produce enough food to feed all of the humans on Earth.

When farmland is tilled, soil is exposed. Without plant roots to hold soil in place, the soil can be carried away by running water and wind. Several centimeters of topsoil may be lost in one year. In some places, it takes more than 1,000 years for new topsoil to develop.

Study Coach

**Sticky-Note Discussion** As you read, use sticky-notes to mark pages in the text where you have a question. Use these questions as part of a discussion with other students.

FOLDABLES™

**C Organizing Information** Make quarter-sheet Foldables to help you organize information about ways land is used and how it can be preserved.

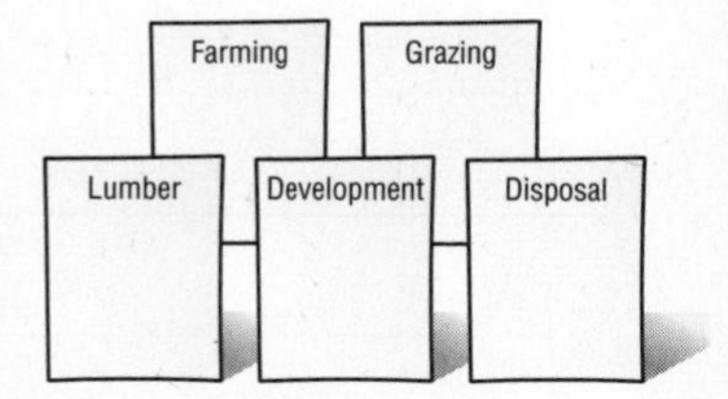

## How can soil erosion be reduced?

Some farmers till, or turn over, the soil before planting a new crop. However, other farmers practice no-till farming. They don't plow the soil from harvest until planting. Instead of tilling, they simply plant seeds between the stubble of last year's vegetation.

Another method of reducing soil loss is contour plowing. Rows are tilled across hills and valleys. When it rains, water and soil are trapped in the plowed rows, reducing soil erosion. Other techniques include planting trees in rows along fields. The trees slow the wind, reducing the amount of soil that can be blown away. Cover crops—crops that are not harvested—also can be planted to reduce erosion. ☑

**Reading Check**

1. **Summarize** What is one method used in farming to reduce soil loss?

______________________

______________________

## How is land used to feed livestock?

Land also is used for feeding livestock. Animals such as cattle eat the vegetation growing on the land. Then the animals are used as food for humans. Many farmers grow corn and hay just for feeding their livestock. These crops provide cattle with a variety of nutrients and can improve the quality of the meat.

## What are forest resources?

According to the Food and Agriculture Organization of the United Nations, about one-fourth of the land on Earth is covered by forest. About 55 percent of this forest is found in developing countries. The other 45 percent occurs in developed countries.

Deforestation is the clearing of forested land for agriculture, grazing, development, or logging. It is estimated that the amount of forested land in the world decreased by 0.24% (94,000 $km^2$) each year between 1990 and 2000. Most of this deforestation occurred in tropical regions.

Deforestation is a serious concern for many people. Many plants and animals that live in tropical rain forests cannot survive in other places. If their homes are destroyed, they may become extinct. Many of these plants might be important for developing new medicines. Scientists currently are researching how deforestation affects plants and animals.

Cutting trees can affect climate as well. Water from tree leaves evaporates into the atmosphere where it can condense to form rain. If many trees are cut down, less water enters the atmosphere and that region receives less rainfall. This is one way humans can affect the water cycle.

**Think it Over**

2. **Explain** How can the cutting of trees affect an area's climate?

______________________

______________________

______________________

## How does development affect land?

Between 1990 and 2000, the number of kilometers of roadways in the United States increased by more than 13 percent. Building highways often leads to more paving as office buildings, stores, and parking lots are constructed.

Paving land prevents water from soaking into the soil. Instead, it runs into sewers and streams. Streams are forced to move a larger volume of water. **Stream discharge** is the volume of water flowing past a certain point per unit of time. If enough rainwater flows into streams, it increases stream discharge. Then streams may flood.

Many communities get their drinking water from underground sources. When land is covered with roads, sidewalks, and parking lots, a reduced amount of rainwater soaks into the ground to refill underground water supplies.

Some communities and businesses have begun to preserve areas rather than pave them. Land is being set aside for environmental protection.

## What are sanitary landfills?

Land also is used to dispose of garbage. About 60 percent of our trash is put into sanitary landfills like the one illustrated below. A **sanitary landfill** is an area where each day's garbage is deposited and covered with soil. The soil keeps the garbage from blowing away and helps decompose some of it. It also reduces the odor that comes from decaying waste.

Sanitary landfills are designed to prevent liquid wastes from seeping into the soil and groundwater below. New sanitary landfills are lined with plastic, concrete, or clay-rich soils that trap the liquid waste. These linings greatly reduce the chance that pollutants will soak into the surrounding soil and groundwater.

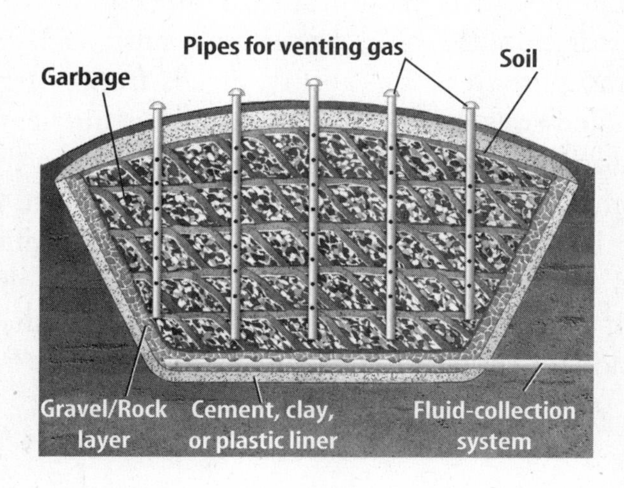


### Think it Over

**3. Think Critically** Why do you think that building a highway in an area leads to more construction in that area?

___

___

### Picture This

**4. Identify** Highlight the part of the landfill that prevents liquid waste from polluting the soil and groundwater.

**Locating New Landfills** Many materials do not decompose in landfills or decompose slowly. So when landfills are filled with garbage, new ones must be built. Locating an acceptable area to build a landfill can be difficult. Type of soil, the depth to groundwater, and neighborhood concerns must be considered. ☑

**Reading Check**

5. **List** reasons why finding a new place to build a landfill is difficult.

________________________

________________________

________________________

________________________

## Hazardous Wastes

Some of the wastes thrown away are dangerous to organisms. Wastes that are poisonous, cause cancer, or can catch fire are called **hazardous wastes.** Hazardous wastes used to be put in landfills with regular household garbage. In the 1980s, new laws made it illegal for industries to put hazardous wastes into sanitary landfills. New technologies have since been developed to help recycle hazardous wastes. Now there is less need to dispose of them.

### How can household hazardous wastes be disposed of properly?

Most individuals throw hazardous materials such as insect sprays, batteries, drain cleaners, bleaches, medicines, and paints in the trash. These substances take years to decay. You can help by disposing of hazardous wastes at special hazardous-waste-collection sites. Contact your local government to find out about collection sites in your area.

### What is phytoremediation?

Hazardous substances from leaking landfills or nearby industries can pollute soil and water. They seep into groundwater and leave toxic substances in the soil. One method for cleaning polluted soil is called phytoremediation (FI toh ruh mee dee AY shun). *Phyto* means "plant" and *remediation* means "to fix a problem." During phytoremediation, the roots of certain plants, such as alfalfa, grasses, and pine trees, can absorb metals from contaminated soils just as they absorb other nutrients. ☑

**Reading Check**

6. **Define** What is phytoremediation?

________________________

________________________

If livestock were to eat the contaminated plants, the harmful metals could end up in your milk or meat. Plants that absorb too much of these metals eventually must be harvested and either composted to recycle the metals or burned. If these plants are destroyed by burning, the ashes left behind still contain hazardous waste and must be disposed of at a hazardous-waste site.

## How are organic pollutants broken down?

Living things can clean up pollutants other than metals. Substances that contain carbon and other elements, such as hydrogen, oxygen, and nitrogen, are called organic compounds. Examples of organic pollutants include gasoline and oil. ☑

Organic pollutants can be broken down into simpler, harmless substances. Plants use some of these harmless substances as nutrients to grow. Some plant roots release enzymes (EN zimez) into the soil. An **enzyme** is a substance that makes a chemical reaction go faster. The enzymes these plants release cause organic pollutants to break down more quickly. Then the plant can use these substances for growth.

**Reading Check**

7. **List** What are some of the elements a substance must contain to be considered an organic compound?

______________________

______________________

______________________

## Natural Preserves

Not all land on Earth is being used to store waste or produce something humans can use. Some land remains mostly uninhabited by humans. National forestlands, grasslands, and national parks in the United States are protected from development. The map below shows some areas of the country that are used for national parks and preserves. Many other countries also set aside land for natural preserves. As the world population continues to grow, the strain on the environment may worsen. Preserving some land in its natural state will benefit future generations.

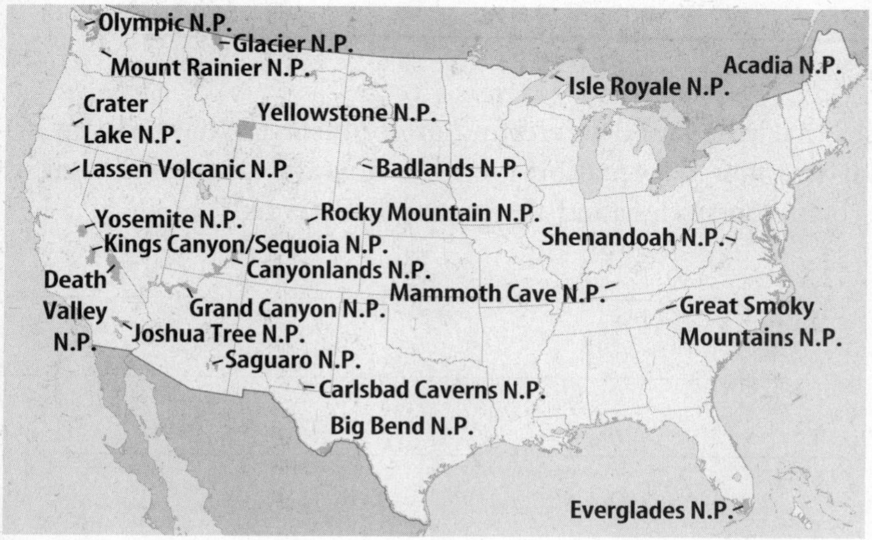

Olympic N.P.
Glacier N.P.
Mount Rainier N.P.
Acadia N.P.
Isle Royale N.P.
Crater Lake N.P.
Yellowstone N.P.
Lassen Volcanic N.P.
Badlands N.P.
Yosemite N.P.
Rocky Mountain N.P.
Kings Canyon/Sequoia N.P.
Shenandoah N.P.
Canyonlands N.P.
Death Valley N.P.
Mammoth Cave N.P.
Grand Canyon N.P.
Great Smoky Mountains N.P.
Joshua Tree N.P.
Saguaro N.P.
Carlsbad Caverns N.P.
Big Bend N.P.
Everglades N.P.

**Think it Over**

8. **Predict** What do you think could happen to the wildlife in national parks if the parks were not protected from development?

______________________

______________________

______________________

## After You Read

### Mini Glossary

**enzyme:** substance that makes a chemical reaction go faster

**hazardous waste:** wastes that are poisonous, cause cancer, or can catch fire

**sanitary landfill:** area where garbage is deposited and covered with soil

**stream discharge:** volume of water flowing past a certain point per unit of time

1. Review the terms and their definitions in the Mini Glossary above. Then write a sentence using one of the terms to explain how humans affect the environment.

_______________________________________________

_______________________________________________

2. Complete the following cause-and-effect chart to show that you understand some of the effects pollutants and overuse of soil can have on land.

| CAUSE OR ACTION | POSSIBLE EFFECT |
|---|---|
| **Farmers till soil.** | |
| **Crops harvested or vegetation dug up for construction** | |
| **Sanitary landfills leak.** | |
| **Hazardous wastes are improperly thrown away.** | |

3. As you read, you marked pages that you found interesting or confusing. What did you find that interested you or that was new information for you? How did discussing your questions with other students help you understand the material presented?

_______________________________________________

_______________________________________________

_______________________________________________

Visit **earth.msscience.com** to access your textbook, interactive games, and projects to help you learn more about land resources.

# Our Impact on Land

## section 3 Conserving Resources

## Before You Read

Do you think humans need to conserve resources? Why or why not? What might you do to help conserve resources?

______________________________________________

______________________________________________

______________________________________________

**What You'll Learn**

- three ways to conserve resources
- the advantages of recycling

## Read to Learn

**Mark the Text**

**Highlight** As you read, highlight the new vocabulary words and definitions to help you organize and understand the information in this section.

### Resource Use

Earth's natural resources are important for making everyday products. Minerals are used to make cars and bikes. Petroleum is used to produce plastics and fuel. If these resources are not used carefully, the environment can be damaged. **Conservation** is the careful use of Earth materials to reduce damage to the environment. Conservation can keep humans from running out of materials, such as certain metals.

### Reduce, Reuse, Recycle

The graph on the next page shows that developed countries, such as the United States, use more natural resources than other regions. Ways to conserve resources include reducing the use of materials and reusing and recycling materials. You can help conserve resources in simple ways. Using both sides of notebook paper and carrying your lunch to school in a non-disposable container help reduce the use of materials. Reusing an item means finding another way to use it instead of throwing it away. Old clothes can be given away or cut into rags for use around your home.

**FOLDABLES™**

**D Find Main Ideas** Make the following three-tab Foldable to help you understand ways you can conserve resources.

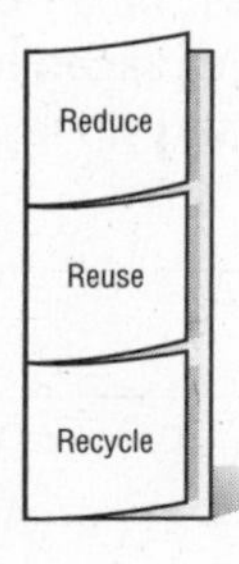

## Applying Math

1. **Use Graphs** How many more liters of oil per person are used in the United States each year than are used in the rest of the world?

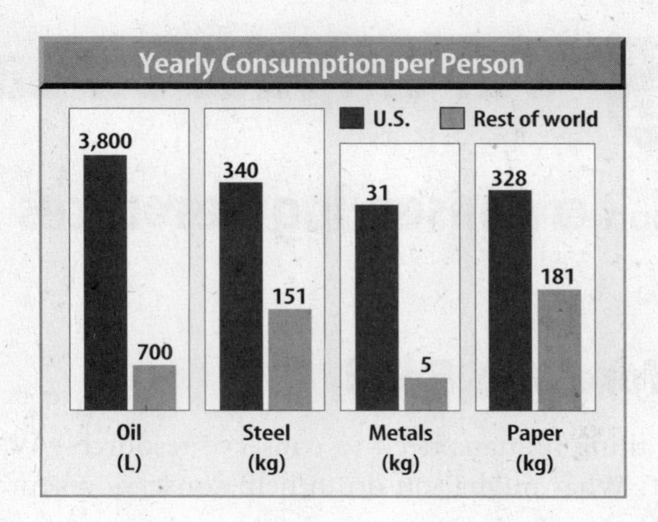


## How can yard waste be reused?

When leaves are raked or grass is cut, the waste is often bagged and thrown away. Instead, those wastes can be composted. **Composting** means piling yard wastes where they can decompose gradually. Decomposed material can provide nutrients for a garden or flowerbed. If everyone in the United States composted, it would reduce the trash put into landfills by 20 percent.

## Reading Check

2. **Identify** Name some common items you could recycle or reuse.

## What materials can be recycled?

Using materials again is called **recycling**. Examples of recyclable wastes are glass, paper, plastic, steel, and tires. Recycling helps conserve Earth's resources, energy, and landfill space. ☑

Paper makes up about 40 percent of the mass of trash. As the graph above shows, Americans throw away huge amounts of paper each year. Recycling this paper would use 58 percent less water and generate 74 percent less air pollution than producing new paper from trees. The graph does not even include newspapers. More than 500,000 trees are cut every week just to print newspapers.

Many businesses have found that recycling can be good for business. For example, some companies can make back some of the money they spend on materials by recycling the waste. Some businesses use scrap materials such as steel to make new products.

These practices save money and reduce the amount of waste sent to landfills. The amount of garbage put in landfills has decreased since 1980. In addition to saving landfill space, reducing, reusing, and recycling also reduces energy use and reduces the amount of raw materials taken from Earth.

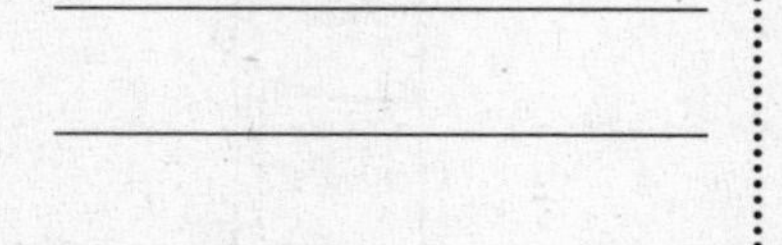


3. **Draw Conclusions** Why has the amount of garbage placed in landfills decreased since 1980?

## What kinds of programs are in use?

Many states or cities have recycling laws. In some places, people who recycle pay lower trash collection fees. In other places, a refundable deposit is made on all drink containers. You may pay a little extra for the drink when you buy it, but you can get that money back when you return the container for recycling.

## Are there costs to recycling?

There are disadvantages to recycling. More people and trucks are needed to haul materials separately from trash. The materials then must be separated at special facilities, as shown in the picture below. A demand for products made from recycled materials must exist. However, the demand for recycled materials is not high because they often cost more.

## What is the outlook for the human population?

It is not likely that the human population will decline or stop growing in the near future. To make up for this, resources must be used wisely. Conserving resources by reducing, reusing, and recycling is an important way that you can make a difference.

### Think it Over

4. **Draw Conclusions** Why do you think recycled materials often are more expensive?

______________________

______________________

______________________

______________________

### Picture This

5. **Think Critically** Why must humans take care of Earth's resources?

______________________

______________________

______________________

______________________

## After You Read

### Mini Glossary

**composting:** piling yard wastes in an area so they can decompose gradually

**conservation:** careful use of Earth materials to reduce damage to the environment

**recycling:** using materials again

1. Review the terms and their definitions in the Mini Glossary above. Then use one of the terms in a sentence explaining its role in caring for the environment.

2. Complete the chart below by listing some benefits and disadvantages of recycling.

| BENEFITS OF RECYCLING | DISADVANTAGES OF RECYCLING |
|---|---|
| Reduces trash in landfills | Must be carried separately from trash |
| | |
| | |
| | |

3. As you read the section, you highlighted the vocabulary words and their definitions. Do you think this helped you remember the definitions of the terms? Why or why not?

**Science Online** Visit **earth.msscience.com** to access your textbook, interactive games, and projects to help you learn more about conserving resources.

# Our Impact on Water and Air

## section 1 Water Pollution

### Before You Read

On the lines below, write some of the ways you use water.

______________________________________________

______________________________________________

**What You'll Learn**

- types of water pollutants and their effects
- ways to reduce water pollution

### Read to Learn

**Mark the Text**

**Highlight** As you read the text, highlight the main idea in each paragraph.

#### Importance of Clean Water

All organisms need water. Plants need water to make food from sunlight. Some animals, such as fish and whales, live in water. Humans and other land animals cannot live without drinking water.

What happens if water isn't clean? Polluted water contains chemicals and organisms that can cause disease and death in many living things. Water also can be polluted with sediments, such as silt and clay.

#### Sources of Water Pollution

Many streams and lakes in the United States are polluted in some way. Even streams that look clear and sparkling may not be safe for drinking.

**Point source pollution** is pollution that enters water from a specific location, such as drainpipes. Pollution from point sources can be controlled or treated before the water flows into a body of water.

Often, no one knows exactly where water pollution comes from. **Nonpoint source pollution** enters a body of water from a large area, such as lawns, construction sites, and roads. Nonpoint sources also include pollutants in rain or snow. Nonpoint source pollution is the largest source of water-quality problems in the United States.

**FOLDABLES™**

**A Compare and Contrast** Make a four-tab Foldable to record information about water pollution.

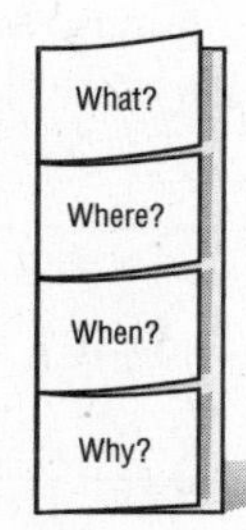

**FOLDABLES™**

**B Cause and Effect** Make a chart Foldable, as shown below, to show the causes and effects of water pollution.

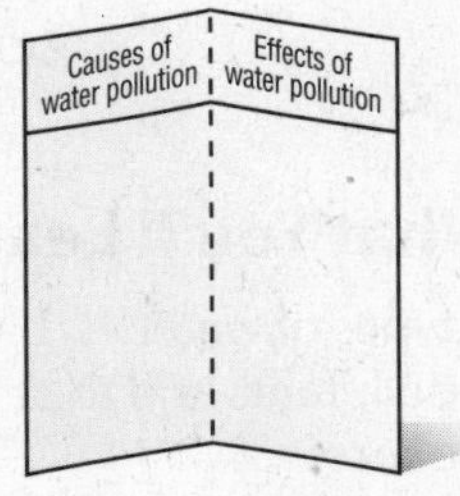


## How does sediment pollute water?

The largest source of water pollution in the United States is sediment. Sediment is pieces of rock and loose soil that are moved by erosion. Although rivers always have carried sediment to oceans, human activities have increased the amount of sediment flowing into rivers, lakes, and oceans. Each year, about 25 billion metric tons of sediment are carried from farm fields to bodies of water. At least 50 billion more tons run off of construction sites, cleared forests, and land used to graze livestock.

Sediment makes water cloudy, blocking the sunlight that underwater plants need to make food. Sediment covers the eggs of organisms that live in the water, preventing them from getting the oxygen they need to develop.

## How do farm and lawn chemicals pollute water?

**Pesticides** are substances that kill pests. Farmers and homeowners use pesticides to kill insects and plants that harm their crops and lawns. Some of these chemicals run off the land and into water, where they may be harmful to people and other organisms.

**Fertilizers** are chemicals that help plants grow. However, rain washes away as much as 25 percent of the fertilizers applied to farms and yards. This fertilizer ends up in ponds, streams, and rivers. Fertilizers contain nitrogen and phosphorus. The algae living in the water use these chemicals to grow and multiply. A lake or pond, such as the one in the figure below, can become choked with algae. When the algae die and decay, oxygen in the lake is used up, and fish and other organisms in the water may die.

### Picture This

1. **Summarize** Use the information in the figure to help explain how fertilizers can cause water pollution.

______________________

______________________

______________________

## What is sewage?

When you flush a toilet or take a shower, the water goes down a drain. This wastewater, called **sewage**, contains human waste, household detergents, and soaps. Human waste contains harmful organisms that can make people sick. In most cities, sewage treatment facilities purify sewage before it enters a body of water.

## Can metals cause water pollution?

Many metals, such as mercury, lead, and nickel, can be poisonous. Lead and mercury in drinking water can harm the nervous system. Yet these metals are part of many useful items, such as paints, stereos, and batteries. Today, environmental laws control the amounts of metals released with wastewater from factories. Because metals remain in the environment for a long time, metals released many years ago still are polluting bodies of water today.

Mining also releases metals into water. For example, in the state of Tennessee, more than 43 percent of all streams and lakes contain metal from mining activities.

**2. Describe** two ways that metals enter bodies of water.

•

## How do oil and gasoline pollute water?

When it rains, oil and gasoline run off roads and parking lots into streams and rivers. These compounds contain pollutants that might cause cancer. Gas stations store gasoline in tanks below the ground. If the tanks rust, they can leak gasoline into nearby soil and groundwater. As little as one gallon of gasoline can make an entire city's water supply unsafe for drinking.

Federal laws passed in 1988 require all new gasoline storage tanks to have a double layer of steel or fiberglass. In addition, all new and old tanks must have equipment to detect spills and must be made of material that will not develop holes. These laws help protect soil and groundwater.

## How does heat pollute water?

When a factory makes a product, heat often is released. Sometimes, water from a nearby river or lake is used to cool factory machines. Then, the heated water is released back into the river or lake. Pollution occurs because hot water contains less oxygen than cool water. Also, organisms that live in water are sensitive to temperature changes. A sudden release of hot water can quickly kill many fish. Water should be cooled in a cooling tower or pond before being released into a river. ☑

**Reading Check**

**3. Think Critically** How can factories cool hot water before releasing it into rivers?

## Reducing Water Pollution

One way to reduce water pollution is to treat water before it enters a body of water. In 1972, the U.S. Congress passed the Water Pollution Control Act. This law provided funds to build sewage treatment facilities and required industries to treat polluted water before releasing it.

The Clean Water Act of 1987 set goals for reducing point source and nonpoint source pollution. The Safe Drinking Water Act of 1996 strengthens health standards for drinking water. ☑

**Reading Check**

4. **Explain** What goals were set by the Clean Water Act?

______________________

______________________

______________________

### How are countries reducing water pollution?

Several countries have worked together to reduce water pollution. The United States and Canada made agreements to reduce pollution in the Great Lakes. Today, the Great Lakes are cleaner than they have been for many years. However, there are still more than 300 human-made chemicals in Lake Erie. The United States and Canada are studying ways to remove those chemicals from the lake.

## How can you help?

The quality of many streams, lakes and rivers has improved. However, there is still much work to be done. You can help by disposing of wastes safely and by conserving water.

### How can wastes be disposed of safely?

When you dispose of household chemicals such as paint or motor oil, don't pour them onto the ground or down the drain. These wastes can move into groundwater or rivers. Instead, read the label on the container for instructions on how to safely dispose of these wastes.

**Think it Over**

5. **Apply** What are two other ways you can help save water?

______________________

______________________

______________________

### How can water be conserved?

You use water every time you flush a toilet, take a bath, or wash dishes. Each person in the United States uses an average of 375 L of water per day. Unless it comes from a home well, this water must be purified before it reaches your home. After you use the water, it must be treated again. Some simple ways to conserve water are to turn off water while brushing your teeth, replace an old toilet with a new one, and to fix leaky faucets.

## After You Read

### Mini Glossary

**fertilizer:** chemical that helps plants and other organisms grow

**nonpoint source pollution:** pollution that enters water from a large area and cannot be traced to a single location

**pesticide:** substance used to keep insects and weeds from destroying crops and lawns

**point source pollution:** pollution that enters water from a specific location and can be controlled or treated before it enters a body of water

**sewage:** wastewater that goes into drains and contains human waste, household detergents, and soaps

1. Review the terms and their definitions in the Mini Glossary. Choose one term and use it to write a sentence telling how water pollution harms people or the environment.

_______________________________________________

_______________________________________________

2. Complete the concept map to show the sources of water pollution.

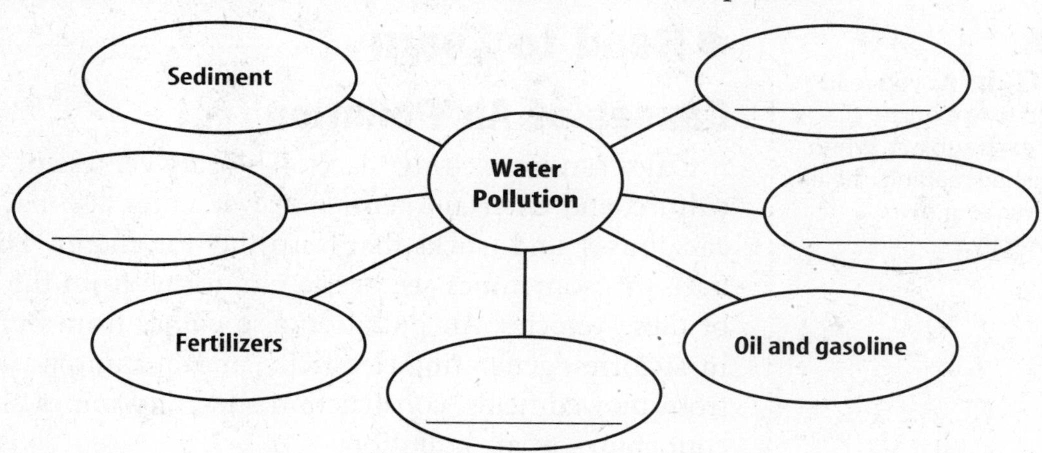


3. As you read this section, you highlighted the main ideas in each paragraph. How did highlighting the main ideas help you understand the information in this section?

_______________________________________________

_______________________________________________

_______________________________________________

Visit **earth.msscience.com** to access your textbook, interactive games, and projects to help you learn more about water pollution.

# Our Impact on Water and Air

## section 2 Air Pollution

**What You'll Learn**
- sources of air pollution
- types of air pollutants and their effects
- ways to reduce air pollution

## Before You Read

Air is just as important to most life on Earth as water is. What are some ways air pollution can harm your health?

______________________________

______________________________

______________________________

Study Coach

**Create-a-Quiz** As you read this section, create a quiz question for each subject. When you have finished reading, see if you can answer your own question correctly.

## Read to Learn

### Causes of Air Pollution

Cities can be exciting places. They are centers of business, culture, and entertainment. However, cities also have many cars, buses, and trucks that burn fuel for energy. The brown haze you sometimes see over a city forms from the exhaust of these vehicles. Air pollution also comes from burning fuel in factories, generating electricity, and burning trash. Dust from plowed fields, construction sites, and mines also contributes to air pollution.

Natural sources may cause air pollution. For example, radon is a gas given off by certain kinds of rock. This gas can seep into basements of homes built on these rocks. Breathing radon can increase the risk of lung cancer. Particles and gases from fires and erupting volcanoes also are natural sources of pollution.

FOLDABLES™

**C Organize Information** Make the following layered-book Foldable to organize information about air pollution.

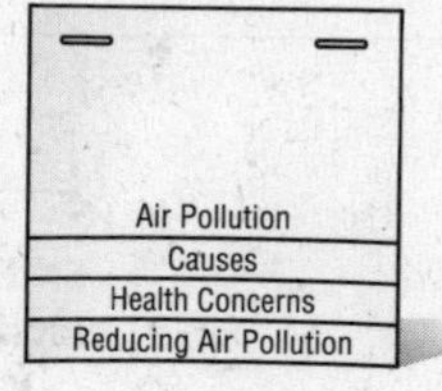


### What is smog?

One type of air pollution found in and around cities is called smog. Major sources of smog include cars, factories, and power plants. A combination of pollutants and natural conditions lead to the formation of smog.

## How does smog form?

A hazy blanket of smog that forms with the help of sunlight is called **photochemical smog.** When gasoline is burned, nitrogen and carbon compounds are released and pollute the air. The figure below shows how these compounds produce smog and ozone. Ozone high in the atmosphere protects you from the Sun's radiation. Ozone near Earth's surface is a major component of smog. It can damage your lungs and harm plants.

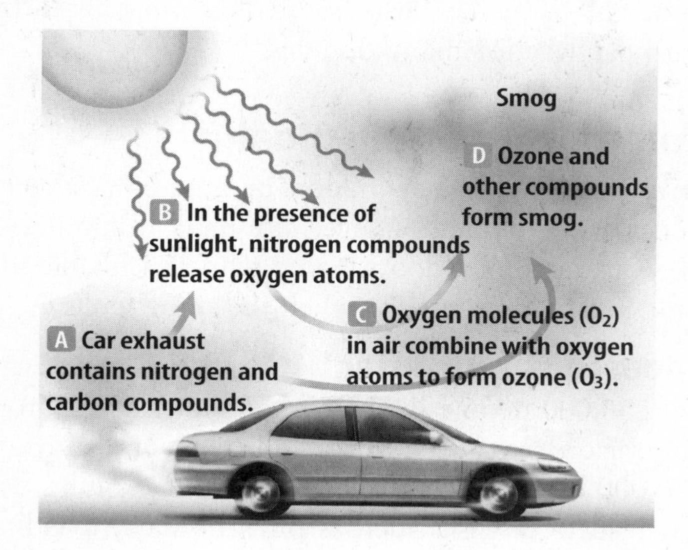


**Picture This**

1. **Identify** What is the source of the oxygen atoms that combine with oxygen in air to form ozone?

## How does nature contribute to smog?

Landforms may help form smog. Mountains can prevent smog from being carried away by winds. The atmosphere also can influence smog formation. When warm air traps cool air near the ground, a temperature inversion occurs. The reduced capacity of the atmosphere to mix materials causes pollutants to accumulate near Earth's surface.

# Acid Rain

Acids form when sulfur oxides from coal-burning power plants and nitrogen oxides from cars combine with moisture in the air. When acidic moisture falls to Earth as rain or snow, it is called **acid rain.**

The **pH scale** measures how much acid a substance contains. A pH of 7 is neutral. Substances with a pH below 7 are **acids.** Substances with a pH above 7 are **bases.**

Natural lakes and streams have a pH between 6 and 8. Acid rain has a pH below 5.6. Acid rain may decrease the pH of streams and lakes, harming some organisms. ☑

**Reading Check**

2. **Define** What is the pH of acid rain?

______________________

## CFCs

About 20 km above Earth's surface is a layer of atmosphere called the ozone layer. Recall that ozone is a molecule made of three oxygen atoms. Ozone in smog near the ground is harmful. However, ozone in the ozone layer helps life on Earth by absorbing some of the Sun's harmful ultraviolet (UV) rays.

Chlorofluorocarbons (CFCs) from air conditioners and refrigerators might be destroying the ozone layer. Each CFC molecule can destroy thousands of ozone molecules. Even though use of CFCs has decreased, CFCs can remain in the upper atmosphere for many decades.

## Air Pollution and Your Health

Have you ever had to stay indoors because of an air or smog pollution alert? When smog levels are high, it is not safe to run, bike, or exercise outdoors. Some schools schedule football games for Saturday afternoons when the smog levels are lower.

### How does dirty air affect health?

About 250,000 people in the United States suffer from breathing disorders caused by air pollution. It also causes about 70,000 deaths each year. Ozone damages lung tissue and may lead to diseases such as pneumonia and asthma. Less severe symptoms caused by ozone pollution include burning eyes, dry throat, and headache. ☑

How do you know if ozone levels in your community are safe? You may have seen an Air Quality Index reported in your newspaper or on television. When the index is at an unhealthy level, people with breathing problems are cautioned to avoid outdoor activities. The table below shows the index along with ways to protect your health when ozone is high.

**Reading Check**

3. **Identify** What is one disease caused by pollution?

______________________

**Air Quality Index**

| Air Quality | Air Quality Index | Protect Your Health |
|---|---|---|
| Good | 0–50 | No health impacts occur. |
| Moderate | 51–100 | People with breathing problems should limit outdoor exercise. |
| Unhealthy for certain people | 101–150 | Everyone, especially children and elderly, should not exercise outside for long periods of time. |
| Unhealthy | 151–200 | People with breathing problems should avoid outdoor activities. |

**Picture This**

4. **Determine** If the Air Quality Index were reported to be at 75, what would the air quality be?

______________________

## How does carbon monoxide affect people?

Carbon monoxide contributes to air pollution. **Carbon monoxide** is a colorless, odorless gas found in car exhaust. This gas can make people ill, even in small amounts, because it replaces oxygen in the blood.

## How does acid rain affect your lungs?

When you breathe the humid air from acid rain, acid is inhaled deep into your lungs. This may cause irritation and reduce your ability to fight respiratory infections. Lungs damaged by acid rain cannot easily move oxygen into blood. This puts stress on your heart.

## How do particles in the air harm the lungs?

Particulate (par TIH kyuh luht) matter also harms the lungs. **Particulate matter** consists of fine particles such as dust, pollen, mold, ash, and soot that are in the air. It comes from forest fires, exhaust from trucks and buses, smoke from factories, or even dust picked up by the wind.

Particulate matter ranges in size from large, visible bits of dust and soil to microscopic particles that form when substances are burned. Small particles are most dangerous because they can travel deeper into the lungs. When particulate matter is breathed in, it can irritate and damage the lungs, causing breathing problems.

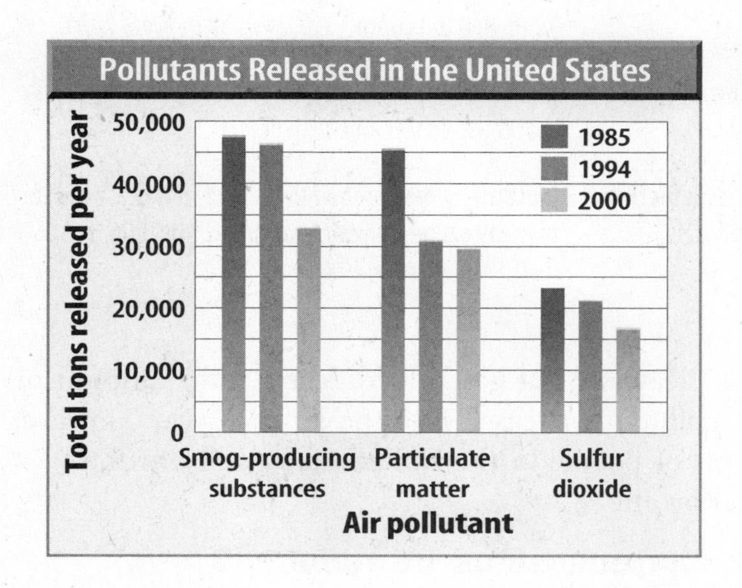


### Think it Over

**5. Infer** What tissues and organs are more likely to be harmed by air pollution?

______________________

______________________

______________________

### Picture This

**6. Analyze** Look at the graph. Describe the overall pattern in the amount of air pollutants released from 1985 to 2000.

______________________

______________________

______________________

## Reducing Air Pollution

Air pollutants cross the borders between states and countries, going wherever the wind carries them. This makes them difficult to control. For example, burning coal in midwestern states causes acid rain in the northeastern United States and Canada. ☑

When states and nations cooperate, pollution problems can be reduced. By 1999, for example, 184 countries had signed an agreement called the Montreal Protocol to stop the manufacture and use of CFCs.

**Reading Check**

7. **Explain** Why do states and countries find air pollution difficult to control?

______

______

______

### How has the United States reduced air pollution?

The United States Congress has passed several laws to protect the air. One was the Clean Air Act of 1990. It helps to reduce air pollution from cars, power plants, and other industries. The Clean Air Act is summarized in the table below.

**Clean Air Regulations**

| | |
|---|---|
| Urban air pollution | All cars manufactured since 1996 must reduce nitrogen oxide emissions by 60 percent and hydrocarbons by 35 percent from their 1990 levels. |
| Acid rain | Sulfur dioxide emissions had to be reduced by 14 million tons from 1990 levels by the year 2000. |
| Airborne toxins | Industries must limit the emission of 200 compounds that cause cancer and birth defects. |
| Ozone-depleting chemicals | Industries were required to immediately cease production of many ozone-depleting substances in 1996. |

**Think it Over**

8. **Predict** Look at the chart on this page. What might people do to reduce air pollutants even further?

______

______

______

The Clean Air Act has helped decrease the amount of some pollutants released into the air each year. However, millions of people in the United States still breathe unhealthy air.

### How can emissions be reduced?

More than 80 percent of sulfur dioxide emissions come from power plants that burn coal with a high sulfur content. When coal is burned, sulfur dioxides combine with moisture to form sulfuric acid, causing acid rain.

**Scrubbers** Sulfur dioxide can be removed by passing the smoke through a scrubber. A **scrubber** causes the gases to react with a limestone and water mixture. Another way to decrease the amount of sulfur dioxide is by burning coal that contains less sulfur. ☑

**Electrostatic Separators** Electric power plants that burn fossil fuels release particulates into the atmosphere. The particulate matter can be removed from smoke with a pollution-control device called an electrostatic separator. This device acts like a magnet for smoke particles. Plates in the separator give the smoke particles a positive charge. The positively charged smoke particles stick to negatively charged plates in the separator.

**Reading Check**

9. **Identify** What substance do scrubbers remove from power plant smoke?

______________________

## How can pollution from cars, trucks, and buses be reduced?

Recent improvements in car design, as well as the use of emission-control devices such as catalytic converters, have reduced car emissions significantly. The gasoline that we use today gives off less pollution than the gas used in the past. Future advances in technology might help to reduce emissions further. This is important because Americans are driving more today than they did in the past. More time spent driving leads to more traffic jams. Cars and trucks produce a lot of pollution when they are stopped in traffic.

## Can you help reduce air pollution?

The Clean Air Act can work only if we all cooperate. Cleaning the air takes money, time, and effort. How might you help reduce air pollution? You might change your lifestyle by walking, riding a bike, or using public transportation to get to a friend's house instead of asking for a car ride. You also can set the thermostat in your home lower in the winter and higher in the summer. ☑

**Reading Check**

10. **Identify** two ways you can reduce air pollution.

______________________

______________________

## After You Read

## Mini Glossary

**acid:** substance with a pH lower than 7

**acid rain:** acidic moisture, with a pH below 5.6, that falls to Earth as rain or snow and can damage forests, harm organisms, and corrode structures

**base:** substance with a pH above 7

**carbon monoxide:** colorless, odorless gas that reduces the oxygen content in the blood, is found in car exhaust, and contributes to air pollution

**particulate matter:** fine solids such as pollen, dust, mold, ash, and soot in the air that can irritate and damage lungs when breathed in

**pH scale:** scale used to measure the amount of acid in a substance

**photochemical smog:** hazy, yellow-brown blanket of smog found over cities that is formed with the help of sunlight, contains ozone near Earth's surface, and can damage lungs and plants

**scrubber:** device that removes sulfur dioxide emissions from coal-burning power plants

1. Review the terms and their definitions in the Mini Glossary. Write one sentence about air pollution using two vocabulary words.

___

___

2. Complete the concept map to show the types of air pollution.

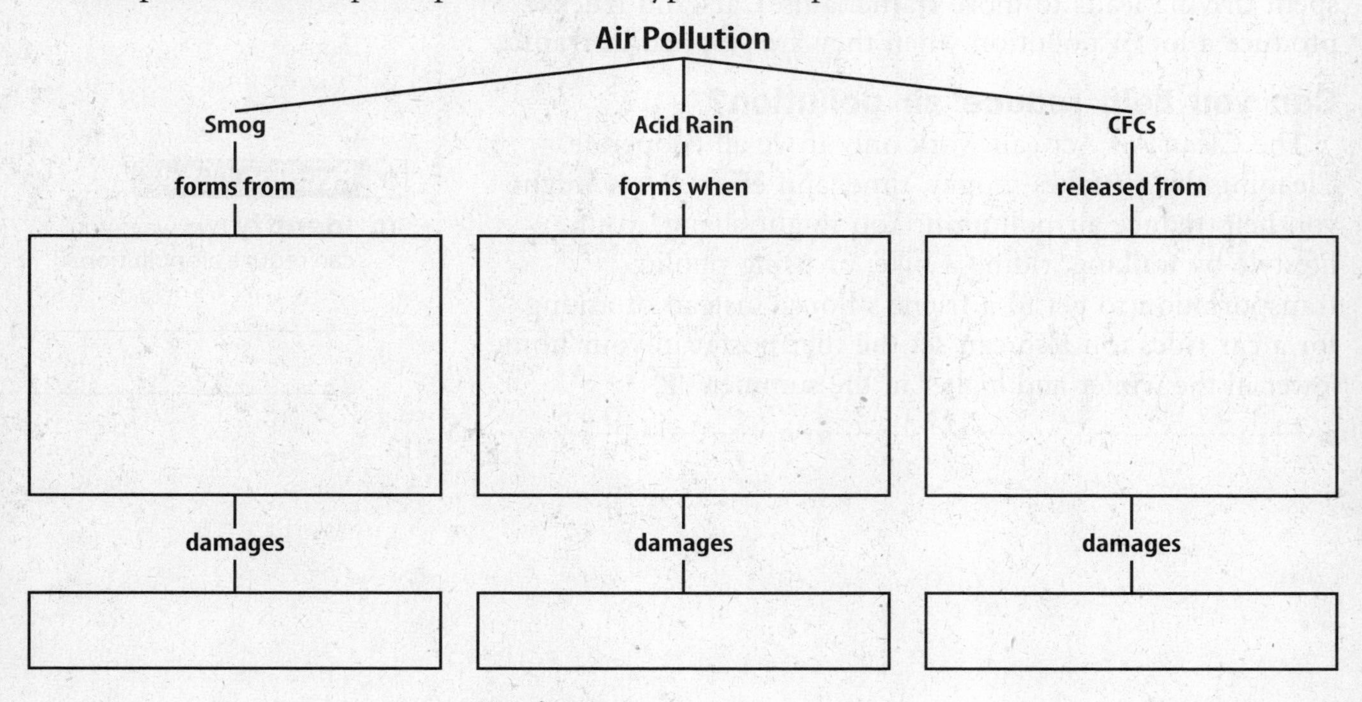


End of Section

**Science Online** Visit **earth.msscience.com** to access your textbook, interactive games, and projects to help you learn more about people and the environment.

# Exploring Space

## section 1 Radiation from Space

## Before You Read

How do we learn about objects in space? What are some of the tools astronomers use?

______________________________________________

______________________________________________

______________________________________________

### What You'll Learn

- what the electromagnetic spectrum is
- the differences between refracting and reflecting telescopes
- the differences between optical and radio telescopes

## Read to Learn

### Electromagnetic Waves

With the help of telescopes, we can see objects in our solar system and far into space. For now, this is the only way to learn about distant parts of the universe. Even if we could travel at the speed of light, it would take many years to travel to the closest star.

### What do you see when you look at a star?

When you look at a star, the light that you see left the star many years ago. Although light travels fast, distances between objects in space are so great that it sometimes takes millions of years for the light to reach Earth.

### What is electromagnetic radiation?

The light and other energy leaving a star are forms of radiation. Radiation is energy that moves from one place to another by electromagnetic waves. Since the radiation has both electric and magnetic properties, it is called electromagnetic radiation. Electromagnetic waves carry energy through empty space as well as through matter. There are electromagnetic waves everywhere around you. When you turn on the radio, the TV, or the microwave, different types of electromagnetic waves surround you.

Study Coach

**Create a Quiz Strategy** As you read, create a five-question quiz about different kinds of telescopes in this section. Switch papers with a classmate and answer each other's questions.

FOLDABLES™

**A Organize Information** Make quarter sheet Foldables to organize information from this section.

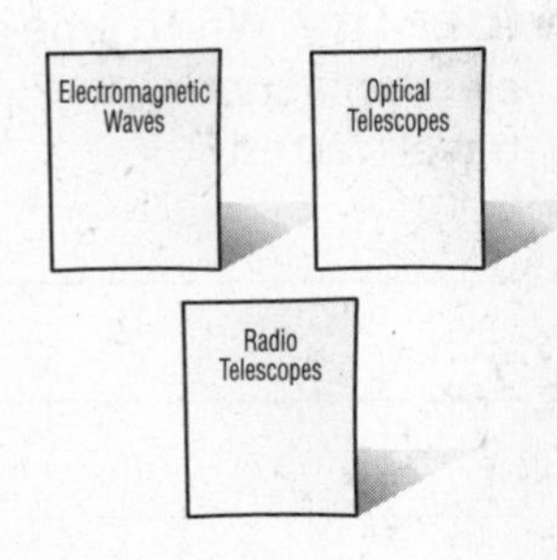

## Where is electromagnetic radiation found?

Electromagnetic radiation is all around you. Two types of electromagnetic waves are radio waves that carry signals to your radio and the light that travels to Earth from the Sun. Other types of electromagnetic waves are gamma rays, X rays, ultraviolet waves, infrared waves, and microwaves. Each of these forms of electromagnetic radiation has a different wavelength.

## What is the electromagnetic spectrum?

The different types of electromagnetic radiation are shown in the electromagnetic spectrum in the figure below. The **electromagnetic spectrum** is the arrangement of the different kinds of electromagnetic radiation according to their wavelengths. Forms of electromagnetic radiation also differ in their frequencies. Frequency is the number of wave crests that pass a given point per unit of time. The shorter the wavelength is, the higher the frequency. The figure below shows the wavelengths and frequencies of some types of electromagnetic radiation.

Microwaves
Infrared radiation
Visible light
Ultraviolet radiation
Frequency (hertz)
$10^9$ $10^{10}$ $10^{11}$ $10^{12}$ $10^{13}$ $10^{14}$ $10^{15}$ $10^{16}$ $10^{1}$
Wavelength (meters)
$10^{-1}$ $10^{-2}$ $10^{-3}$ $10^{-4}$ $10^{-5}$ $10^{-6}$ $10^{-7}$ $10^{-8}$ $10$

## How fast do electromagnetic waves travel?

Even though electromagnetic waves have different wavelengths, they all travel at the same speed. All electromagnetic waves travel at the speed of light, or 300,000 km/s. Stars give off visible light and other electromagnetic waves. It can take millions of years for some stars' light waves to reach Earth because the universe is so large. The light you see when you look at a star left the star many years ago.

Scientists can learn about the source of the electromagnetic radiation by studying a star's light waves. One tool that scientists use to study electromagnetic radiation in space is a telescope. A telescope is an instrument that magnifies, or enlarges, images of distant objects. There are different kinds of telescopes.

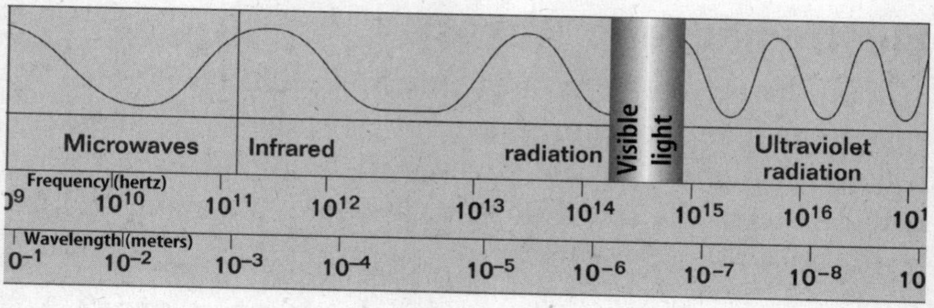

### Picture This

**1. Interpret Scientific Illustrations** Which type of electromagnetic radiation has a longer wavelength: infrared or ultraviolet?

______________________

### Think it Over

**2. Identify** Which type of electromagnetic wave travels fastest?

______________________

______________________

______________________

## Optical Telescopes

An optical telescope collects visible light, which is a form of electromagnetic radiation, to produce magnified images of objects. The telescope collects light using either an objective lens or mirror. The objective lens or mirror then forms an image at the focal point of the telescope. The focal point is where light that is bent by the lens or reflected by the mirror comes together to form an image. The image is magnified by another lens, the eyepiece. There are two types of optical telescopes.

**Refracting Telescope** A **refracting telescope** collects light using convex lenses. Convex lenses curve outward, like the surface of a ball. Light from an object passes through the convex objective lens. The lens bends the light to form an image at the focal point. The eyepiece magnifies the image. The illustration below on the left shows how a refracting telescope works.

**Reflecting Telescope** A **reflecting telescope** collects light using a concave mirror. The concave mirror is curved inward, like the inside of a bowl. When light strikes the mirror, the mirror reflects, or bounces, light to the focal point where it forms an image. Sometimes a smaller mirror is used to reflect light into the eyepiece lens, where it is magnified for viewing. The illustration below on the right shows how a reflecting telescope works.

**Picture This**

3. **Compare and Contrast** Name one difference between a refracting telescope and a reflecting telescope.

______________________

______________________

**Refracting Telescope**

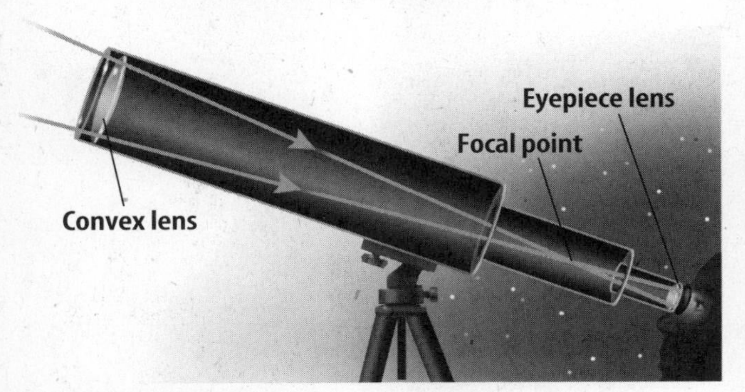


In a refracting telescope, a convex lens focuses light to form an image at the focal point.

**Reflecting Telescope**

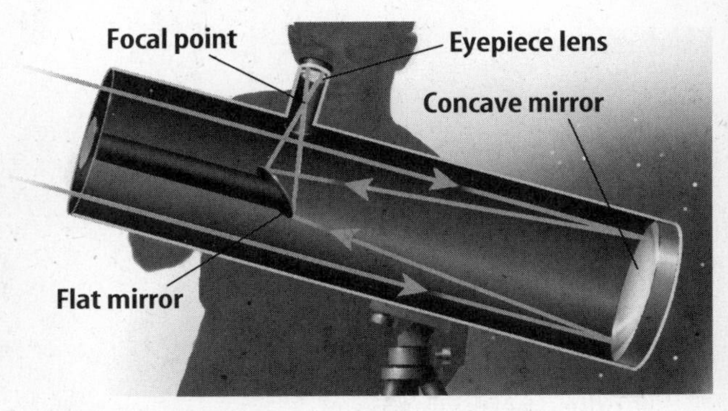


In a reflecting telescope, a concave mirror focuses light to form an image at the focal point.

## How are optical telescopes used?

Most optical telescopes used by astronomers—scientists who study space—are housed in buildings called **observatories**. Observatories have dome-shaped roofs that open. This allows astronomers to view the night sky. ☑

**Reading Check**

**4. Describe** What is an observatory?

______________________

______________________

## What is the *Hubble Space Telescope*?

Not all optical telescopes are housed in observatories. The *Hubble Space Telescope* is a large reflecting telescope that orbits Earth. It was launched in 1990 by the space shuttle *Discovery*. Earth's atmosphere can absorb and distort energy received from space. Since *Hubble* was placed outside this atmosphere, scientists expected it to produce clear pictures. However, a mistake was made when the telescope's largest mirror was shaped. It did not produce clear images. In 1993, a set of small mirrors was installed to correct the faulty images. Two more missions to service *Hubble* happened in 1997 and 1999—as shown in the photograph. In 1999, *Hubble* sent back clear images of a large cluster of galaxies known as Abell 2218.

NASA

**Picture This**

**5. Explain** Why does the *Hubble Space Telescope* produce better images of space than telescopes on Earth?

______________________

______________________

## How do reflecting telescopes work?

In the early 1600s, the Italian scientist Galileo Galilei aimed a small telescope at the stars. Since then, telescopes have been greatly improved. Today, large reflecting telescopes use mirrors to direct light and to magnify images. These mirrors are several meters wide and are extremely hard to build. Instead of constructing one large mirror, some telescopes have mirrors that are constructed out of many small mirrors that are pieced together. The mirrors of the twin Keck telescopes in Hawaii are 10 meters wide and were made in this way.

## How do optical telescopes work?

The most recent advances in optical telescopes involve active and adaptive optics. Active optics use a computer to correct for changes in temperature, mirror distortions, and bad viewing conditions. Adaptive optics are even more advanced. Adaptive optics use lasers to probe the atmosphere for air turbulence and send back information to a computer. The computer then makes adjustments to the telescope's mirror. Telescope images are clearer when corrections for air turbulence, temperature changes, and mirror-shape changes are made.

# Radio Telescopes

Stars and other objects in space give off different kinds of electromagnetic energy. One example of that energy is the radio wave. Radio waves are a kind of long-wavelength energy in the electromagnetic spectrum. A **radio telescope** is used to collect and record radio waves that travel through space. Unlike visible light, radio waves pass freely through Earth's atmosphere. Because of this, radio telescopes are useful 24 hours a day under most weather conditions. ☑

Radio waves reaching Earth's surface strike the large, concave dish of a radio telescope. This dish reflects the waves to a focal point where a receiver is located. The information gathered allows scientists to detect objects in space, to map the universe, and to search for signs of intelligent life on other planets.

### Think it Over

6. **Think Critically** Why should scientists continue to try to improve telescopes?

______________________

______________________

### Reading Check

7. **Explain** Why don't radio telescopes need to be above Earth's atmosphere in order to collect data from space?

______________________

______________________

______________________

## After You Read

### Mini Glossary

**electromagnetic spectrum:** arrangement of electromagnetic waves according to their wavelengths

**observatory:** building that can house an optical telescope; often has a dome-shaped roof that can open for viewing

**radio telescope:** telescope that collects and records radio waves that travel through space; can be used day or night and in most weather conditions

**reflecting telescope:** optical telescope that collects light using a concave mirror to reflect light and form an image at the focal point

**refracting telescope:** optical telescope that collects light using convex lenses to bend light and form an image at the focal point

**1.** Review the terms and their definitions above. Choose a term and write a sentence that shows you understand the meaning of that term.

______________________________________________

______________________________________________

______________________________________________

**2.** Fill in the missing information in the table.

| | Optical Telescope | Radio Telescope |
|---|---|---|
| Type of electromagnetic radiation collected | ______________ | Radio Waves |
| Parts used to collect electromagnetic radiation | ______ and mirrors, eyepieces | Curved dish and ______ |
| Affected by atmosphere? | ______________ | ______________ |

**3.** Did you answer all the questions on the quiz you took? How did the *Create a Quiz* strategy help you remember what you read?

______________________________________________

______________________________________________

______________________________________________

Visit **earth.msscience.com** to access your textbook, interactive games, and projects to help you learn more about radiation from space.

# chapter 22 Exploring Space

## section 2 Early Space Missions

## Before You Read

What have humans sent into space? Why were those things sent into space?

______________________________________________

______________________________________________

______________________________________________

### What You'll Learn

- about natural and artificial satellites
- the difference between artificial satellites and space probes
- about the history of the race to the moon

## Read to Learn

### The First Missions into Space

Astronomers have used telescopes to learn a lot about the Moon and the planets. However, astronomers want to gain more knowledge by sending humans to these places or by sending spacecraft where humans can't go.

**Mark the Text**

**Identify the Main Point** Highlight the main point in each paragraph. Use a different color to highlight a detail or an example that helps explain the main point.

### How do spacecraft travel?

Spacecraft must travel faster than 11 km/s to break free of Earth's gravity and enter Earth's orbit. They can do this with special engines called rockets. **Rockets** are engines that carry their own fuel and have everything they need for the burning of fuel. They don't require air to carry out the process. Therefore, they can work in space, which has no air.

### What are the different types of rockets?

The simplest rocket engine is made of a burning chamber and a nozzle. A more complicated rocket has more than one burning chamber. There are two types of rockets. The difference between them is what fuel, or propellant, they burn.

**Solid-Propellant Rockets** Solid-propellant rockets are simpler in design. However, they cannot be shut down and restarted after they are ignited.

**FOLDABLES™**

**B Organizing Information** Make a two-tab Foldable to help you organize information about satellites.

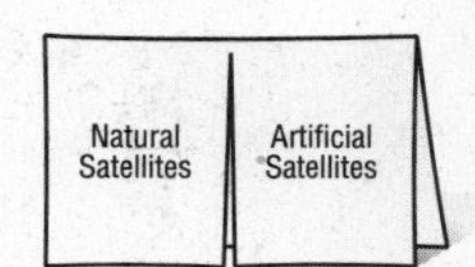

**Liquid-Propellant Rockets** Liquid-propellant rockets can be shut down after they are ignited and restarted. For this reason, they are used for long-term space missions. Scientists on Earth send signals to stop and restart the spacecraft's engines in order to change the spacecraft's direction. Liquid propellant rockets powered many space probes, including the two *Voyagers* and *Galileo.* In the photograph below, a liquid-propellant rocket stands on the launchpad. ☑

**Reading Check**

1. **Identify** What kind of rocket was used to power the two *Voyager* space probes?

______________________

NASA

**Picture This**

2. **Describe** What is the advantage of the liquid-propellant rocket in the photo over a solid-propellant rocket?

______________________

______________________

## How are fuels used to launch rockets?

Solid-propellant rockets use a rubber-like fuel. The fuel contains its own oxidizer. The burning chamber of a rocket is a tube that has a nozzle at one end. As the solid propellant burns, hot gases exert pressure on all inner surfaces of the tube. The tube pushes back on the gas except at the nozzle where hot gases escape. Thrust builds up and pushes the rocket forward.

Liquid-propellant rockets use a liquid fuel and an oxidizer, such as liquid oxygen stored in separate tanks. To ignite the rocket, the oxidizer is mixed with the liquid fuel in the burning chamber. As the mixture burns, forces are exerted and the rocket is propelled forward.

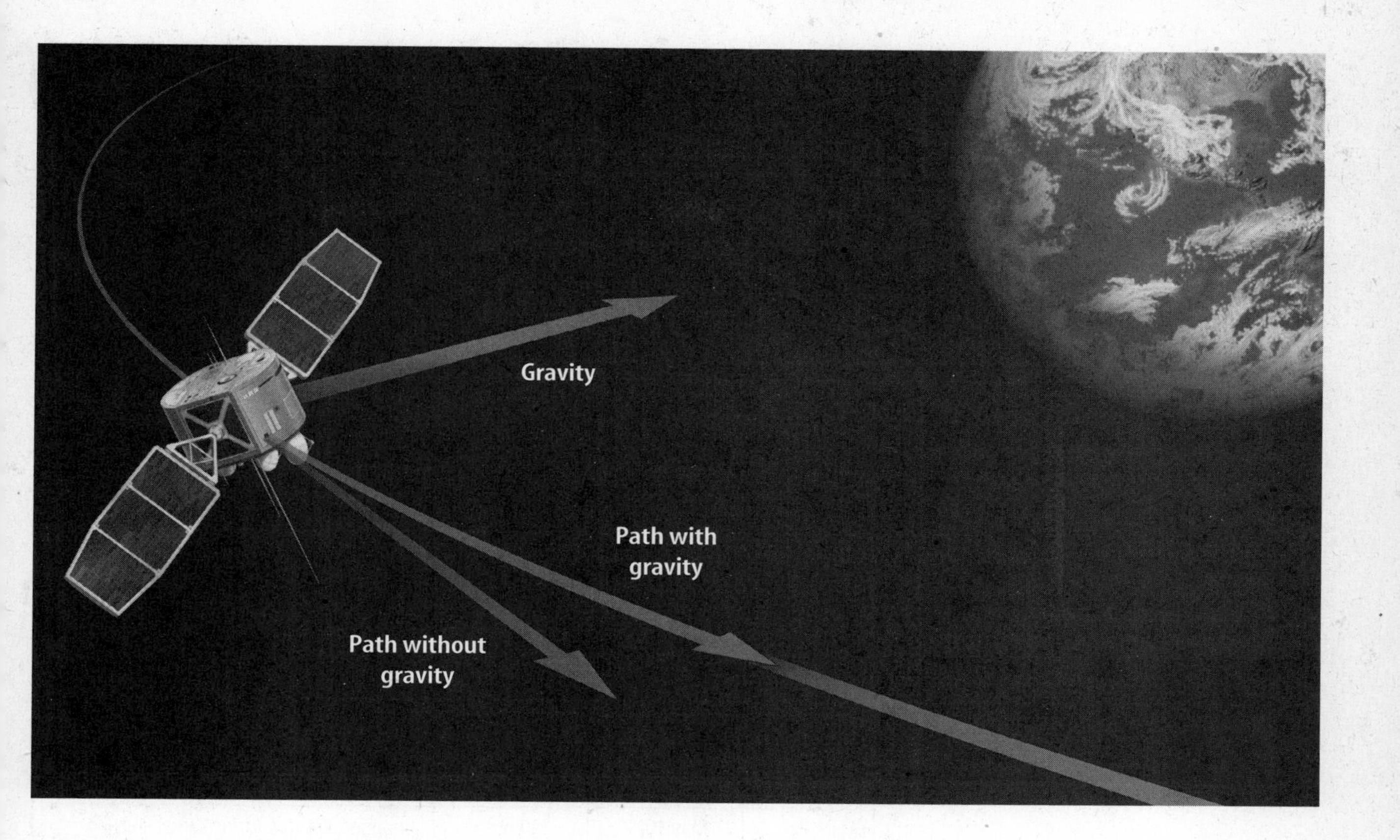


## What are satellites?

The space age began in 1957 when the former Soviet Union used a rocket to send *Sputnik I* into space. *Sputnik I* was the first artificial satellite. A **satellite** is any object that revolves around another object. The Moon, which circles Earth, is an example of a natural satellite.

When an object enters space, it travels in a straight line unless a force makes it change direction. Earth's gravity is such a force. Earth's gravity pulls a satellite toward Earth. The illustration above shows the effect of gravity on a satellite's path. The result of the satellite traveling forward while at the same time being pulled toward Earth is a curved path. This curved path is called an **orbit**. *Sputnik I* orbited Earth for 57 days before gravity pulled it back into Earth's atmosphere, where it burned up.

## What are satellites used for?

Today, thousands of artificial satellites orbit Earth. They have many uses. Communications satellites transmit radio and television programs to locations around the world. Other satellites collect scientific data. Weather satellites constantly monitor Earth's global weather patterns.

### Picture This

3. **Interpret Scientific Illustrations** Where would the satellite go if Earth's gravity did not exist?

______________________

______________________

______________________

## Space Probes

Rockets also carry instruments into space to collect data. A **space probe** is an instrument that gathers information and sends it back to Earth. Space probes travel into and beyond the solar system, carrying cameras and other equipment to collect data. They carry radio transmitters and receivers to communicate with scientists on Earth. The table shows some of the space probes launched by the National Aeronautics and Space Administration (NASA).

### Think it Over

4. **Compare** What did the *Mariner 2* and *Magellan* space missions have in common?

_______________

_______________

| Some Early Space Missions | | | |
|---|---|---|---|
| Mission Name | Date Launched | Destination | Data Obtained |
| Mariner 2 | August 1962 | Venus | verified high temperatures in Venus's atmosphere |
| Pioneer 10 | March 1972 | Jupiter | sent back photos of Jupiter—first probe to encounter an outer planet |
| Viking 1 | August 1975 | Mars | orbiter mapped the surface of Mars; lander searched for life on Mars |
| Magellan | May 1989 | Venus | mapped Venus's surface and returned data on the composition of Venus's atmosphere |

## Where are other important space probes?

*Voyager 1* and *Voyager 2* were launched in 1977. *Voyager I* flew past Jupiter and Saturn. *Voyager 2* flew past Jupiter, Saturn, Uranus, and Neptune. The objective of both probes is to explore beyond the solar system. Scientists expect both probes to send data back to Earth for at least 20 more years.

*Galileo*, launched in 1989, reached Jupiter in 1995. A smaller probe was released from *Galileo* into Jupiter's violent atmosphere. The small probe collected information about Jupiter's makeup, temperature, and pressure. *Galileo* also gathered information about Jupiter's moons, rings, and magnetic fields. The data from *Galileo* show that there may be an ocean of water under the surface of Europa, one of Jupiter's moons. *Galileo* also took photographs of a powerful volcanic vent on Io, another one of Jupiter's moons.

### Applying Math

5. **Calculate** How many years did it take the probe *Galileo* to reach Jupiter?

_______________

## Moon Quest

*Sputnik I* only sent out a beeping sound as it orbited Earth. But people soon realized that sending a human into space was not far off.

The former Soviet Union sent the first human into space in 1961. Cosmonaut Yuri A. Gagarin orbited Earth and returned safely. President John F. Kennedy called for the United States to send humans to the Moon and return them safely to Earth. He wanted to do this by the end of the 1960s. The race for space had begun.

The U.S. program to reach the Moon began with **Project Mercury.** The goal of *Project Mercury* was to orbit a piloted spacecraft and to bring it back safely. On May 5, 1961, Alan B. Shepard became the first U.S. citizen in space. In 1962, John Glenn became the first U.S. citizen to orbit Earth. ☑

## What was Project Gemini?

The next step in reaching the Moon was **Project Gemini.** There were two astronauts on every *Gemini* mission. *Gemini* spacecraft were larger than *Mercury* spacecraft. On one mission, astronauts met and connected with another spacecraft that was in orbit. The *Gemini* program also studied the effects of space travel on the human body.

Scientists also sent robotic probes to learn about the Moon. These probes did not carry humans. The probe *Ranger* proved that a spacecraft could be sent to the Moon. In 1966, *Surveyor* landed on the Moon's surface, proving that the surface could support spacecraft and humans. *Lunar Orbiter* took pictures of the Moon's surface to help choose future landing sites.

## When did humans first walk on the Moon?

The final stage of the U.S. program to reach the Moon was **Project Apollo.** On July 20, 1969, the spacecraft *Apollo 11* landed on the Moon's surface. Neil Armstrong was the first human to set foot on the Moon. Edwin Aldrin walked on the Moon with Armstrong while Michael Collins remained in the Command Module. There were a total of six *Apollo* landings. Astronauts brought more than 2,000 samples of Moon rock and soil back to Earth before the program ended in 1972.

**Reading Check**

6. **Summarize** What was the goal of *Project Mercury*?

_______________

_______________

_______________

**Think it Over**

7. **Sequence** Write the year in which each event occurred.

_______ The first human is sent into space.

_______ The first human orbits Earth.

_______ The first human walks on the Moon.

## After You Read

### Mini Glossary

**orbit:** a curved path around a star or planet

**Project Apollo:** the final stage in the U.S. program to reach the Moon; the first person walked on the Moon on July 20, 1969

**Project Gemini:** the second stage in the U.S. program to reach the Moon in which an astronaut team connected with another spacecraft that was in orbit

**Project Mercury:** the first stage in the U.S. program to reach the Moon, in which a spacecraft with an astronaut orbited Earth and returned safely

**rocket:** engine that can work in space and burns liquid or solid fuel

**satellite:** any object that revolves around another object

**space probe:** an instrument that is carried into space, collects data, and sends the data back to Earth

**1.** Review the terms and their definitions in the Mini Glossary. Use the term *orbit* to describe a satellite.

______________________________________________

**2.** Complete the following diagram to describe early U.S. space missions. Use the terms: *Earth, Gemini, Apollo, Mercury,* and *the Moon.*

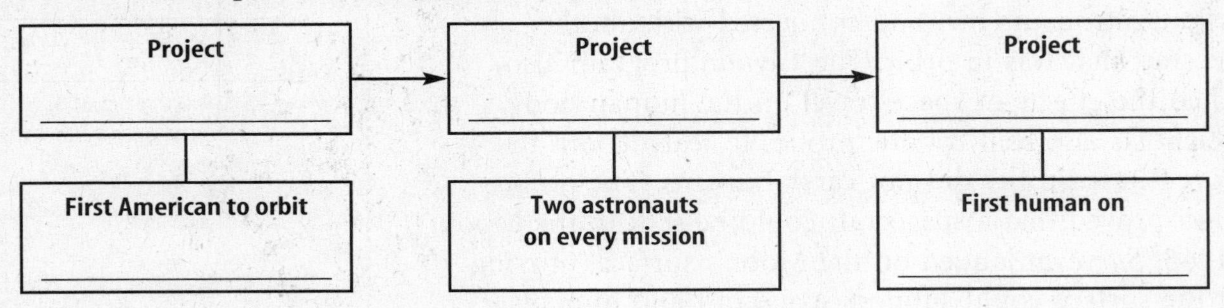


**3.** Choose one of the question headings in the Read to Learn section. Write the question in the box. Then answer it in your own words.

**Write your question here.**

______________________________________________

______________________________________________

Visit **earth.msscience.com** to access your textbook, interactive games, and projects to help you learn more about early space missions.

# chapter 22 Exploring Space

## section 3 Current and Future Space Missions

### Before You Read

Do you have questions about the other planets? Write three questions about the planets on the lines below.

_______________________________________________

_______________________________________________

_______________________________________________

**What You'll Learn**

- about the space shuttle
- about orbital space stations
- about plans for future space missions
- about the application of space technology to everyday life

### Read to Learn

#### The Space Shuttle

NASA's early rockets cost millions of dollars and could be used only once. They were used to launch a small capsule holding astronauts into orbit.

NASA realized it would be less expensive and less wasteful to reuse resources. The space shuttle was created. The **space shuttle** is a reusable spacecraft that carries astronauts, satellites, and other materials to and from space.

At launch, the space shuttle stands on end and is connected to an external liquid-fuel tank and two solid-fuel booster rockets. When the shuttle reaches an altitude of about 45 km, the emptied, solid-fuel booster rockets drop off and parachute back to Earth. The rockets are recovered and reused. The liquid-fuel tank also separates and falls to Earth but is not recovered.

#### What happens on the space shuttle?

In space, the shuttle orbits Earth. Astronauts conduct scientific experiments, such as how space flight affects the human body. They also launch, repair, and retrieve satellites. When the mission is complete, the shuttle glides back to Earth and lands like an airplane.

Study Coach

**Outline** Outline the facts you learn about current and future space missions. For each mission, include a question that scientists hope to answer.

FOLDABLES™

**C Find Main Ideas** Create a six-tab Foldable to summarize the main ideas from the section.

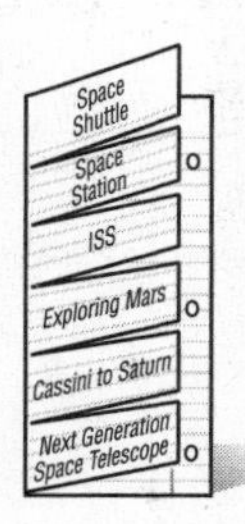

## Space Stations

Astronauts can spend only a short time living in the space shuttle. Its living area is small, and the crew needs more room to live, exercise, and work. A **space station** has living quarters, work and exercise areas, and all the equipment and life support systems that humans need in order to live and work in space.

In 1973, the United States launched the space station *Skylab.* Crews of astronauts spent up to 84 days there. They performed scientific experiments and collected data on the effects on humans of living in space. In 1979, the empty *Skylab* fell out of orbit and burned up as it entered Earth's atmosphere.

The former Soviet Union launched the space station *Mir.* Crews from the former Soviet Union spent more time on board *Mir* than crews from any other country.

**Think it Over**

1. **Infer** Why can't long-term projects be done on the space shuttle?

______________________

______________________

## Cooperation in Space

In 1995, the United States and Russia began an era of cooperation and trust in exploring space. One American and two Russians were launched into space together on a Russian spacecraft. Then a Russian traveled into space on an American shuttle. There were many missions involving space shuttles docking at *Mir.* Each was an important step toward building and operating the new *International Space Station.*

### What is the *International Space Station*?

The *International Space Station (ISS)* will be a permanent laboratory in space designed for long-term research projects. Some of the research will be used to improve medicines and the treatment of many diseases. Sixteen nations are working together to build sections of the *ISS.* The sections will then be carried into space where the *ISS* will be constructed. The space shuttle and Russian rockets will transport the sections. The illustration below shows the proposed design for the *ISS.* ☑

**Reading Check**

2. **Identify** What are some ways that the *ISS* will be used?

______________________

______________________

______________________

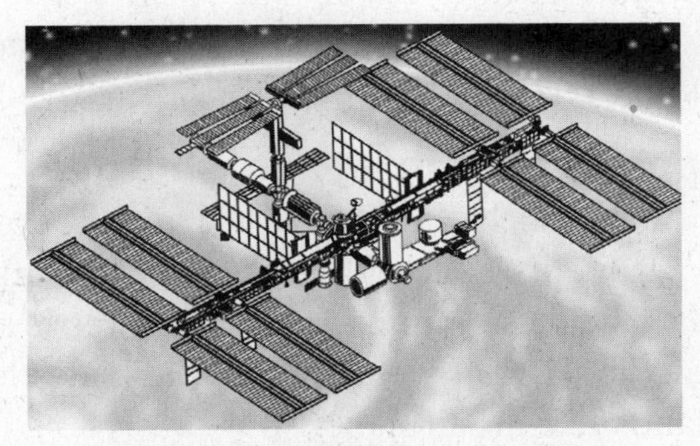

**Proposed *International Space Station***

## How is the *ISS* being constructed?

The *ISS* is being built in phases. Phase One involved the space shuttle *Mir* docking missions. Phase Two began in 1998 when the first *ISS* module, or unit, was put into orbit. Another unit was attached, and the first crew of three people went to live and work on the station. Phase Two ended in 2001 when a U.S. laboratory was added. Labs from Japan, Europe, and Russia will follow. It will take 47 launches to finish the *ISS* by 2006, its scheduled completion date. NASA plans to station seven-person crews onboard for several months at a time.

## Exploring Mars

In 1996, two Mars missions were launched, *Mars Global Surveyor* and *Mars Pathfinder. Surveyor* orbited Mars. It took high-quality photos of the planet's surface. *Pathfinder* landed on Mars. Rockets and a parachute slowed its fall. *Pathfinder* carried scientific instruments to study the surface, including a remote-controlled robot rover called *Sojourner.* ☑

In 2002, the spacecraft *Mars Odyssey* began to map Mars. The information it gathered proved that soil contained frozen water on one part of Mars. In 2003, the twin robot rovers *Spirit* and *Opportunity* were launched from Earth to explore the surface of Mars. They will study the rocks, soils, and water on Mars. In 2008, a rover called *Phoenix* will be sent to dig over a meter into the surface.

**Reading Check**

**3. Identify** What was the name of the probe launched in 1996 that landed on Mars?

______________________

**Think it Over**

**4. Infer** How could digging in Martian soil help scientists learn about Mars?

______________________

______________________

## New Millennium Program

NASA has plans for future space missions. The New Millennium Program (NMP) will develop equipment to be sent into the solar system.

## Exploring the Moon

The *Lunar Prospector* spacecraft was launched in 1998. For one year it orbited the Moon, mapped it, and collected data. The data showed that there might be ice in craters at the Moon's poles. At the end of the mission, *Prospector* was crashed, on purpose, into a lunar crater. Scientists used special telescopes to look for water vapor that might have been tossed up when the spacecraft hit. They didn't find any water. But they believe that water ice is there. This water would be useful if a colony is ever built on the Moon.

## *Cassini*

In October 1997, NASA launched the space probe *Cassini.* This probe's destination is Saturn. When it lands, the space probe will explore Saturn and surrounding areas for four years. One part of its mission is to deliver the European Space Agency's *Huygens* probe to Saturn's largest moon, Titan. Some scientists theorize that Titan's atmosphere may be similar to the atmosphere of early Earth. ☑

**Reading Check**

5. **Describe** What do scientists hope to learn about Saturn's moon Titan from *Huygens*?

______________________

______________________

## What will the new space telescopes be like?

Not all missions involve sending astronauts or probes into space. Plans are being made to launch a new space telescope that is capable of observing the first stars and galaxies in the universe. The *James Webb Space Telescope*, shown in the figure below, will be the successor to the *Hubble Space Telescope.* As part of the Origins project, it will provide scientists with the chance to study how galaxies evolved, how stars produce elements, and how stars and planets are formed. To accomplish these tasks, the telescope will have to see objects that are 400 times fainter than any objects seen by telescopes on Earth. NASA hopes to launch the *James Webb Space Telescope* as early as 2010.

**Picture This**

6. **Think Critically** Why might scientists want to learn how galaxies evolved?

______________________

______________________

______________________

NASA

## What are some benefits of space technology?

Research done for the space programs is also used to solve problems on Earth. It has led to better ways to detect and treat heart disease. It has helped doctors create a way to find eye problems in infants. Knowledge gained from shuttle research has helped scientists develop cochlear implants. These tiny ear devices have helped thousands of deaf people hear.

Space technology can even help catch criminals and prevent accidents. Scientists developed a way to sharpen images they got from space. Now police use this same method to read blurry photos of license plates, as shown in the picture below.

Police cars and ambulances use an instrument developed in space research. As an emergency vehicle approaches a traffic light, the instrument changes the signal so other cars have time to stop safely. Global Positioning System (GPS) technology uses satellites to determine location on Earth's surface.

The Cover Story/CORBIS

### Think it Over

**7. Think Critically** What are some ways space technology benefits people on Earth?

______________________

______________________

______________________

### Picture This

**8. Draw Conclusions** How might the use of space technology help police enforce the law?

______________________

______________________

______________________

## After You Read

### Mini Glossary

**space shuttle:** reusable spacecraft that carries astronauts, satellites, and other materials to and from space

**space station:** a structure with living quarters, work and exercise areas, and equipment and life support systems for humans to live and work in space

1. Review the terms and their definitions in the Mini Glossary. Write a sentence to explain how the space shuttle is used to build the space station.

_______________________________________________

_______________________________________________

2. Complete the diagram to review what you learned about the *International Space Station.*

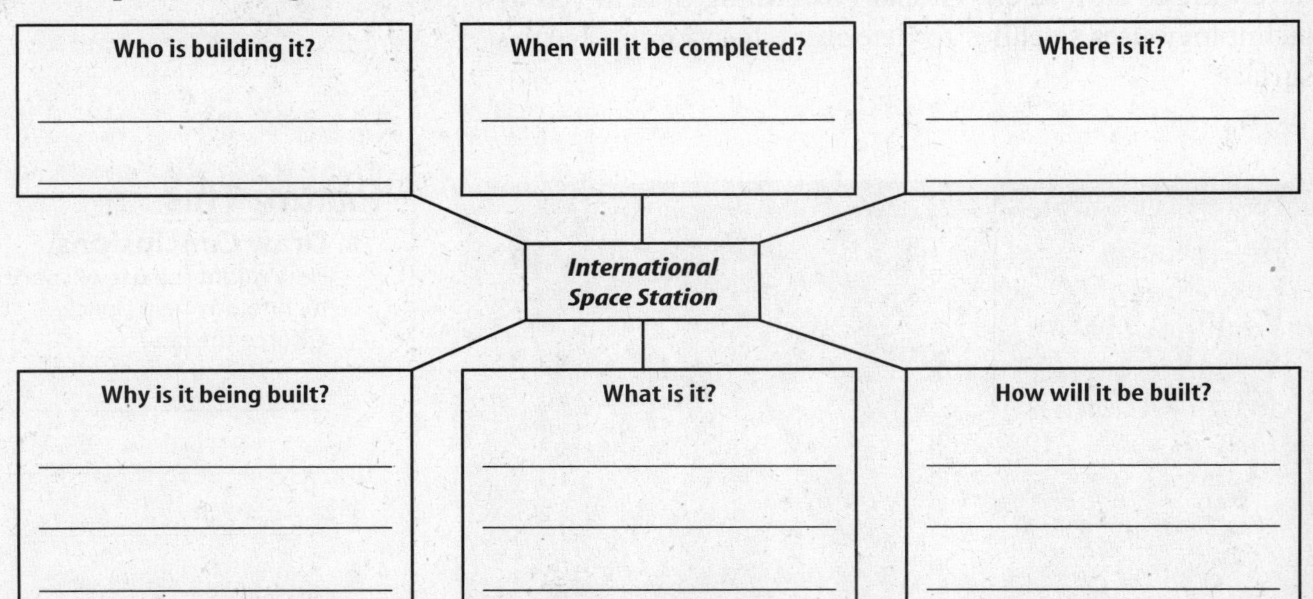


3. Describe two ways that space research has helped people on Earth.

_______________________________________________

_______________________________________________

Visit **earth.msscience.com** to access your textbook, interactive games, and projects to help you learn more about current and future space missions.

# The Sun-Earth-Moon System

## section 1 Earth

### Before You Read

What do you already know about Earth's shape, its size, and how it moves? Write what you know on the lines below.

______________________________________________

______________________________________________

______________________________________________

**What You'll Learn**

- Earth's shape, size, and movements
- the difference between rotation and revolution
- what causes the seasons

### Read to Learn

**Study Coach**

**Make a Sketch** As you read, draw your own sketches to help you understand and remember new information.

## Properties of Earth

In the morning, the Sun rises in the east. It moves across the sky during the day. Finally, the Sun sets in the west. Is the Sun moving—or are you?

People once thought that Earth was a flat object at the center of the universe. They believed that the Sun went around Earth in a big circle each day. Now, most people know that Earth is not flat, and the Sun only looks like it is moving around Earth. Scientists have discovered that Earth spins and that Earth moves around the Sun. It is the spinning motion of Earth that makes it look like the Sun is moving across the sky.

### What is Earth's shape?

Basketballs, tennis balls, and Earth have something in common. They are all round, three-dimensional objects called **spheres** (SFIHRZ). The distance from the center of a sphere to any point on the surface is the same.

Aristotle, a Greek astronomer and philosopher who lived around 350 B.C., observed that Earth made a curved shadow on the Moon during an eclipse. His observations led him to think that Earth was a sphere.

**FOLDABLES™**

**A Find Main Ideas** Make the following six-tab Foldable to identify and record the main ideas about Earth.

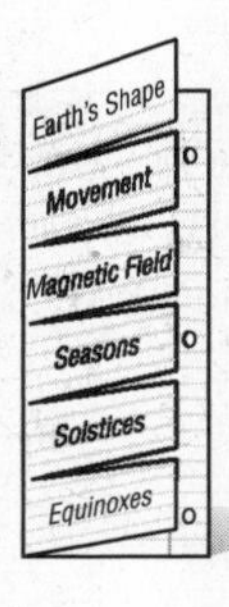

## How do we know Earth is a sphere?

Today, we have observations from astronauts and pictures from artificial satellites and space probes to show us Earth's shape. Now we also know Earth is not a perfect sphere. It bulges at the equator and is somewhat flat at the poles. The table below shows the differences in Earth's diameter at the equator and from pole to pole.

**Applying Math**

1. **Solve a One-Step Equation** On the table, find the numbers for Diameter (equator) and for Diameter (pole to pole). Subtract the number for Diameter (pole to pole) from the number for Diameter (equator). What does this tell you about Earth's shape?

______

______

______

| Physical Properties of Earth | |
|---|---|
| Diameter (pole to pole) | 12,714 km |
| Diameter (equator) | 12,756 km |
| Circumference (poles) | 40,008 km |
| Circumference (equator) | 40,075 km |
| Mass | $5.98 \times 10^{24}$ kg |
| Average density | 5.52 g/cm$^3$ |
| Average distance to the Sun | 149,600,000 km |
| Period of rotation (1 day) | 23 h, 56 min |
| Period of revolution (1 year) | 365 days, 6 h, 9 min |

## Does Earth spin?

Earth spins like a top. The imaginary center line around which Earth spins is called Earth's **axis.** The poles are at the north and the south ends of Earth's axis. The spinning of Earth on its axis is called **rotation.**

Earth's rotation causes day and night. As Earth rotates, your area of Earth faces toward the Sun in the morning and away from the Sun at night. Earth rotates once each day. A rotation takes about 24 hours.

## Magnetic Field

Scientists hypothesize that the movement of material inside Earth's core, along with Earth's rotation, generates a magnetic field. Like a bar magnet, Earth has opposite north and south magnetic poles. Earth's magnetic field protects you from harmful radiation. It does this by trapping many charged particles that reach Earth from the Sun. ☑

**Reading Check**

2. **Explain** How is Earth like a bar magnet?

______

______

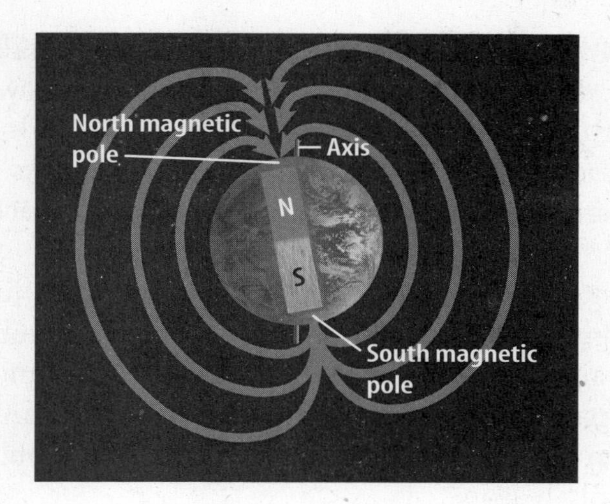


## What is Earth's magnetic axis?

When a compass needle points north, you are seeing proof of Earth's magnetic field. The line that joins Earth's north and south magnetic poles is called its magnetic axis. As shown in the figure above, the magnetic axis does not line up with Earth's rotational axis. In fact, the location of the magnetic axis changes slowly over time. A compass whose needle points north will lead you to Earth's magnetic north pole, not the rotational north pole.

## What causes changing seasons?

Flowers bloom as the days get warmer. The Sun appears higher in the sky, and daylight lasts longer. Spring seems like a fresh, new beginning. What causes these changes?

### Does Earth's orbit cause seasons?

Recall that Earth's rotation causes day and night. Another movement of Earth is called revolution. **Revolution** is Earth's orbit, or the path of Earth, as it goes around the Sun. It takes a year for Earth to orbit the Sun. ☑

The shape of Earth's path around the Sun is an **ellipse** (ee LIHPS)—a long, curved shape, similar to a stretched-out circle. The Sun is not located in the center of the ellipse but is a little toward one end. Earth is closest to the Sun around January 3, and farthest from the Sun around July 4.

Although Earth's orbit takes it nearer and farther from the Sun, the change in distance is small and does not cause seasons. If Earth's distance from the Sun caused the seasons, January—when the Earth is nearest to the Sun—would have the warmest days. This is not the case, however, in the northern hemisphere.

**Picture This**

3. **Draw** lines through Earth's magnetic axis and Earth's rotational axis to show they do not line up.

**Reading Check**

4. **Define** What is Earth's revolution?

______________________

______________________

## Does Earth's tilted axis cause seasons?

Earth's axis is tilted 23.5 degrees from a line drawn perpendicular to the plane of its orbit. It is this tilt that causes seasons. The tilt explains why Earth receives such a different amount of solar energy from place to place during the year.

In the northern hemisphere, summer begins in June and ends in September. This is when the northern hemisphere is tilted toward the Sun. During summer, there are more hours of sunlight—or solar energy. Longer periods of sunlight are one reason that summer is warmer than winter, but this is not the only reason.

**Think it Over**

**5. Infer** In the winter, are daylight hours longer or shorter than in summer?

______________________

______________________

## How does Earth's tilt affect solar radiation?

Earth's tilt causes the Sun's radiation to strike the hemispheres at different angles. Sunlight strikes the hemisphere tilted toward the Sun at an angle closer to 90 degrees than the hemisphere tilted away. Thus the hemisphere tilted toward the Sun receives more solar radiation than the hemisphere tilted away from the Sun.

Summer occurs in the hemisphere tilted toward the Sun, when its radiation strikes Earth at a high angle and for longer periods of time. The hemisphere receiving less radiation experiences winter.

**Think it Over**

**6. Infer** If it is winter in the northern hemisphere, which hemisphere is getting more of the Sun's radiation?

______________________

______________________

# Solstices

The **solstice** is the day when the Sun reaches its greatest distance north or south of the equator. In the northern hemisphere, the summer solstice occurs on June 21 or 22, and the winter solstice occurs on December 21 or 22. The position of Earth in relation to the Sun at different times of the year is shown in the figure on the next page. In the southern hemisphere, the winter solstice is in June and the summer solstice is in December.

Summer solstice is the longest period of daylight of the year. From the summer solstice to the winter solstice, the number of daylight hours keeps decreasing. The winter solstice is the shortest period of daylight of the year. Then the number of daylight hours begins increasing again.

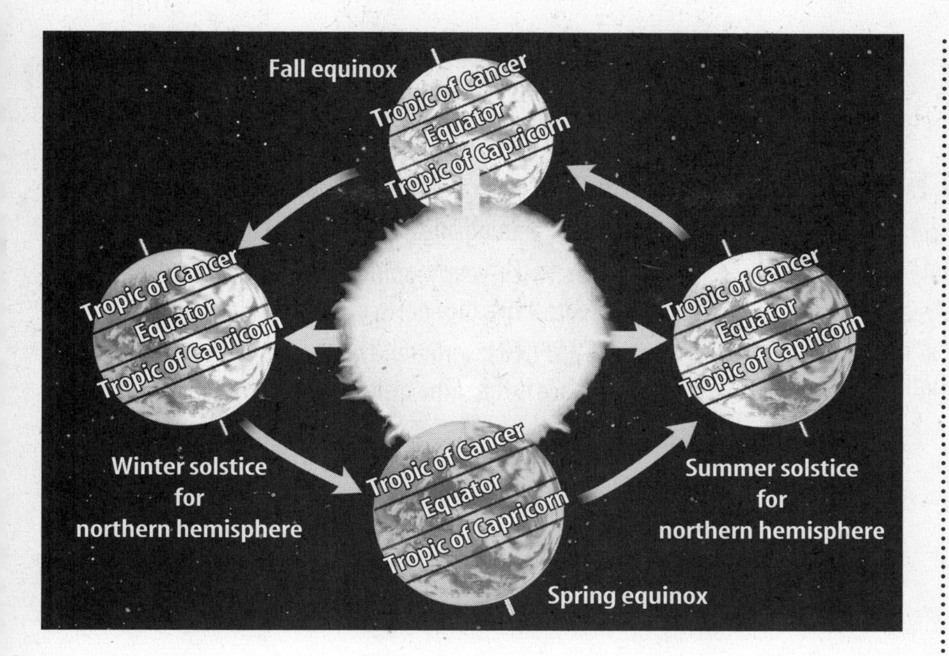


## Equinoxes

An **equinox** (EE kwuh nahks) occurs when the Sun is directly above Earth's equator. The tilt of Earth's axis means that the Sun's position relative to the equator is constantly changing. Most of the time, the Sun is either north or south of the equator. But two times a year the Sun is directly over the equator. This results in the spring and fall equinoxes. At an equinox, the Sun strikes the equator at the highest possible angle, 90 degrees. This can be seen in the figure above. ☑

During an equinox, neither the northern hemisphere nor the southern hemisphere is tilted toward the Sun. The number of daylight hours and nighttime hours is nearly equal all over the world.

In the northern hemisphere, the spring equinox occurs on March 20 or 21, and the fall equinox occurs on September 22 or 23. In the southern hemisphere, the spring equinox occurs in September, while the fall equinox occurs in March.

### Picture This

7. **Explain** Use the figure to explain over which part of Earth the Sun is located during the winter solstice in the northern hemisphere.

______________________

______________________

### Reading Check

8. **Define** When does an equinox occur?

______________________

______________________

## After You Read

### Mini Glossary

**axis:** imaginary center line around which Earth spins

**ellipse (ee LIHPS):** elongated, closed curve that described Earth's yearlong orbit around the sun

**equinox (EE kwuh nahks):** twice-yearly time—each spring and fall—when the Sun is directly over the equator and the number of daylight and nighttime hours are equal worldwide

**revolution:** Earth's yearlong elliptical orbit around the Sun

**rotation:** spinning of Earth on its axis

**solstice:** twice-yearly point at which the Sun reaches its greatest distance north or south of the equator

**sphere (SFIHR):** a round, three-dimensional object whose surface is the same distance from its center at all points

1. Review the terms and their definitions in the Mini Glossary. Write a sentence or two about the effects of Earth's rotation and its tilted axis.

___

___

___

2. Complete the table by labeling the statements true or false.

| Earth's Properties and Seasons | True or False? |
|---|---|
| Earth's shape is a slightly flattened sphere. | ___ |
| Earth's seasons are caused by its tilt. | ___ |
| The shape of Earth's orbit is a circle. | ___ |
| The shape of Earth's orbit is an ellipse. | ___ |
| After the summer solstice, daylight hours increase. | ___ |
| During a solstice, the Sun is at its farthest point north or south of the equator. | ___ |

Visit **earth.msscience.com** to access your textbook, interactive games, and projects to help you learn more about Earth.

# The Sun-Earth-Moon System

## section 2 The Moon—Earth's Satellite

### Before You Read

What do you already know about the moon? List physical characteristics or phases of the moon on the lines below. Check your information as you read the section.

_______________________________________________

_______________________________________________

_______________________________________________

**What You'll Learn**
- the phases of the Moon
- why solar and lunar eclipses occur
- the Moon's physical characteristics

### Read to Learn

#### Motions of the Moon

The Moon's movements are similar to Earth's movements. Just as Earth rotates on its axis, the Moon rotates on its axis. Earth revolves around the Sun, while the Moon revolves around Earth. The Moon's revolution around Earth is responsible for the changes in the Moon's appearance.

If the Moon rotates on its axis, why can't you see it spin around in space? The Moon's rotation takes 27.3 days—the same amount of time it takes to revolve once around Earth. Because these two motions take the same amount of time, the same side of the Moon always faces Earth. So, even though the Moon rotates on its axis, the same side is always visible from Earth.

#### What lights the Moon?

The surface of the Moon reflects the light of the Sun. Just as half of Earth experiences day as the other half experiences night, half of the Moon is lighted while the other half is dark. As the Moon revolves around Earth, different portions of its lighted side can be seen. This is why the Moon appears to change form or shape.

Study Coach

**Create a Quiz** As you read the text, create a quiz question for each subject. When you have finished reading, see if you can answer your own questions correctly.

FOLDABLES™

**B Classify** Make the following six-tab Foldable to identify the main ideas about Earth's Moon.

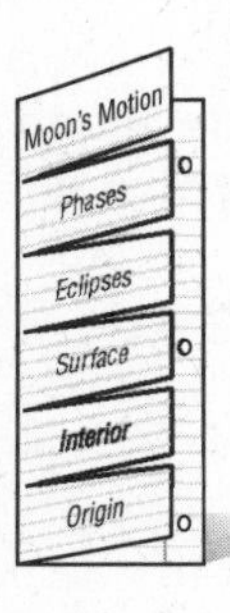

## Phases of the Moon

**Moon phases** are the different ways the Moon appears from Earth. The phase of the Moon depends on the relative positions of the Moon, Earth, and the Sun, as shown in the figure below.

A **new moon** occurs when the Moon is between Earth and the Sun. During a new moon, the lighted half of the Moon is facing the Sun and the dark side of the Moon faces Earth. Even though the Moon is in the sky, it cannot be seen. A new moon rises and sets in the sky at the same time as the Sun.

**Moon Phases**

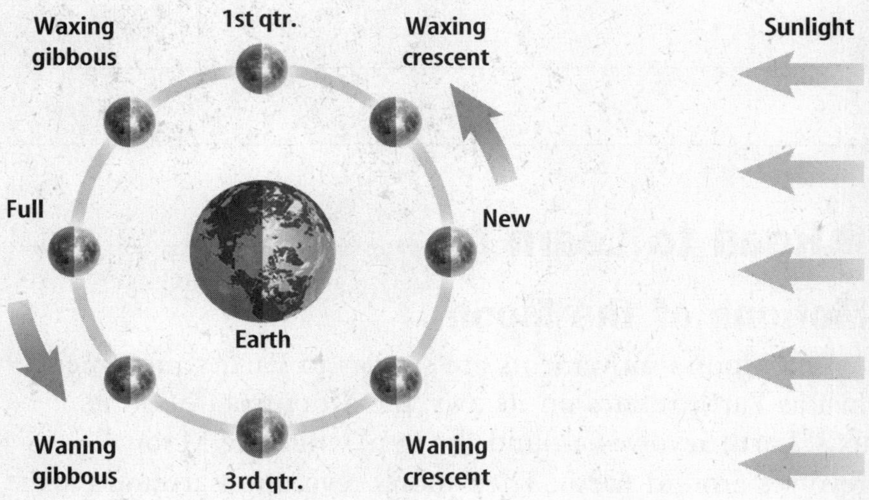

Waxing gibbous
1st qtr.
Waxing crescent
Sunlight
Full
New
Earth
Waning gibbous
3rd qtr.
Waning crescent

**Picture This**

1. **Identify** Highlight the lighted side of each figure of the Moon.

**Waxing Phases** After the new moon, the phases begin waxing. **Waxing** means that more of the lighted half of the Moon can be seen each night. About 24 h after a new moon, a thin slice of the Moon can be seen. This phase is called the waxing crescent. About a week after a new moon, you can see half of the lighted side of the Moon, or about one quarter of the Moon's surface. This is the first quarter phase. ☑

**Reading Check**

2. **Apply** What is happening when the moon is in its waxing phases?

_______________

_______________

The phases continue to wax. When more than one quarter of the Moon's surface is visible, it is called waxing gibbous. *Gibbous* is the Latin word for "humpbacked." A **full moon** occurs when all of the Moon's surface that faces Earth reflects light.

**Waning Phases** After the full moon, the phases are said to be waning. **Waning** means that you can see less and less of the lighted half of the Moon each night. About 24 h after a full moon, you begin to see the waning gibbous moon. About a week after a full moon, you can again see half of the lighted side of the Moon, or one quarter of the Moon's surface. This is the third-quarter phase. As the waning phases continue, you see less and less of the Moon. The last of the waning phases is the waning crescent, when just a small slice of the Moon is visible. This takes place just before another new moon.

It takes about 29.5 days for the Moon to complete its cycle of phases. Recall that it takes about 27.3 days for the Moon to revolve around Earth. The difference in the numbers is due to Earth's revolution. It takes about two extra days for the Sun, Earth, and the Moon to return to their same relative positions.

**Applying Math**

3. **Calculate** About how many times does the moon complete its cycle of phases around Earth in one year?

## Eclipses

Imagine living 10,000 years ago. You are gathering nuts and berries when, without warning, the Sun disappears. The darkness lasts only a short time, and the Sun soon returns to full brightness. You know something strange has happened, but you don't know why or how. It will be almost 8,000 years before anyone can explain what you just experienced.

The event just described was a total solar eclipse (ih KLIPS). Today, most people know what causes eclipses. What causes the day to become night and then change back into day?

### What causes an eclipse?

The revolution of the Moon around Earth causes eclipses. Eclipses take place when Earth blocks light from reaching the Moon, or when the Moon blocks light from reaching a part of Earth. Sometimes, during a new moon, the Moon's shadow falls on Earth. This causes a solar eclipse. During a full moon, Earth may cast a shadow on the Moon. This causes a lunar eclipse.

An eclipse can take place only when the Sun, the Moon, and Earth are lined up perfectly. Because the Moon's orbit is not in the same plane as Earth's orbit around the Sun, lunar eclipses take place only a few times each year.

**Think it Over**

4. **Infer** What is between the Sun and the Moon during a lunar eclipse?

## What is an eclipse of the Sun?

A **solar eclipse** occurs when the Moon moves directly between the Sun and Earth and casts its shadow over part of Earth. A solar eclipse is shown in the figure below. Depending on where you are on Earth, you may be in a total eclipse or a partial eclipse. Only a small area of Earth is part of the total solar eclipse during the eclipse event.

The darkest portion of the Moon's shadow on Earth is called the umbra (UM bruh). A person standing within the umbra experiences a total solar eclipse. During a total solar eclipse, the only part of the Sun that is visible is a white glow around the edge of the eclipsing Moon.

Surrounding the umbra is a lighter shadow on Earth's surface. This lighter shadow is called the penumbra (puh NUM bruh). Those who are standing in the penumbra experience a partial solar eclipse. **WARNING:** *Regardless of which eclipse you view, never look directly at the Sun. The light can permanently damage your eyes.*

**Picture This**

**5. Label** On the diagram, label the umbra and the penumbra.

**Solar Eclipse**

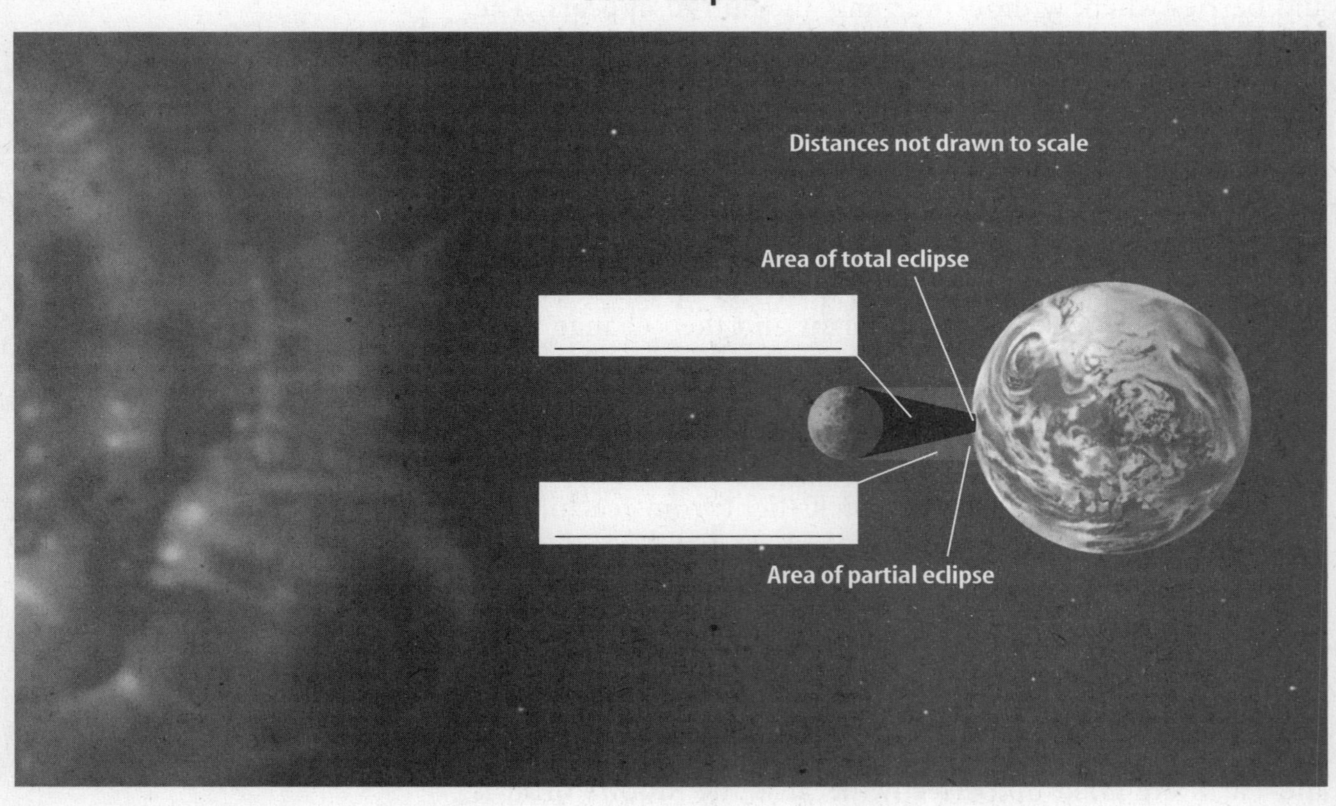

## Lunar Eclipse

### What is an eclipse of the Moon?

When Earth moves directly between the Sun and the Moon and its shadow falls on the Moon, a **lunar eclipse** occurs. A lunar eclipse begins when the Moon moves into Earth's penumbra. As the Moon continues to move, it enters Earth's umbra, and you can see a curved shadow on the Moon's surface. As the Moon moves completely into Earth's umbra, it goes dark. This is a total lunar eclipse. A total lunar eclipse is shown in the figure above. Sometimes sunlight bent through Earth's atmosphere will cause the eclipsed Moon to have a reddish appearance.

A partial lunar eclipse occurs when only a portion of the Moon moves into Earth's shadow. Then, the rest of the Moon is in Earth's penumbra and still gets some direct sunlight. When the Moon is totally within Earth's penumbra, it is called a penumbral lunar eclipse. It is difficult to tell when a penumbral lunar eclipse happens because some sunlight continues to fall on the side of the Moon facing Earth.

**Picture This**

6. **Label** On the diagram, label the umbra and penumbra.

## During which lunar phase do eclipses occur?

Lunar eclipses do not happen every month. Lunar eclipses happen only during the full moon phase. ☑

A total lunar eclipse can be seen by anyone on the nighttime side of Earth as long as the Moon is not hidden by clouds. Only a few people get to witness a total solar eclipse, however. Only those in the small area where the Moon's umbra strikes Earth can witness it.

**Reading Check**

7. **Explain** When do lunar eclipses occur?

______________________

______________________

## The Moon's Surface

When you look at the Moon, you can see many depressions called craters. Meteorites, asteroids, and comets striking the Moon's surface created most of these craters. When the objects struck the Moon, cracks may have formed in the Moon's crust, allowing lava to reach the surface and fill up the large craters. Dark, flat regions formed as the lava spread. These regions are called **maria** (MAHR ee uh). ☑

The igneous rocks of the maria are 3 billion to 4 billion years old. So far, they are the youngest rocks to be found on the Moon. This shows that craters formed after the Moon's surface had cooled. The maria formed early while molten rock still remained in the Moon's interior. The Moon must once have been as geologically active as Earth is today. As the Moon cooled, the interior separated into distinct layers.

**Reading Check**

8. **Define** What are maria?

______________________

______________________

## The Moon's Origin

Before the Apollo space missions in the 1960s and 1970s, there were three leading theories about the origin of the Moon. One theory was that the Moon was captured by Earth's gravity. Another stated that the Moon and Earth condensed from the same cloud of dust and gas. An alternative theory proposed that Earth ejected molten material that became the Moon.

### What is the impact theory?

The data gathered by the Apollo missions led many scientists to support a new theory. This theory, called the impact theory, states that the Moon formed billions of years ago from condensing gas and debris thrown off when Earth collided with a Mars-sized object. The blast that resulted ejected material from both objects into space. A ring of gas and debris formed around Earth. Finally, particles in that ring joined together to form the Moon.

**Think it Over**

9. **Recognize Cause and Effect** What was a result of the data gathered in the Apollo missions?

______________________

______________________

## Inside the Moon

Just as scientists study earthquakes to gather information about Earth's interior, scientists study moonquakes to understand the structure of the Moon. The information scientists gather from moonquakes has helped them make several possible models of the Moon' interior. One model is shown in the figure below. In it, the Moon's crust is about 60 km thick on the side facing Earth. On the side facing away from Earth, the Moon's crust is thought to be about 150 km thick. Under the crust, another solid layer, the mantle, may be 1,000 km deep. A zone of the mantle where the rock is partly melted may extend even farther down. Below this mantle, there may be a solid, iron-rich core.

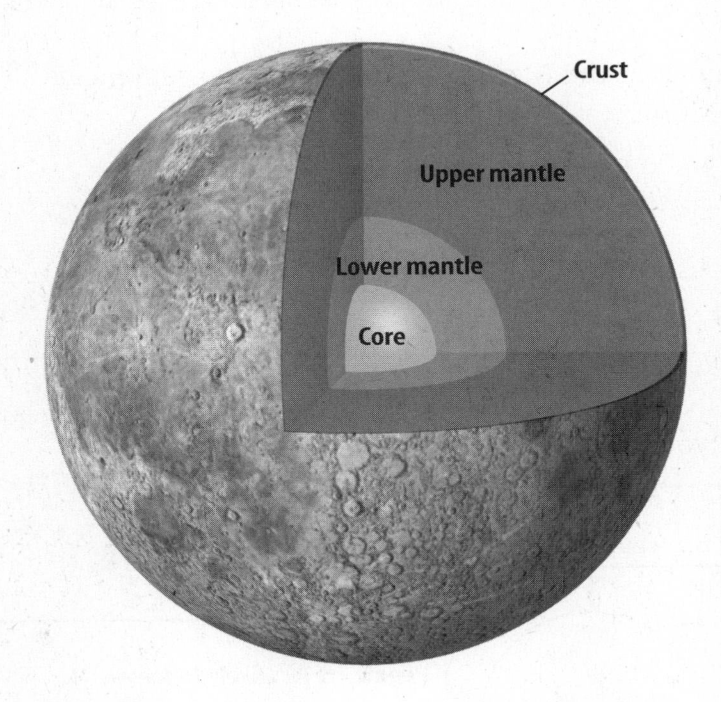


## What has been learned about the Moon in history?

Much has been learned about the Moon and Earth by studying the Moon's phases and eclipses. Earth and the Moon are in motion around the Sun. From studying the curved shadow that Earth casts on the Moon, early scientists learned that Earth is a sphere. When Galileo first used his telescope to look at the Moon, he saw that it was not smooth but had craters and maria. Today, scientists study rocks collected from the Moon. By doing so, they hope to learn more about Earth.

**Applying Math**

**10. Calculate** About what is the difference in thickness between the Moon's crust on the side facing Earth and the crust facing away from Earth?

______________________

**Picture This**

**11. Interpret Scientific Illustrations** List the layers of the Moon in order from the interior to the surface.

______________________

______________________

______________________

______________________

# After You Read

## Mini Glossary

**full moon:** phase that occurs when all of the Moon's surface facing Earth reflects light

**lunar eclipse:** occurs when Earth passes directly between the Sun and the Moon and Earth's shadow falls on the Moon

**maria (MAHR ee uh):** dark-colored, relatively flat regions of the Moon formed when ancient lava reached the surface and filled craters on the Moon's surface

**moon phase:** change in appearance of the Moon as viewed from the Earth, due to the relative positions of the Moon, Earth, and the Sun

**new moon:** moon phase that occurs when the Moon is between Earth and the Sun, at which point the Moon cannot be seen because its lighted half is facing the Sun and its dark side faces Earth

**solar eclipse:** occurs when the Moon passes directly between the Sun and Earth and casts a shadow over part of Earth

**waning:** describes phases that occur after a full moon, as the visible lighted side of the Moon grows smaller

**waxing:** describes phases following a new moon, as more of the Moon's lighted side becomes visible

1. Review the terms and their definitions in the Mini Glossary. Write two sentences explaining different phases of the Moon.

_______________________________________________

_______________________________________________

_______________________________________________

2. Fill in the concept map with what you know about eclipses.

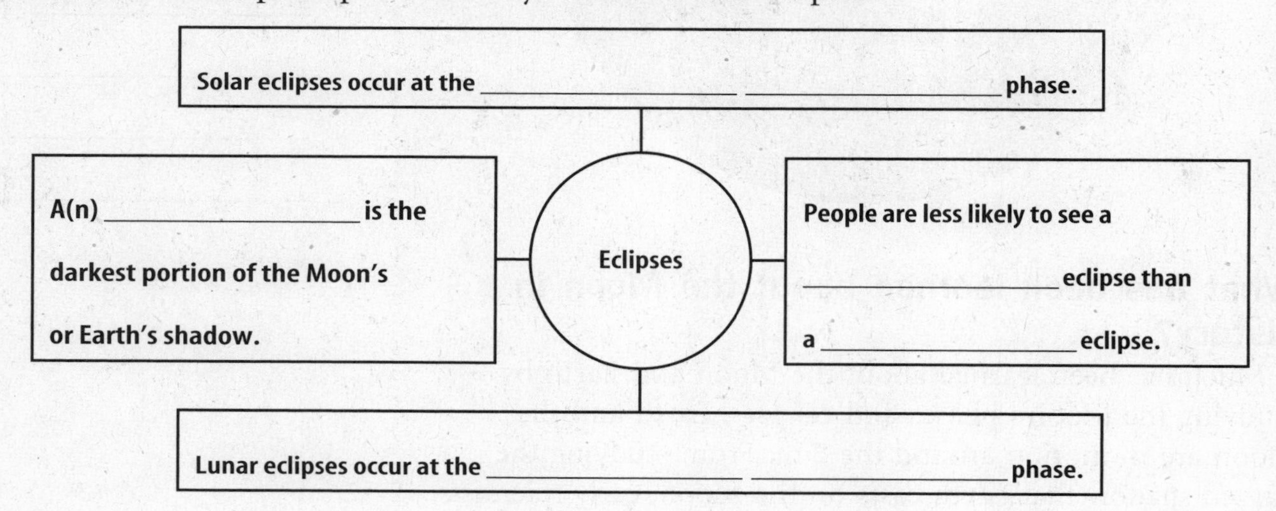

Visit **earth.msscience.com** to access your textbook, interactive games, and projects to help you learn more about Earth's satellite, the Moon.

# The Sun-Earth-Moon System

## section 3 Exploring Earth's Moon

### Before You Read

People have always been curious about the Moon. What would you like to know about the Moon? In the space below, write some questions you have about the Moon.

______________________________________________

______________________________________________

______________________________________________

**What You'll Learn**

- recent discoveries about the Moon
- facts that might affect future space travel to the Moon

### Read to Learn

#### Missions to the Moon

For centuries, scientists have tried to discover what the Moon is made of and how it formed. In 1959, the former Soviet Union launched the first *Luna* spacecraft. This spacecraft made it possible to study the Moon up close.

Two years later, the United States began a similar space program. The United States launched the first *Ranger* spacecraft and a series of *Lunar Orbiters.* The spacecraft in these early missions took detailed photographs of the Moon.

The *Surveyor* spacecraft were the next step. The *Surveyor* spacecraft were designed to take more detailed photographs and to actually land on the Moon. Five of these spacecraft landed on the Moon's surface and analyzed lunar soil. The goal of the *Surveyor* program was to gather information about the Moon that would allow astronauts to land there one day.

In 1969, the astronauts of *Apollo 11* landed on the Moon. Between 1969 and 1972, when the *Apollo* missions ended, 12 U.S. astronauts had walked on the Moon.

Mark the Text

**Identify the Main Point** Highlight the main point in each paragraph. Using a different color, highlight an example that helps explain the main point.

FOLDABLES™

**C Organize Information** Make the following 2-tab Foldable to organize information about Moon missions and Moon mapping.

| Moon Missions | Moon Mapping |
| --- | --- |

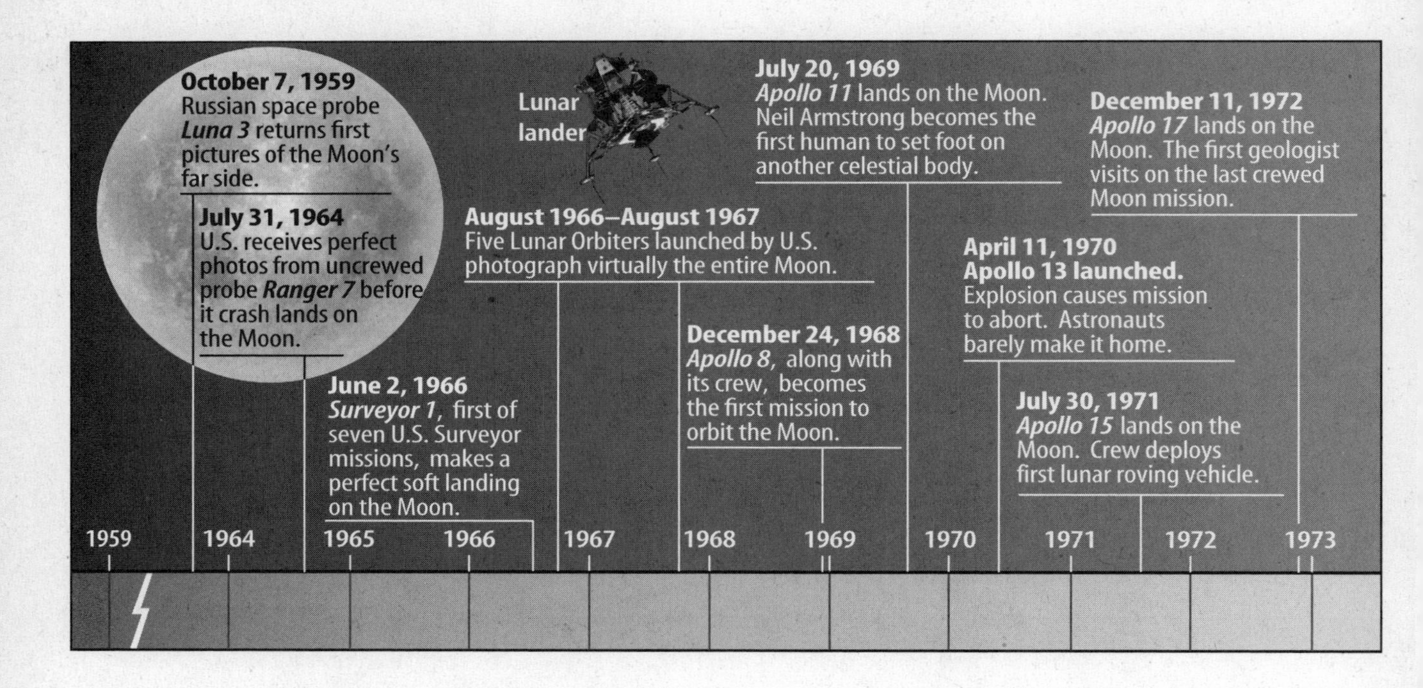


Picture This

**1. Use Tables** Which Apollo mission deployed the first lunar roving vehicle?

_______________

## Is the Moon being studied today?

The time line above shows important events in the exploration of the Moon. But, there is still much to learn about the Moon. The United States has started to study the Moon again. In 1994, the spacecraft *Clementine* was placed into lunar orbit. *Clementine's* purpose was to conduct a survey of the Moon's surface. An important part of the study was to collect data on the mineral content of Moon rocks. While in orbit around the Moon, *Clementine* also mapped features on the Moon's surface, including huge impact basins.

## What is an impact basin?

An **impact basin,** or impact crater, is a depression left behind when a meteorite or other object strikes the Moon. The South Pole-Aitken Basin is the oldest impact basin that has been identified so far.

Impact basins like the South Pole-Aitken Basin are very interesting to scientists. Because this deep crater is located at one of the poles, the Sun's rays never reach the bottom of the crater. Therefore, the bottom of the crater is always in shadow. The temperatures there are extremely cold. Scientists hypothesize that if a comet collided with the Moon, ice could have been deposited there. Some of that ice might still be found in the shadows at the bottom of the crater. In fact, *Clementine* sent information that showed the presence of water, just as scientists had hypothesized.

FOLDABLES™

**D Ask Questions** Make the following Foldable to identify what you already know and what you want to know about the *Clementine* and *Lunar Prospector* missions.

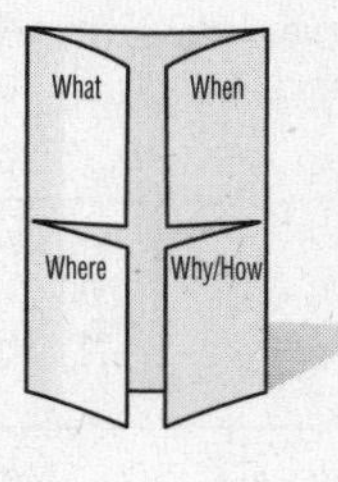

## Mapping the Moon

Photographs taken by *Clementine* were used to create detailed maps of the Moon's surface. Data from *Clementine* showed that the Moon's crust did not have the same thickness all over the Moon. The crust on the side of the Moon that faces Earth is much thinner than the crust on the far side. Additional information showed that the Moon's crust is thinnest under impact basins.

### What is the *Lunar Prospector*?

In 1998, NASA sent the small *Lunar Prospector* spacecraft into orbit around the Moon. For one year the spacecraft circled the Moon from one pole to the other. It flew around the Moon once every two hours.

The *Lunar Prospector* collected data that confirmed that the Moon has a small, iron-rich core at its center. This finding supports the impact theory of how the Moon was formed. The small core is a result of a small amount of iron that could have blasted away from Earth. ☑

### Where is there ice on the Moon?

In addition to photographing the surface, *Lunar Prospector* carried instruments that gathered information for mapping the Moon. The maps were of the Moon's gravity, its magnetic field, and how much and where certain elements were found in the Moon's crust. Scientists finally had data from the entire surface of the Moon, rather than just the areas around the Moon's equator.

The *Lunar Prospector* confirmed that ice was present in deep craters at both poles of the Moon. Using data from *Lunar Prospector*, scientists made maps that show the location of ice at each pole. At first scientists thought that ice crystals were mixed with lunar soil. More recent information suggests that the ice deposit may be in the form of more compact deposits. ☑

**Reading Check**

2. **Describe** What is in Moon's core that supports the impact theory?

______________________

**Reading Check**

3. **Identify** Where is there ice on the Moon's surface?

______________________

______________________

## After You Read

### Mini Glossary

**impact basin:** a depression left on the surface of the Moon caused by an object striking its surface

1. Review the term and its definition in the Mini Glossary. Write a sentence explaining what causes impact basins to form.

2. Complete the chart to review missions that gathered information about the Moon.

| Spacecraft | Mission |
|---|---|
| *Ranger and Lunar Orbiters* | To photograph the Moon |
| *Surveyor* | |
| *Clementine* | |
| *Lunar Prospector* | |

3. How did highlighting help you read this section? Reread the sentences you highlighted in the text. Now that you have read the entire section, do you think you highlighted the right sentences? Make any corrections you think would help you.

Visit **earth.msscience.com** to access your textbook, interactive games, and projects to help you learn more about exploring Earth's moon.

# The Solar System

## section 1 The Solar System

### Before You Read

Name the planets in the solar system that you already know.

_______________________________________________

_______________________________________________

**What You'll Learn**

- past and present ideas about the solar system
- how the solar system formed
- how the Sun's gravity holds planets in orbit

### Read to Learn

#### Ideas About the Solar System

Based on their observations, early humans believed the Sun and planets moved around Earth. Today, people understand that Earth and the other planets and objects in the solar system orbit, or move around, the Sun.

**Earth-Centered Model** Early Greek scientists thought the planets, the Sun, the Moon, and the stars rotated around Earth. This is called the Earth-centered model of the solar system. It included Earth, the Moon, the Sun, five planets—Mercury, Venus, Mars, Jupiter, and Saturn—and the stars.

**Sun-Centered Model** In 1543, Nicholas Copernicus published his model of the solar system. He stated that Earth and the other planets revolved around the Sun and that the Moon revolved around Earth. He explained that the Sun and the planets only looked like they were moving around Earth because Earth rotates. This is the Sun-centered model of the solar system.

Galileo Galilei used his telescope to observe that Venus went through a full cycle of phases like the Moon's. Also, Venus looked smaller when its phase was near full. This could only be explained if Venus were orbiting the Sun, not Earth. Galileo concluded that the Sun is the center of the solar system.

Study Coach

**Ask Questions** As you read, write down your questions. Use the questions to find out more about topics that are not clear, or topics that are particularly interesting.

FOLDABLES™

**A Find Main Ideas** Make the following two-tab Foldable to help you identify the main ideas about past and present views on the solar system.

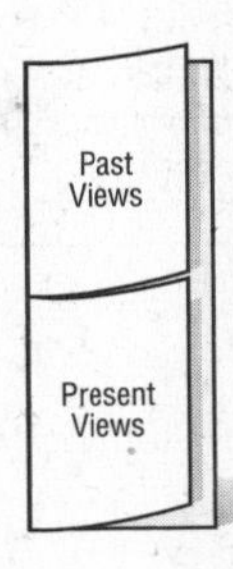

## What is the modern view of the solar system?

Today, we know that the **solar system** is made up of nine planets, including Earth, and many smaller objects that orbit the Sun. The Sun and the position of the nine planets relative to the Sun are shown in the figure on this page and the next page. The solar system also includes a huge amount of space that stretches out in all directions from the Sun.

The Sun contains 99.86 percent of the mass in the solar system. Therefore, the Sun has a lot of gravity. The Sun's gravity is strong enough to hold the planets and other objects in their orbits. ☑

**Reading Check**

1. **Explain** What force holds the planets in their orbits?

______________________

______________________

# How the Solar System Formed

Scientists hypothesize that the solar system formed more than 4.6 billion years ago. They have found clues that it may have formed from a cloud of gas, ice, and dust. Over time, this cloud pulled together to form a large, tightly packed, spinning disk. The center of the disk heated up to about 10 million degrees Celsius, and the reaction known as nuclear fusion began. That is how the star, the Sun, formed at the center of the solar system.

## How did the planets form?

Not all of the gas, ice, and dust was pulled into the center of the spinning disk to form the Sun. Some matter collided and stuck together to form planets and asteroids. The nine planets of the solar system are divided into two groups, the inner planets and the outer planets.

**Picture This**

2. **Interpret Scientific Illustrations** Which planet is closest to the Sun? Which planet is farthest from the Sun? Which is the third planet from the Sun?

______________________

______________________

______________________

## What are the nine planets?

The inner planets of the solar system—Mercury, Venus, Earth, and Mars—are small, rocky planets with iron cores. The outer planets are Jupiter, Saturn, Uranus, Neptune, and Pluto. Except for Pluto, the outer planets are much larger than the inner planets. They are made up mostly of lighter substances, including hydrogen, helium, methane, and ammonia. ☑

These light substances are not found in great quantities in the inner planets. The high temperatures closer to the Sun turned these substances to gas. They could not cool enough to form solids.

**Reading Check**

3. **Identify** Name the inner planets.

______________________

______________________

______________________

______________________

## Motions of the Planets

When Nicholas Copernicus developed his Sun-centered model of the solar system, he thought the orbits of the planets were circles. In the early 1600s, Johannes Kepler discovered that the orbits of the planets are oval shaped, or elliptical. He also found that the Sun's position in the orbits is slightly off-center.

Kepler discovered that the planets orbit the Sun at different speeds. Planets closer to the Sun travel faster than planets farther away from the Sun. The outer planets also have longer distances to travel and take much longer to orbit the Sun than the inner planets.

**Think it Over**

4. **Infer** Which planet takes longer to orbit the Sun—Mars or Neptune?

______________________

## After You Read

### Mini Glossary

**solar system:** system of nine planets, including Earth, and many smaller objects that orbit the Sun

1. Review the term and its definition in the Mini Glossary. On the lines below, write something you have learned about the solar system.

_______________________________________________

_______________________________________________

_______________________________________________

2. Complete the chart that shows how the solar system may have formed.

| |
|---|
| **1. The solar system formed from a cloud of ____________, ____________, and ____________.** |
| **2. The cloud condensed to form a(n) ____________.** |
| **3. ____________ formed first. It was at the center of the new solar system.** |
| **4. The other material in the solar system collided and formed ____________ planets.** |
| **5. The inner planets are ____________, ____________, ____________, and ____________.** **The outer planets are ____________, ____________, ____________, ____________, and ____________.** |

3. Review the questions you wrote as you read this section. What resources could you use to find answers to your questions? Did the questions you write help you understand the information?

_______________________________________________

_______________________________________________

End of Section

Science Online Visit **earth.msscience.com** to access your textbook, interactive games, and projects to help you learn more about the solar system.

# The Solar System

## section 2 The Inner Planets

### Before You Read

What do you know about Mercury and Venus? What would you like to know about these inner planets?

______________________________________________

______________________________________________

______________________________________________

**What You'll Learn**

- facts about the inner planets
- what each inner planet is like
- compare and contrast Venus and Earth

### Read to Learn

**Study Coach**

**Make Flash Cards** Make four flash cards to help you study this section. On one side of the card, write the name of an inner planet. On the other side, write facts about that planet.

#### Inner Planets

Today, people know a great deal about the solar system. Scientists use telescopes to study the planets both from Earth and from space. They also use space probes to study the solar system. Much of the information you will read in this section was gathered by space probes.

#### Mercury

**Mercury** is the planet closest to the Sun. The spacecraft *Mariner 10* sent pictures of Mercury to Earth in 1974 and 1975. Scientists learned that Mercury, like Earth's Moon, has many craters. But unlike the Moon, Mercury has cliffs as high as 3 km on its surface. These cliffs might have formed when the crust of the planet broke as the core of the planet was cooling and shrinking.

Scientists learned that Mercury has a weak magnetic field. This shows that Mercury has an iron core, the same as Earth. Some scientists think that Mercury's crust solidified while the iron core was still hot and liquid. As the core became more solid, it became smaller. The cliffs resulted from breaks in Mercury's crust caused by the shrinking of the core.

**FOLDABLES™**

**B Compare and Contrast** Make the following Foldable to understand how the inner planets are similar and different.

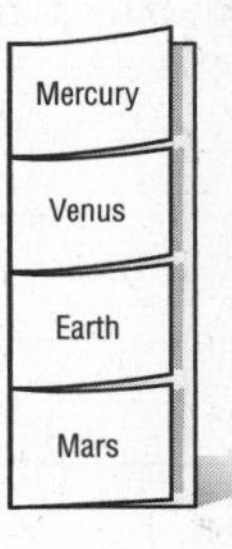

## Does Mercury have an atmosphere?

Mercury has no true atmosphere. This is because Mercury has a low gravitational pull and high temperatures during the day. Most gases that could form an atmosphere escape into space. Earth-based observations have found traces of sodium and potassium around the planet. However, these atoms probably come from rocks in Mercury's crust. Therefore, Mercury has no true atmosphere. This lack of atmosphere and its nearness to the Sun cause Mercury to have great extremes in temperature. Mercury's temperature can reach as high as 425°C during the day, and it can fall to as low as −170°C at night. A picture of Mercury and some facts about the planet are shown below. ☑

**Reading Check**

1. **Describe** What is the range in temperature on Mercury?

___________________________

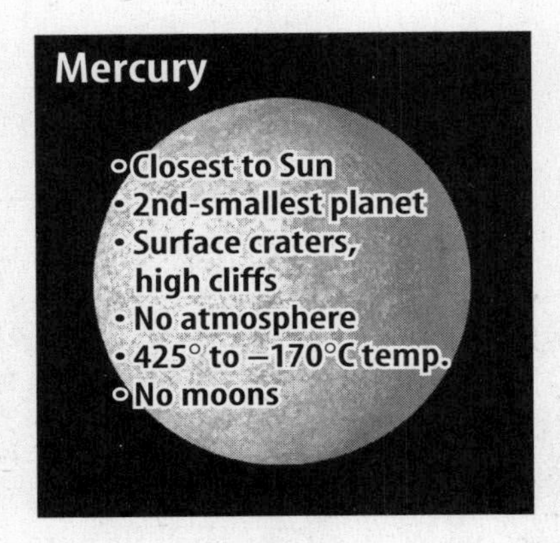


## Venus

**Venus** is the second planet from the Sun. Venus is sometimes called Earth's twin because its size and mass are similar to Earth's. When *Mariner 2* flew past Venus in 1962, the satellite sent back information about Venus's atmosphere and rotation. From 1990 to 1994, the U.S. *Magellan* probe used radar to make detailed maps of Venus's surface. A picture of Venus and some facts about the planet are shown below.

**Think it Over**

2. **Identify** What are some physical characteristics of Venus?

___________________________

___________________________

___________________________

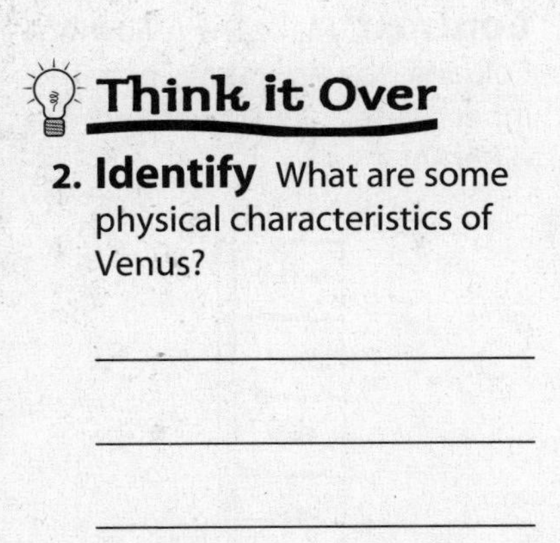

## How hot is it on Venus?

The thick clouds on Venus block most of the Sun's light from reaching the planet's surface. The clouds and carbon dioxide gas in the atmosphere trap heat from the Sun. Temperatures on the surface of Venus range from 450°C to 475°C.

## Earth

**Earth** is the third planet from the Sun. It is about 150 million km from the Sun, or one astronomical unit (AU). Earth is the only planet in the solar system that has large amounts of liquid water. More than 70 percent of Earth's surface is covered by liquid water. Earth is also the only planet that supports life. Earth's atmosphere protects life forms from the Sun's harmful radiation. The atmosphere also causes most meteors to burn up before they reach the surface of the planet. A picture of Earth and some facts about the planet are shown below. ☑

**Reading Check**

3. **Apply** How does Earth's atmosphere help support life?

_______________

_______________

_______________

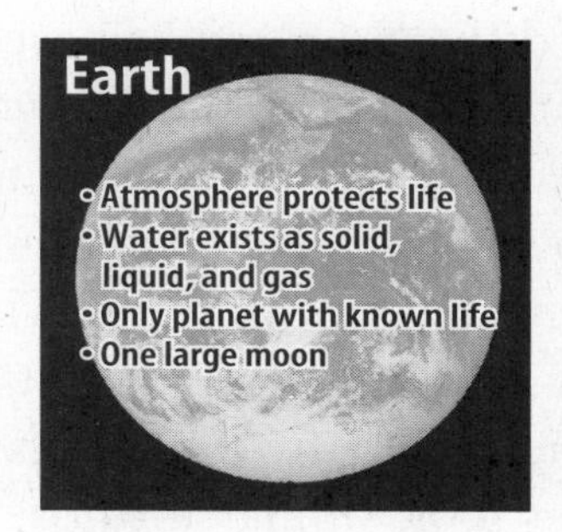


## Mars

**Mars** is the fourth planet from the Sun. It is called the red planet. Its red color is caused by iron oxide in the soil. Polar ice caps on Mars can be seen through telescopes from Earth. The ice caps are made of frozen water covered by a layer of frozen carbon dioxide. A picture of Mars and some facts about the planet are shown below.

## What have scientists learned from missions to Mars?

Several spacecraft have made missions to Mars. From these missions, scientists have learned that there are long channels on the planet that might have been carved by flowing water. The largest known volcano in the solar system is on Mars. It is called Olympus Mons. It is probably not an active volcano. There are also large valleys in the Martian crust. ☑

**Reading Check**

4. **Infer** Is the volcano Mt. Saint Helens on Earth larger than Olympus Mons on Mars?

______________________

## What did the *Viking* probes do?

The *Viking 1* and *2* probes arrived at Mars in 1976. Each probe had two parts—an orbiter and a lander. The orbiters remained in space. They took photographs of the entire surface of Mars. The landers touched down on the surface of Mars. They carried equipment to search for signs of life on the planet. No conclusive evidence of life was found on Mars. ☑

**Reading Check**

5. **Explain** What were the Viking 1 and 2 probes looking for on Mars?

______________________

______________________

## How were *Pathfinder, Global Surveyor,* and *Odyssey* used?

The *Mars Pathfinder* analyzed Martian rock and soil. These data indicated that iron might have reached the surface of Mars from underground. *Global Surveyor* took pictures that showed features like gullies that could have been formed by flowing water. *Mars Odyssey* had instruments that detected frozen water. The water forms a layer of frost under a thin layer of soil. It is possible that volcanic activity might melt frost beneath the Martian surface. The features look similar to those formed by flash floods on Earth, such as on Mount St. Helens. You can see how they compare in the figure below.

**Picture This**

6. **List** the features that the gullies on Mars and on Mount St. Helens have in common.

______________________

______________________

______________________

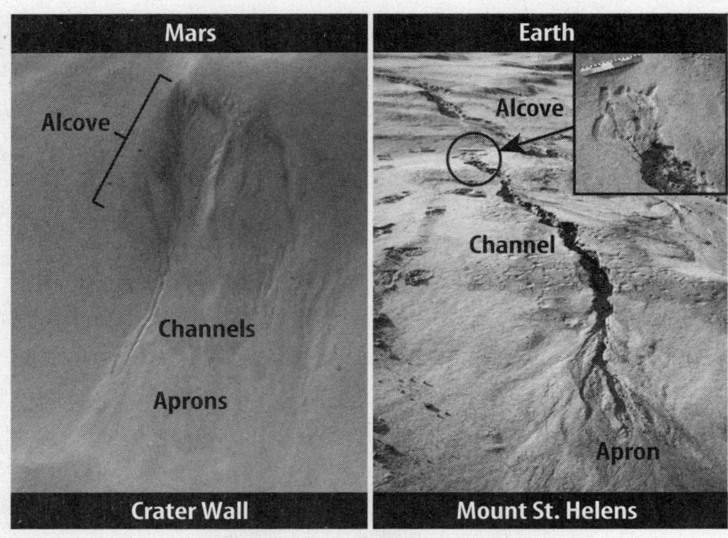


NASA/JPL/Malin Space Science Systems

## What makes up Mars's atmosphere?

Mars's atmosphere is much thinner than Earth's atmosphere. It is made up mostly of carbon dioxide with some nitrogen and argon. Temperatures on the surface of Mars can be as high as 35°C and as low as −125°C. The change in temperature between day and night causes strong winds, which in turn cause global dust storms. This information is important if humans ever explore Mars. ☑

## Are there seasons on Mars?

Mars's axis is tilted 25°, which is close to Earth's tilt of 23.5°. So, Mars has seasons as it orbits the Sun. The polar ice caps on Mars change with the season. During winter, carbon dioxide freezes at the poles. The polar ice caps get larger. During summer, the carbon dioxide ice changes to gas. The ice caps get smaller. It is winter at one pole when it is summer at the other pole. The color of the ice caps and other areas on Mars also changes with the seasons. This is due to the movement of dust and sand during dust storms.

## Does Mars have moons?

Mars has two small moons—Phobos and Deimos. Phobos orbits Mars once every 7 hours. It has a large crater and chains of smaller craters. Deimos orbits Mars once every 31 hours. It is farther away from Mars's surface. Its surface looks smoother than that of Phobos. Its craters have partially filled with soil and rock.

**Reading Check**

7. **Recognize Cause and Effect** What is the result of extreme change in day and night temperatures on Mars?

**Think it Over**

8. **Compare** Why does Mars have seasons?

## After You Read

### Mini Glossary

**Earth:** third planet from the Sun; has plenty of liquid water and an atmosphere that protects life

**Mars:** fourth planet from the Sun; has polar ice caps and a reddish appearance caused by iron oxide in the soil

**Mercury:** planet closest to the Sun; does not have a true atmosphere; has a surface with many craters and high cliffs

**Venus:** second planet from the Sun; similar to Earth in mass and size; has thick clouds

1. Review the terms and their definitions in the Mini Glossary. Write something interesting you learned about Mars, Venus, or Mercury.

_______________________________________________

_______________________________________________

2. Complete the table to organize the information from this section.

| THE INNER PLANETS | | | |
|---|---|---|---|
| | ORDER FROM SUN | ATMOSPHERE | TEMPERATURES |
| MERCURY | Closest | ________ | Highs: 425°C Lows: −170°C |
| VENUS | ________ | Heavy clouds<br>Carbon dioxide gas | Highs: ________ Lows: ________ |
| EARTH | 3rd | ________<br>________ | Not given |
| MARS | ________ | Mostly carbon dioxide<br>Some nitrogen and argon | ________ |

3. Review the flash cards you made. How did this help you learn the content of the section? How could you use the flash cards to prepare for a test on the inner planets?

_______________________________________________

_______________________________________________

Visit **earth.msscience.com** to access your textbook, interactive games, and projects to help you learn more about the inner planets.

# The Solar System

## section 3 The Outer Planets

### Before You Read

What do you know about the outer planets Jupiter, Uranus, Saturn, Neptune, or Pluto? What would you like to learn?

_______________________________________________

_______________________________________________

_______________________________________________

**What You'll Learn**

- facts about the outer planets: Jupiter, Uranus, Saturn, Neptune, and Pluto
- how Pluto is different from other planets

### Read to Learn

## Outer Planets

*Voyager, Galileo,* and *Cassini* were not the first space probes to explore the outer planets. However, much new information about the outer planets has come from these probes.

Study Coach

**Make Flash Cards** Make five flash cards to help you study this section. On one side of each card, write the name of one of the outer planets. On the other side, write facts about that planet.

## Jupiter

**Jupiter** is the fifth planet from the Sun. It is the largest planet in the solar system. Data from space probes show that Jupiter has faint rings around it made of dust. Io, one of Jupiter's moons, has active volcanoes.

### What is Jupiter's atmosphere made of?

Jupiter is made up mostly of hydrogen and helium with some ammonia, methane, and water vapor. Scientists hypothesize that the atmosphere of hydrogen and helium gas changes to liquid hydrogen and helium toward the middle of the planet. Below this liquid layer may be a rocky core that is probably different from any rock on Earth.

Jupiter's atmosphere has bands of white, red, brown, and tan clouds. Storms of swirling gas have been observed on the planet. The **Great Red Spot** is the most spectacular of these storms.

FOLDABLES™

**C Compare and Contrast** Make the following Foldable to help you understand how the outer planets are similar and different.

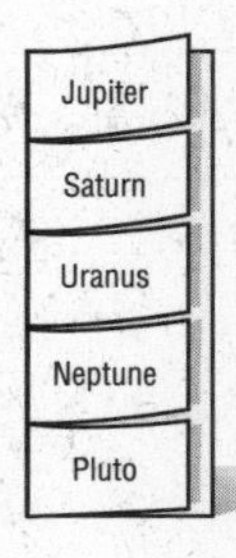

## How many moons orbit Jupiter?

At least 61 moons orbit Jupiter. In 1610, the astronomer Galileo Galilei was the first person to see the four largest moons. Io (I oh) is the large moon closest to Jupiter.

Jupiter's gravity and the gravity of the next large moon, Europa, pull on Io. This force heats up Io. The result is that Io has the most active volcanoes in the entire solar system.

Europa is made up mostly of rock. It has a thick crust of ice. Under the ice there might be a deep ocean. If this ocean does exist, it would be one of the few places in the solar system with large quantities of liquid water. The next moon is Ganymede. Ganymede is the largest moon in the solar system—larger than the planet Mercury. Callisto, the last of Jupiter's large moons, is made up mostly of ice and rock. Callisto is another place in the solar system where there may be a large quantity of water. Pictures of Jupiter and Callisto, as well as some facts about Jupiter, are shown below. ☑

**Reading Check**

1. **Define** What are Io and Callisto?

______________________

______________________

# Saturn

**Saturn** is the sixth planet from the Sun. It is the second-largest planet in the solar system. Saturn is the least dense planet in the solar system.

## What is Saturn's atmosphere like?

Saturn is similar to Jupiter. Both planets are large and made up mostly of gas. Saturn has a thick outer atmosphere made up mostly of hydrogen and helium. Deeper within the atmosphere the gases change to liquid. Below its atmosphere and liquid layers, Saturn might have a small, rocky core.

**Picture This**

2. **Describe** List four facts that describe Jupiter.

______________________

______________________

______________________

______________________

Jupiter's moon Callisto

JPL

## What are Saturn's rings and moons like?

Each of Saturn's large rings is made up of thousands of thin rings. These are made of ice and rock particles. Some particles are as tiny as a speck of dust, and some are tens of meters across. Saturn has the most complex ring system in the solar system.

At least 31 moons orbit Saturn. The planet's gravity holds them in their orbits. Titan is the largest of Saturn's moons. It is larger than the planet Mercury. A picture of Saturn and some facts about the planet are shown below.

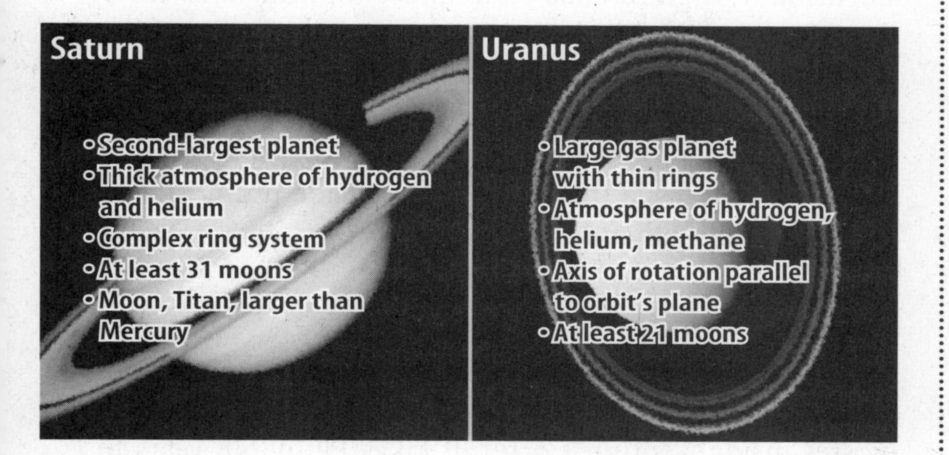


## Uranus

<u>**Uranus**</u> (YOOR uh nus) is the seventh planet from the Sun. It is a large planet and also is made up mostly of gas. Thin, dark rings surround the equator. Scientists know that Uranus has at least 21 moons. Its largest moon, Titania, has many craters and deep valleys.

## What are the characteristics of Uranus?

The atmosphere of Uranus is made up of hydrogen, helium, and some methane. Methane gives the planet a bluish-green color. A few clouds and storms can be seen on Uranus. There may be liquid water under its atmosphere.

Uranus has an unusual rotation. It is tilted on its side. The axes of rotation of the other planets, except Pluto, are nearly perpendicular to the planes of their orbits. Uranus's axis of rotation is nearly parallel to the plane of its orbit. Some scientists believe that a collision may have caused Uranus to tip over in this way. A picture of Uranus and some facts about the planet are shown above. ☑

### Think it Over

3. **Compare and Contrast** Describe two ways that Saturn and Uranus are different.

_______________

_______________

_______________

### ✓ Reading Check

4. **Recognize Cause and Effect** What do scientists believe may have caused Uranus to tilt on its axis?

_______________

_______________

## Neptune

**Neptune** is usually the eighth planet from the Sun. However, part of Pluto's orbit crosses inside Neptune's orbit. From 1979 until 1999, Pluto was closer to the Sun than Neptune was.

### Think it Over

**5. Recognize Cause and Effect** What gas causes Uranus and Neptune to have a bluish-green color?

**a.** hydrogen
**b.** methane
**c.** helium
**d.** carbon dioxide

### What characteristics does Neptune have?

Neptune's atmosphere is similar to Uranus's atmosphere. Methane gives the atmosphere of Neptune its bluish-green color, just as it does for Uranus. Neptune has dark-colored storms similar to the Great Red Spot on Jupiter. These storms and bright clouds form and disappear. This shows that Neptune's atmosphere is active and changes rapidly.

There may be a layer of liquid water under Neptune's atmosphere. The planet probably has a rocky core. Neptune has at least 11 moons and several rings. Neptune's largest moon, Triton, has a thin atmosphere made up mostly of nitrogen and methane.

## Pluto

**Pluto** is the smallest planet in the solar system. It is the ninth planet from the Sun. However, at times Pluto's orbit crosses inside Neptune's orbit. It takes Pluto 248 years to orbit the Sun.

Pluto is very different from the other outer planets. It has a thin atmosphere, and all the other outer planets have very thick atmospheres. Pluto has a solid, icy-rock surface. All the other outer planets are believed to have liquid layers around rocky cores.

### Think it Over

**6. Infer** Could Pluto support life?

______________________

______________________

______________________

______________________

### Does Pluto have a moon?

Pluto has one moon, called Charon. Pluto and Charon are so close in size that some scientists consider them to be two planets. Many scientists think that Pluto and Charon are actually part of an area called the Kuiper Belt. The Kuiper Belt is a large disk of icy objects that lies beyond the orbit of Neptune.

# After You Read

## Mini Glossary

**Great Red Spot:** giant, high-pressure storm in Jupiter's atmosphere

**Jupiter:** largest planet, and fifth planet from the Sun; has an atmosphere made up mostly of hydrogen and helium

**Neptune:** usually the eighth planet from the Sun; is large, gaseous, and bluish-green in color

**Pluto:** usually the ninth planet from the Sun; has a solid icy-rock surface and a single moon, Charon

**Saturn:** second-largest and sixth planet from the Sun; has a complex ring system, at least 31 moons, and a thick atmosphere made mostly of hydrogen and helium

**Uranus (YOOR uh nus):** seventh planet from the Sun; is large and gaseous, has a distinct bluish-green color.

**1.** Review the terms and their definitions in the Mini Glossary. Choose an outer planet and write a sentence that tells something you learned about it.

______________________________________________

______________________________________________

______________________________________________

**2.** Complete the table below to organize the information from this section.

| THE OUTER PLANETS | | | |
|---|---|---|---|
| | ORDER FROM THE SUN | ATMOSPHERE | MOONS |
| Jupiter | 5th | __________ | __________, including __________ |
| Saturn | __________ | Thick; hydrogen and helium | At least 31, including Titan |
| Uranus | 7th | __________ | __________, including __________ |
| Neptune | __________ | Thick methane | __________, including __________ |
| Pluto | Usually, 9th | __________ | 1, Charon |

Visit **earth.msscience.com** to access your textbook, interactive games, and projects to help you learn more about the outer planets.

**End of Section**

# The Solar System

## section 4 Other Objects in the Solar System

### What You'll Learn

- how comets change when they near the Sun
- the differences among comets, meteoroids, and asteroids

### Before You Read

Look up into the sky on a clear night. There are many objects you can see in addition to the Moon. What do you think these objects are? What would you like to know about them?

---

---

---

Mark the Text

**Highlight** Highlight the descriptions of comets, meteors, and asteroids as you read about them in this section.

FOLDABLES™

**D Organize Information** Make the following three-tab Foldable to help you organize information about comets, meteoroids, and asteroids.

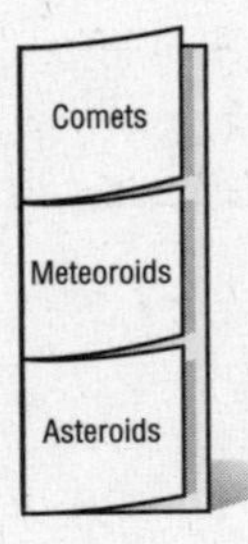


### Read to Learn

## Comets

Planets and moons are not the only objects in the solar system. Comets, meteoroids, and asteroids are other important objects that orbit the Sun.

You may have heard of Halley's Comet. A **comet** is made up of dust and pieces of rock mixed with frozen water, methane, and ammonia. Halley's Comet was last seen from Earth in 1986. It takes Halley's Comet 76 years to orbit the Sun. Astronomer Jan Oort suggested that billions of comets surround the solar system. This cloud of comets, called the Oort Cloud, is located beyond the orbit of Pluto.

### What is the structure of a comet?

A comet is a mass of frozen ice and rock similar to a large, dirty snowball. As a comet approaches the Sun, the Sun's heat turns the ice to gas. This releases dust and bits of rock which form a bright cloud, or coma, around the nucleus, or solid part, of the comet. The solar wind pushes on the gas and dust to form tails that point away from the Sun.

## Meteoroids, Meteors, and Meteorites

After many trips around the Sun, most of the ice in a comet's nucleus has evaporated. The comet is now just rocks and dust, spread out within the original comet's orbit. These objects are called meteoroids. A meteoroid that enters Earth's atmosphere and burns up is called a **meteor.** Another term for a meteor is a shooting star.

Whenever Earth passes through the old orbit of a comet, small pieces of rock and dust enter Earth's atmosphere. The event is called a meteor shower. A **meteorite** is a large meteoroid that does not burn up completely in Earth's atmosphere and strikes Earth. Most meteorites are probably the remains from asteroid collisions or broken-up comets. Others come from the Moon and Mars. ☑

**Reading Check**

1. **Define** What is a meteor shower?

______________________

______________________

## Asteroids

An **asteroid** is a piece of rock made up of material like that which formed the planets. Most asteroids are located in an area between the orbits of Mars and Jupiter called the asteroid belt as shown in the figure. Other asteroids are scattered throughout the solar system.

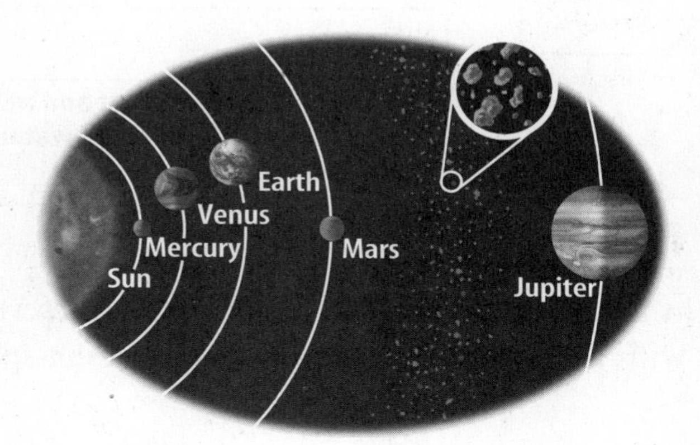


**Picture This**

2. **Interpret Scientific Illustrations** Between the orbits of which planets is the asteroid belt located?

______________________

### What else do we know about asteroids?

Some asteroids are tiny. Others measure hundreds of kilometers. The first asteroid ever discovered, Ceres, is the largest. It measures 940 km in diameter.

Comets, asteroids, and most meteorites were formed early in the history of the solar system. Scientists study these space objects to learn what the solar system might have been like long ago. Understanding this could help scientists better understand how Earth formed.

**Think it Over**

3. **Explain** What is important about studying objects in space?

______________________

______________________

______________________

## After You Read

### Mini Glossary

**asteroid:** a piece of rock made up of material similar to that which formed the planets

**comet:** space object made of dust and rock particles mixed with frozen water, methane, and ammonia

**meteor:** a meteoroid that burns up in Earth's atmosphere

**meteorite:** a meteoroid that strikes Earth's surface

1. Review the terms and their definitions in the Mini Glossary. Write a sentence to tell what the Oort cloud is.

_______________________________________________

_______________________________________________

2. Complete the concept chart with the correct words from the Mini Glossary.

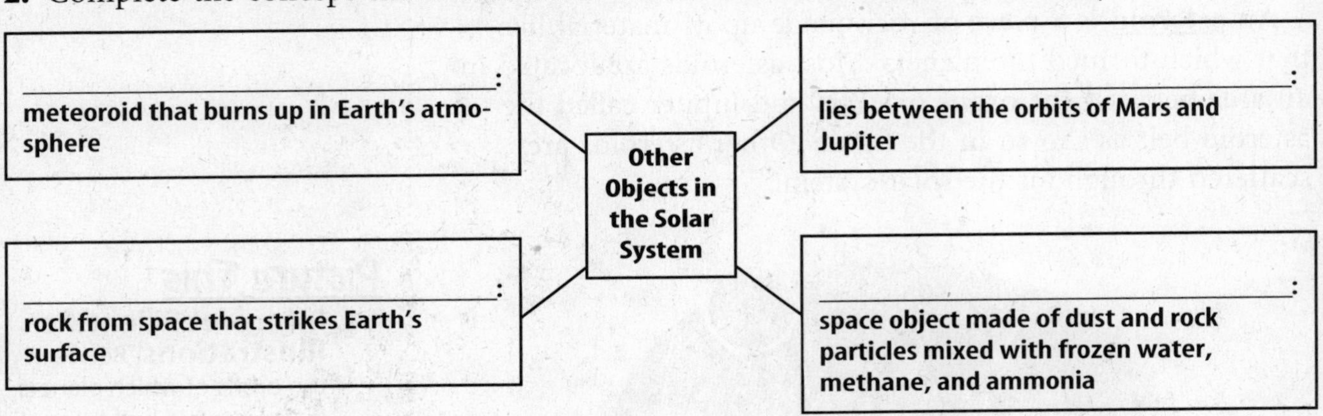


3. Reread the sentences you highlighted in the text. Did this strategy help you describe comets, meteors, and asteroids? Work with a partner and take turns describing space objects to each other.

_______________________________________________

_______________________________________________

_______________________________________________

# Stars and Galaxies

## section 1 Stars

## Before You Read

Describe the sky on a cloudless, moonless night. What would you see? Write the names of any stars you know about.

_______________________________________________

_______________________________________________

_______________________________________________

**What You'll Learn**

- about constellations
- the difference between absolute magnitude and apparent magnitude

## Read to Learn

### Constellations

It's fun to look at clouds and find animals, faces, and objects. It takes more imagination to play this game with stars. Ancient Greeks, Romans, and other people who lived long ago found patterns, or shapes, made by stars in the night sky. These star patterns are called **constellations** (kahn stuh LAY shuns). In these star patterns, they saw characters, animals, and objects from stories they knew well.

From Earth, a constellation looks like spots of light arranged in a particular shape against the night sky. However, the stars in a constellation often have no relationship to each other in space.

### What are some common constellations?

Modern astronomy divides the sky into 88 constellations. Many of these were named by early astronomers. The Big Dipper is part of the constellation Ursa Major. The two stars at the front of the Big Dipper point to the star Polaris. Polaris is often called the North Star. That is because Polaris is almost directly over Earth's north pole. Polaris is located at the end of the Little Dipper in the constellation Ursa Minor. See the figure on the next page for the locations of Polaris, the Big Dipper, and the Little Dipper.

Study Coach

**Identify What You Know** Create a K-W-L chart for this chapter. Write what you already know about stars, what you want to know, and what you learn as you read this section.

FOLDABLES™

**A Record Data** For this section, create a Foldable to record important facts, notes, and new vocabulary about stars.

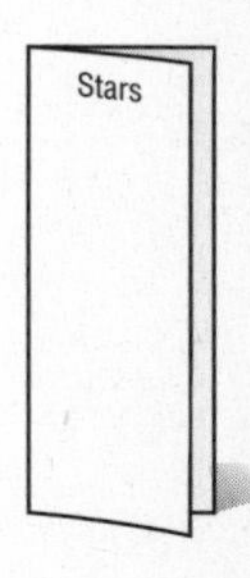

## Picture This

**1. Interpret Diagram** Do the stars appear to rotate clockwise or counter-clockwise around Polaris?

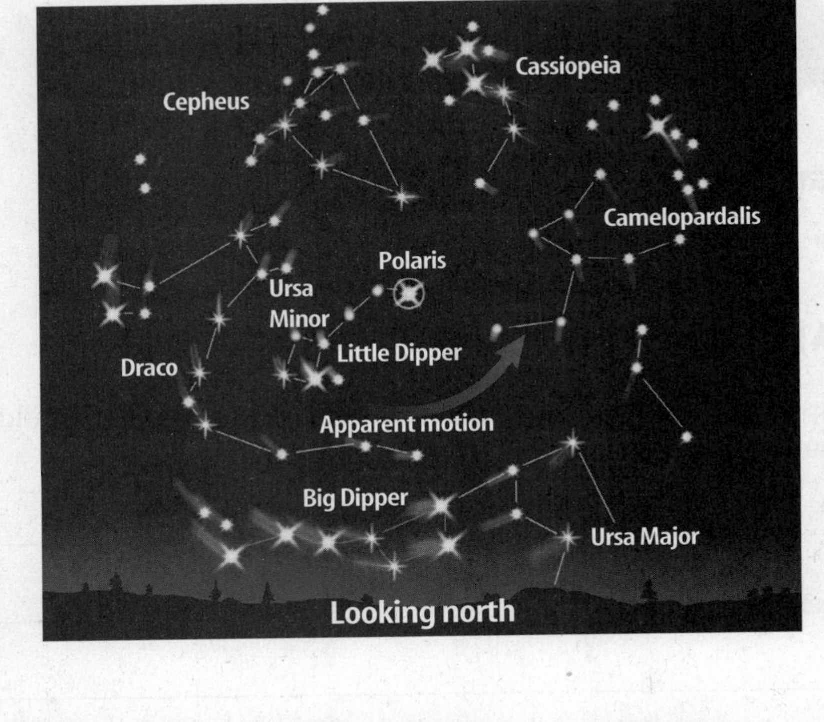


## Why do constellations appear to move?

You may have noticed that stars appear to move during the night. Constellations in the northern sky appear to circle around Polaris. Because of this, they are called circumpolar constellations. They appear to move because Earth is moving.

The figure above shows the circumpolar constellations rotating around Polaris. Because of their unique position, you can see the circumpolar constellations all year long. Other constellations, like Orion, can only be seen in certain seasons. In the summer, Orion can't be seen north of the equator because the northern hemisphere faces Orion during the day. ☑

**2. Identify** Which constellations can be seen all year?

## Absolute and Apparent Magnitudes

When you look at constellations, you'll notice that some stars are brighter than others. Sometimes stars look brighter than others because they're closer to Earth.

There are two ways to describe a star's brightness. The **absolute magnitude** (MAG nuh tewd) of a star is the amount of light it gives off. The **apparent magnitude** is the amount of light that reaches Earth, or how bright it looks. A star that is dim can look bright in the sky if it's close to Earth. A star that is bright can appear dim if it's far away. For example, Rigel is a brighter star than Sirius, but Sirius appears brighter because it is 100 times closer to Earth than Rigel is.

## Measurement In Space

One way scientists measure the distance between Earth and a nearby star is to measure parallax (PER uh laks). Parallax is what makes an object seem to change its position when you look at it from two different positions. Stretch your arm out in front of you and look at your thumb with one eye closed. Now open your eye and close your other eye and look at your thumb. Your thumb looks like it has moved, even though it has not. That apparent shift is parallax. Try it again, but with your thumb closer to your face. What did you see? Your thumb appears to move when it is closer to your eyes. The nearer an object is, the greater its parallax. ☑

### How is parallax measured?

Astronomers measure the parallax of a nearby star to see how far away it is from Earth. Astronomers observe the same star at two different times of the year. Astronomers look at how the star seems to change positions compared with stars that are farther away. Then they use the angle of the parallax and the size of Earth's orbit to calculate the distance of the star from Earth.

Space is so enormous that scientists need a special way to describe distances. Distances between stars and galaxies are measured in light-years. A **light-year** is the distance that light travels in one year. Light travels 300,000 km/s.

**Reading Check**

3. **Determine** Which would have a greater parallax—an object close to you or one that is far away?

______________________

______________________

## Properties of Stars

The color of a star indicates its temperature. For example, hot stars are a blue-white color. Stars that have a medium temperature, like the Sun, are yellow. A cooler star looks orange or red.

Astronomers use an instrument called a spectroscope to learn what a star is made of. The spectroscope spreads light out into a band of colors which might include dark lines. These dark lines stand for elements in a star's atmosphere. These patterns of lines help astronomers identify the elements in a star's atmosphere. ☑

**Reading Check**

4. **Identify** What do the dark lines in the band of colors produced by a spectroscope represent?

______________________

______________________

## After You Read

### Mini Glossary

**absolute magnitude (MAG nuh tewd):** the amount of light that a star gives off

**apparent magnitude:** the amount of a star's light that reaches Earth

**constellation (kahn stuh LAY shun):** a group of stars that forms a pattern in the night sky

**light-year:** the distance that light travels in one year

1. Review the terms and their definitions in the Mini Glossary. Write a sentence to explain why two stars can have the same absolute magnitude but may have different apparent magnitudes.

_______________________________________________

_______________________________________________

2. Complete the diagram to explain what you learned about stars.

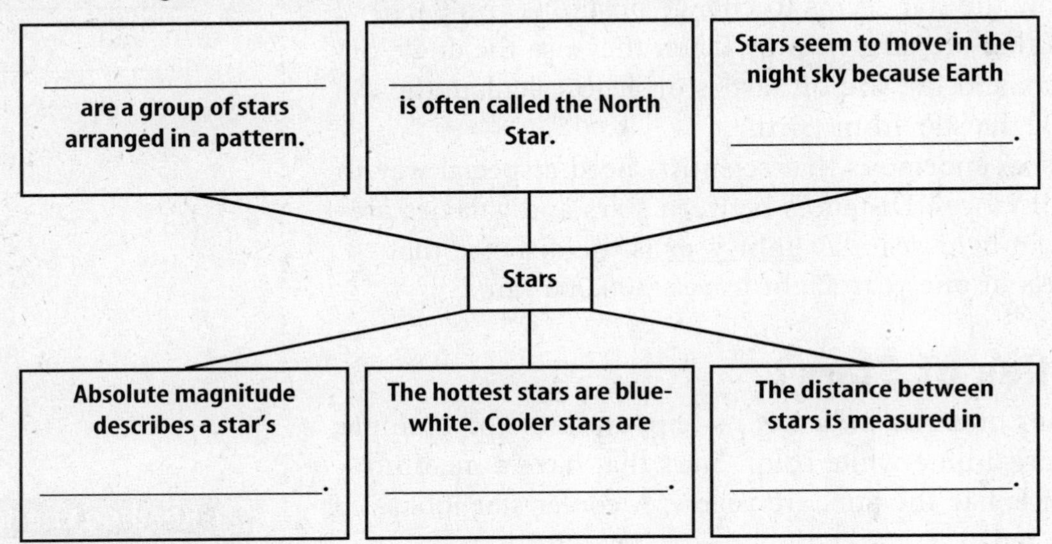


3. Look back at the K-W-L chart you made as you read this section. Did you add to what you already knew? Did you learn what you wanted to know? Did the K-W-L chart help you to understand what you read?

_______________________________________________

_______________________________________________

_______________________________________________

Visit **earth.msscience.com** to access your textbook, interactive games, and projects to help you learn more about stars.

# Stars and Galaxies

## section 2 The Sun

### Before You Read

What comes to mind when you think about the Sun? Brainstorm some words and write them below.

_______________________________________________

_______________________________________________

**What You'll Learn**

- the Sun is the closest star to Earth
- the structure of the Sun
- the features of the Sun, such as sunspots and solar flares

### Read to Learn

#### The Sun's Layers

The Sun is an ordinary star and is the center of our solar system. It is also the closest star to Earth. Almost all life on Earth depends on energy from the Sun.

Like other stars, the Sun is an enormous ball of gas that produces energy in its core, or center. This energy is produced by fusing hydrogen into helium. This energy travels outward to the Sun's atmosphere. The energy is given off as light and heat.

**Mark the Text**

**Underline** the different properties of the Sun as you read.

#### The Sun's Atmosphere

The Sun is made up of different layers. The lowest layer of the Sun's atmosphere is the **photosphere** (FOH tuh sfihr). This is the layer that gives off the light we see from Earth. The photosphere is often called the surface of the Sun. Temperatures there are about 6,000 K. The layer above the photosphere is called the **chromosphere** (KROH muh sfihr). This layer is about 2,000 km thick. There is a change of zone between 2,000 km and 10,000 km above the photosphere. Above this zone is the outer layer of the Sun's atmosphere. This outer layer is called the **corona** (kuh ROH nuh). The corona is the largest layer of the Sun's atmosphere. It reaches millions of kilometers into space. The illustration on the next page shows the different layers of the Sun.

**FOLDABLES™**

**B Take Notes** Create a Foldable to record the main ideas about the Sun. Include information about the Sun's layers, atmosphere, and surface features.

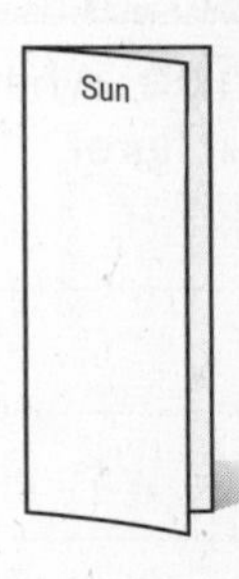

## Think it Over

1. **List** Number the parts of the Sun's atmosphere shown below, with 1 being the innermost layer and 4 being the outermost layer.

_____ chromosphere

_____ corona

_____ core

photosphere

### The Sun's Atmosphere

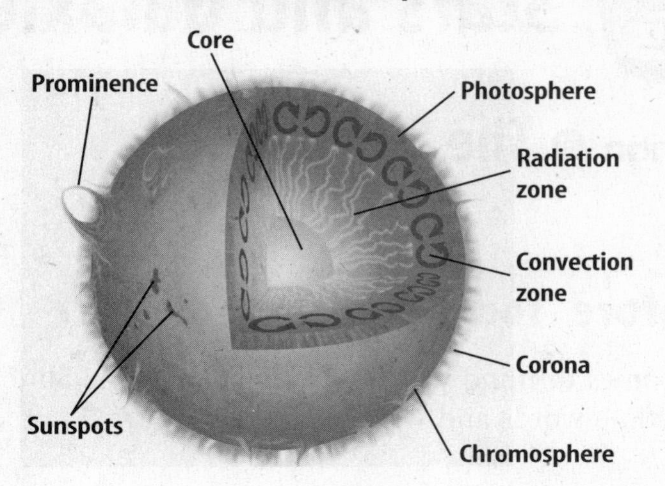


## Surface Features

From our point of view on Earth, the Sun's surface looks smooth. But the Sun's surface has many features. Among them are sunspots, prominences, flares, and CMEs.

### What is a sunspot?

**Sunspots** are areas of the Sun's surface that appear dark. Sunspots look this way because they are cooler than the area around them. Scientists have been studying sunspots for hundreds of years. They have observed the way that sunspots move. The fact that sunspots move has led scientists to determine that the Sun rotates. However, the Sun does not rotate like Earth does. The Sun rotates faster at its equator than at its poles. Sunspots near the equator take about 25 days to rotate once. Near the poles, sunspots take about 35 days.

Sunspots are not permanent features on the Sun. They appear and disappear over days, weeks, or months. The number of sunspots increases and decreases in a regular cycle of time. About every 10 or 11 years, there is a period of many large sunspots. In between those times, there are fewer sunspots.

**Reading Check**

2. **Describe** What happens in a solar flare?

_____________________

_____________________

_____________________

### What are prominences and solar flares?

Sunspots are related to other features on the Sun's surface. Sunspots and strong magnetic fields are found together on the Sun. The magnetic fields might cause prominences, which are huge arching columns of gas.

The gases near a sunspot may suddenly brighten and rapidly shoot outward. This is called a solar flare. ☑

## What is a CME?

When large amounts of electrically-charged gas shoot out from the Sun's corona, the event is called a CME. CME stands for coronal mass ejection.

CMEs present little danger to life on Earth, but they do affect our planet. CMEs can damage satellites. They can cause radio interference. Near the poles, they can produce a display of shifting colorful lights in the night sky. These displays tend to occur at Earth's poles. One such display of lights is called the Aurora borealis, or northern lights. The picture below shows the Aurora borealis.

**3. Infer** Why do you think the Aurora borealis is also known as the northern lights?

______________________________

______________________________

## The Sun—An Average Star

The Sun is an average star. It is middle-aged and its absolute magnitude is about average. The Sun shines with a yellow light. Although the Sun is an average star, it is much closer to Earth than other stars. Light from the Sun reaches Earth in about eight minutes. Light from other stars takes many years to reach Earth. ☑

The Sun is unusual in one way. It is not close to any other stars. Most stars are found in groups of two or more stars that orbit each other. Stars can also be held together by each other's gravity. This kind of group is a star cluster. Most star clusters are far from the solar system. They might be visible as a fuzzy bright patch in the night sky.

**Reading Check**

**4. Identify** How long does it take for the light from the Sun to reach Earth?

______________________________

## After You Read

### Mini Glossary

**chromosphere (KROH muh sfihr):** one of the middle layers of the Sun's atmosphere

**corona (kuh ROH nuh):** the top, largest layer of the Sun's atmosphere

**photosphere (FOH tuh sfirh):** the lowest layer of the Sun's atmosphere; gives off light

**sunspot:** an area on the Sun's surface that is cooler and less bright than surrounding areas

1. Review the terms and their definitions in the Mini Glossary. Write a sentence using three terms to describe the Sun's atmosphere.

_______________________________________________

_______________________________________________

_______________________________________________

2. Complete the chart to show how the Sun is like other stars and different from other stars.

| The Sun vs. Other Stars | |
|---|---|
| **Similarities** | **Differences** |
| It is a huge ball of ____________. | Its light reaches Earth in ____________. |
| It produces energy in its ____________. | Life on ____________ depends on it. |
| It has an ____________ that has different layers. One is the corona. | It is not close to other ____________. |

3. Look at the list of words you brainstormed to describe the Sun before you read this section. What words would you add to this list? Look at the text you underlined to describe the Sun. Now look at your new list. What was the most surprising thing you learned about the Sun?

_______________________________________________

_______________________________________________

_______________________________________________

Visit **earth.msscience.com** to access your textbook, interactive games, and projects to help you learn more about the Sun.

## section 3 Evolution of Stars

### Before You Read

What makes one star different from another? Do you think the Sun is the same as other stars? Write your ideas on the lines below.

______________________________

______________________________

______________________________

**What You'll Learn**

- how stars are sorted into groups
- ways the Sun is the same as other types of stars
- ways the Sun is different from other types of stars
- how stars develop

### Read to Learn

## Classifying Stars

When you look at the night sky, all stars might look about the same. However, they're very different. They vary in age and size. They vary in temperature and brightness as well. These features led scientists to organize stars into categories, or groups.

### How is a star's temperature related to its brightness?

In the early 1900s, two scientists named Ejnar Hertzsprung and Henry Russell noticed that hotter stars are usually brighter. In other words, stars with higher temperatures have brighter absolute magnitudes.

### How do scientists show this relationship?

Hertzsprung and Russell developed a graph to show this relationship. You can see this graph on the next page. The temperatures are at the bottom. Absolute magnitude goes up the left side. A graph that shows this relationship between a star's temperature and its brightness is called a Hertzsprung-Russell diagram, or an H-R diagram.

Study Coach

**Make Flash Cards** to help you record new vocabulary words. Write the word on one side of the flash card and a brief definition on the other side.

FOLDABLES™

**C** Create a Foldable as shown below about evolution of stars. Label the three columns Star Classification, Star Temperature and Color, and How a Star Evolves.

| Star Classification | Star Temp. and Color | How a Star Evolves |
|---|---|---|
| | | |

## Picture This

**1. Complete the Diagram** Color hot, bright stars blue. Color cool, dim stars red. Color in-between stars yellow. Read to find out how to color dwarfs and giants.

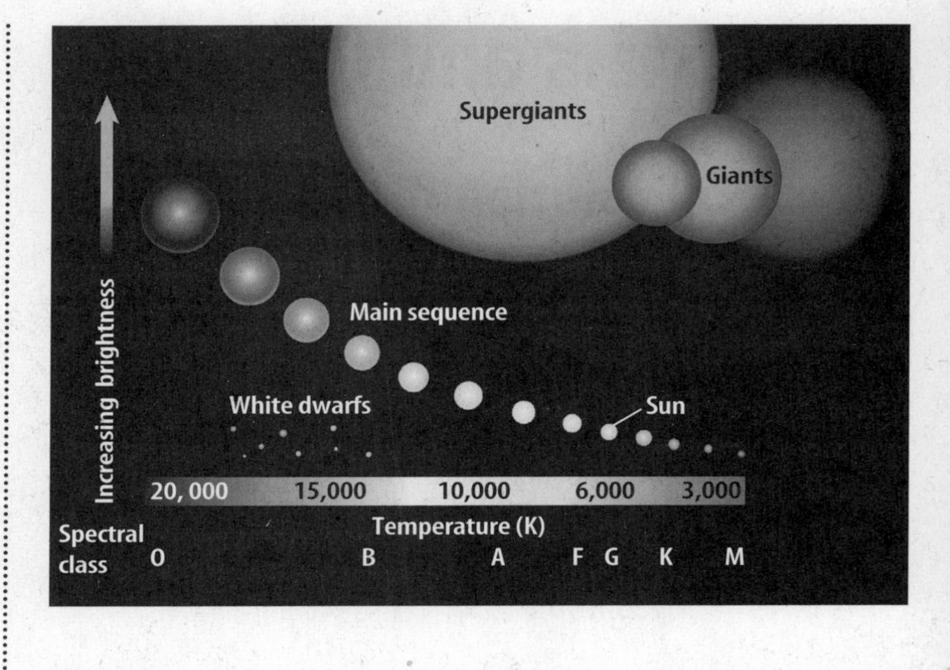


### What is the main sequence?

The H-R diagram above shows the connection between a star's temperature and its brightness. As you can see, most stars seem to fit into a band that runs from the upper left to the lower right. This band is called the main sequence. Hot, blue, bright stars begin at 20,000 K and continue to about 15,000 K. Cool, red, dim stars range from 5,000 K to 3,000 K. Yellow stars, like the Sun, are in between.

### What are dwarfs?

About 90 percent of all stars are main sequence stars. Most of these are small, red stars found in the lower right of the H-R diagram. Some of the stars that are not in the main sequence are hot, but they are not bright. These small stars are called white dwarfs, although they are usually blue in color. White dwarfs are found on the lower left of the H-R diagram. ☑

**Reading Check**

**2. Identify** What are small stars that are hot but not bright called?

_______________

### What are giants?

Other stars are very bright, but they are not hot. These large stars are called giants or red giants, because they are usually red in color. They're found on the upper right of the H-R diagram. The largest giants are called supergiants. These stars can be hundreds of times bigger than the Sun and thousands of times brighter.

## How do stars shine?

For centuries, people have wondered what stars were made of and what made them shine. Over time, people realized the Sun had been shining for billions of years. What material could burn for so long?

### What process creates the light that reaches Earth?

In the 1930s, scientists made an important discovery about atoms. Scientists observed that the nuclei, or centers, of atoms reacted with one another. They hypothesized that the center of the Sun was hot enough to cause hydrogen atoms to fuse, or link together, and form another kind of atom—helium atoms. This reaction, called fusion, releases huge amounts of energy. Much of this energy is released as different kinds of light. A very small part of this light comes to Earth. ☑

**Reading Check**

3. **Identify** What is the name of the process in which hydrogen is converted to helium?

______________________

## Evolution of Stars

The H-R diagram explained a lot about stars. However, scientists wondered why some stars didn't fit in the main sequence. Scientists also wondered what happened when a star used up its hydrogen fuel. Now, there are theories about how stars evolve, or change over time. These theories also explain what makes stars different from one another, and what happens when a star "dies."

When a star uses up its hydrogen, that star is no longer in the main sequence. This can take less than 1 million years for the brightest stars. It can take billions of years for the dimmest stars. The Sun has a main sequence life span of about 10 billion years. Half of its life is still in the future.

**Applying Math**

4. **Calculate** About how many years are left in the Sun's main sequence life span?

______________________

### How are stars formed?

Stars begin as a large cloud of gas and dust called a **nebula** (NEB yuh luh). The pull of gravity between the particles of gas and dust causes the nebula to contract, or shrink. The nebula can break apart into smaller and smaller pieces. Each piece eventually might collapse to form a star.

The particles in the smaller pieces of nebula move closer together. This causes temperatures in each piece to rise. When the temperature in the core of a piece of nebula reaches 10 million K, fusion begins. Energy is released from the core and travels outward. Now the object is a star.

## What is a giant?

After a star is formed, the heat created by fusion creates outward pressure. Without this pressure, the star would collapse from its own gravity. The star becomes a main sequence star. It continues to use its hydrogen fuel. The different stages in the life of a star are shown in the illustration on this page and the next page.

When hydrogen in the core of the star runs out, the core contracts and temperatures inside the star increase. The outer layers of the star expand and cool. In this late stage in its life cycle, a star is called a **giant.**

As the core contracts, its temperature continues to rise. By the time it reaches 100 million K, the star is huge. Its outer layers are much cooler than when it was a main sequence star. In about 5 billion years, the Sun will become a giant.

### Think it Over

**5. Infer** What is the relationship between how much hydrogen a star has and the star's temperature?

________________________

________________________

## What is a white dwarf?

The star's core contracts even more after it uses much of its helium and the outer layers escape into space. This leaves only the hot, dense core. At this stage in a star's life cycle, it is about the size of Earth. It is called a **white dwarf.** In time, the white dwarf will cool and stop giving off light.

## What are supergiants and supernovas?

The length time it takes for a star to go through its stages of life depends on its mass. The stages happen more quickly and more violently in stars that are more than eight times more massive than the Sun. In massive stars, the core heats up to much higher temperatures. Heavier and heavier elements form in the core. The star expands into a **supergiant.** Finally, iron forms in the core. Iron can't release energy through fusion. The core collapses violently. This sends a shock wave outward through the star. The outer part of the star explodes. This produces a kind of star called a supernova. A supernova can be millions of times brighter than the original star was.

## What is a neutron star?

What happens next depends on the size of the supernova's collapsed core. If the collapsed core is between 1.4 and 3 times as massive as the Sun, the core shrinks until it is only about 20 km in diameter. In this dense core, there are only neutrons. This kind of star is called a **neutron star**. Because the star is so dense, one teaspoonful of a neutron star would weigh more than 600 million metric tons on Earth.

## What is a black hole?

The core of some supernovas is more than three times more massive than the Sun. Nothing can stop the core's collapse in these supernovas. All of the core's mass collapses to a point. The gravity near this point is so strong that not even light can escape from it. Because light cannot escape from this region, it is called a **black hole**. If you could shine a light into a black hole, the light would disappear into it. However, a black hole is not like a vacuum cleaner. It does not pull in faraway objects. Stars and planets can orbit around a black hole, as long as they are far enough away.

## Where does a nebula's matter come from?

You learned that a star begins as a nebula. Where does the matter, or gas and dust, come from to form the nebula? Some of it was once in other stars. A star ejects large amounts of matter during the course of its life. Some of this matter becomes part of a nebula. It can develop into new stars. The matter in stars is recycled many times.

The matter that is created in the cores of stars and during supernova explosions is also recycled. Elements such as carbon and iron can become parts of new stars. Spectrographs of the Sun show that it contains some carbon, iron, and other heavy elements. However, the Sun is too young to have formed these elements itself. The Sun condensed from material that was created in stars that died long ago.

Some elements condense to form planets and other objects. In fact, your body contains many atoms that were formed in the cores of ancient stars.

### Think it Over

6. **Infer** If the collapsed core of a supernova is 2.4 times as massive as the Sun, what will it become next?

_______________

## After You Read

### Mini Glossary

**black hole:** the final stage in the evolution of a very massive star, where the core collapses to a point that its gravity is so strong that not even light can escape

**giant:** a late stage in the life of a low-mass star, when the core contracts but its outer layers expand and cool; a large, bright, cool star

**nebula (NEB yuh luh):** a large cloud of gas and dust where stars are formed

**neutron star:** a very dense core of a collapsed star that can shrink to about 20 km in diameter and contains only neutrons

**supergiant:** late stage in the life cycle of a massive star in which the core heats up and the star expands; a large, very bright star

**white dwarf:** a late stage in the life cycle of a low-mass star; formed when its outer layers escape into space, leaving behind a hot, dense core; a small, dim, hot star

1. Review the terms and their definitions in the Mini Glossary. Write a sentence to compare a white dwarf and a giant.

___

___

___

2. Fill in the blanks to review what you have learned about the life of a massive star.

A massive star forms in a ________. The star burns hydrogen fuel as a main ________ star. The core heats up. The star expands and cools into a ________. The star then explodes as a ________. Depending on its mass, it will then become either a ________ or a ________.

3. Could you use the flash cards you created to describe how the Sun developed? What information was helpful? What other information should have been on the cards?

___

___

___

Visit **earth.msscience.com** to access your textbook, interactive games, and projects to help you learn more about the evolution of stars.

# Stars and Galaxies

## section 4 Galaxies and the Universe

### Before You Read

Imagine that someone on the other side of the universe wanted to send you mail. How might you give someone an address for Earth?

______________________________________________

______________________________________________

**What You'll Learn**

- the Sun's position in the Milky Way Galaxy
- what forces affect our solar system
- what forces affect other galaxies

### Read to Learn

#### Galaxies

How can you describe the location of Earth? We are in the solar system. The solar system is in a galaxy called the Milky Way. A **galaxy** is a large group of stars, gas, and dust held together by gravity.

There are many other galaxies. Every galaxy has the same elements, forces, and types of energy that are found in our solar system.

You learned that stars are grouped together in galaxies. In the same way, galaxies are grouped into clusters. The Milky Way is part of a cluster called the Local Group. The Local Group is made up of about 45 galaxies in different sizes and shapes. There are three major types of galaxies.

**Mark the Text**

**Highlight** the main point in each paragraph. Use a different color to highlight a detail or example that helps explain the main point.

#### What are the three major types of galaxies?

Spiral galaxies have spiral arms that wind outward from the center. The arms are made up of bright stars, dust, and gas. The Milky Way galaxy is a spiral galaxy.

Elliptical (ih LIHP tih kul) galaxies are a common type of galaxy. They are shaped like large, three-dimensional ellipses.

Irregular galaxies include all the galaxies that don't fit into the other two groups. These galaxies have many different shapes.

**FOLDABLES™**

**D Summarize** Create a three-tab Foldable to summarize the main ideas from the section.

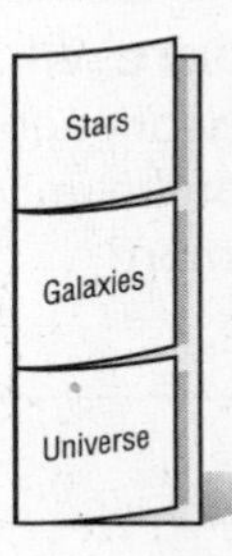

## The Milky Way Galaxy

There might be one trillion stars in the Milky Way. It is about 100,000 light-years across. Find the Sun in the image of the Milky Way below. It is about 26,000 light-years from the galaxy's center in one of the spiral arms. In the galaxy, all stars orbit around a central region, or core. It takes about 225 million years for the Sun to orbit the center of the Milky Way.

Scientists put the Milky Way into the spiral galaxy group. However, it's difficult to know the exact shape because we can't look at the galaxy from the outside. You can't see the shape of the Milky Way because the location of our solar system is in one of its spiral arms. But you can see the Milky Way stretching across the sky. It looks like a dusty band of dim light. All the stars you can see in the night sky are part of the Milky Way. Like many other galaxies, the Milky Way has a black hole at its center.



**Picture This**

1. **Explain** Why can't you see the shape of the Milky Way?

______________________

______________________

______________________

## Origin of the Universe

Scientists have offered different models, or ideas, for how the universe began. One model is the steady state theory. It suggests that the universe always has been the same as it is now. The universe expands and new matter is created. This keeps the density of the universe in a steady state.

A second model is the oscillating (AH sih lay ting) model. This model states that the universe formed and then it expanded, or grew larger. Over time, the rate of growth slowed down. Then the universe began to contract, or shrink. Then the whole process began again. In other words, it oscillates back and forth in size.

A third model is called the big bang theory. This theory states that the universe began with a big bang and has been expanding ever since.

**Think it Over**

2. **Compare** What do the three theories about the origin of the universe have in common?

______________________

______________________

## Expansion of the Universe

Think of the sound of a whistle on a passing train. The pitch of the whistle rises as the train moves closer. Then the pitch of the whistle drops as the train moves away. This happens because the sound waves coming from the whistle are compressed, or shortened, as the train gets closer. This effect is called the Doppler (DAH plur) shift. ☑

### Does the Doppler shift affect light?

The Doppler shift happens with light too. Like sound, light moves in waves. If a star is moving toward Earth, the light waves are shortened. If a star is moving away from Earth, the light waves are stretched out. Blue-violet light waves are shorter than red light waves. Scientists can identify blue-violet light from stars moving toward Earth. When a star is moving away from Earth, the light shifts toward red. This is called a red shift.

### How do we know the universe is expanding?

In 1929, Edwin Hubble noticed a red shift in the light from galaxies outside the Local Group. This meant the galaxies are moving away. If all galaxies outside the Local Group are moving away from Earth, then the entire universe must be expanding.

## The Big Bang Theory

The **big bang theory** is the leading theory about how the universe formed. It states that the universe began about 13.7 billion years ago. There was a huge explosion. In less than a second, the universe grew from the size of a pinhead to 2,000 times the size of the Sun. Even today, galaxies are still moving away from this explosion.

Scientists don't know if the universe will expand forever or stop expanding. If there is enough matter in the universe, gravity might stop the expansion. Then the universe would contract until everything came back to a single point. But studies show the universe is expanding faster, not slower. Scientists are still trying to figure out what will happen to the universe.

**Reading Check**

3. **Apply** You hear a police siren in the distance. If the siren's pitch is getting higher, is the police car coming closer or moving away?

_______________

**Think it Over**

4. **Draw Conclusions** Why do you think the leading theory about how the universe formed is called the big bang theory?

_______________

_______________

# After You Read

## Mini Glossary

**big bang theory:** the theory that the universe began about 13.7 billion years ago with a huge explosion and has been expanding ever since

**galaxy:** a large group of stars, dust, and gas held together by gravity

1. Review the terms and their definitions in the Mini Glossary. Write a sentence using the terms *big bang theory* and *galaxy.*

2. Complete the diagram to show how Earth fits into the Universe. Use the following terms: Milky Way, Solar System, and Local Group.

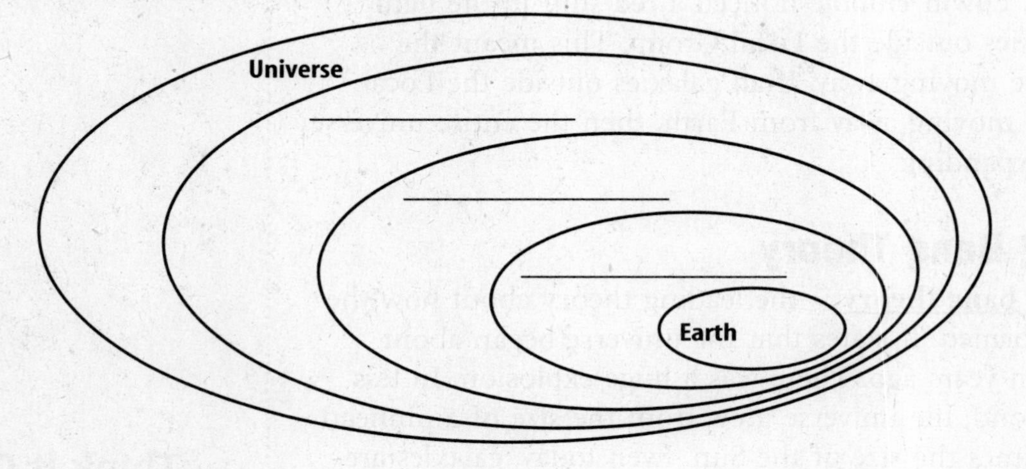


3. Look at your highlighted text about the Milky Way. Write a short description of the Milky Way that includes three details. Did highlighted text help you write your description? What other strategy could have helped you keep track of details about the Milky Way?

ScienceOnline Visit **earth.msscience.com** to access your textbook, interactive games, and projects to help you learn more about galaxies and the universe.